High-Temperature Superconductivity in Cuprates

Fundamental Theories of Physics

An International Book Series on The Fundamental Theories of Physics: Their Clarification, Development and Application

Volume 125

High-Temperature Superconductivity in Cuprates

The Nonlinear Mechanism and Tunneling Measurements

by

Andrei Mourachkine

Université Libre de Bruxelles, Brussels, Belgium

SPRINGER-SCIENCE+BUSINESS MEDIA, B.V.

A C.I.P. Catalogue record for this book is available from the Library of Congress.

ISBN 978-90-481-6098-3 ISBN 978-0-306-48063-8 (eBook)
DOI 10.1007/978-0-306-48063-8

Printed on acid-free paper

To the young generation of scientists

The important thing in science is not so much to obtain new facts as to discover new ways of thinking about them.

—William Lawrence Bragg

Contents

Preface

"The Frontiers of Knowledge (to coin a phrase) are always on the move. Today's discovery will tomorrow be part of the mental furniture of every research worker. By the end of next week it will be in every course of graduate lectures. Within the month there will be a clamour to have it in the undergraduate curriculum. Next year, I do believe, it will seem so commonplace that it may be assumed to be known by every schoolboy.

"The process of advancing the line of settlements, and cultivating and civilizing the new territory, takes place in stages. The original papers are published, to the delight of their authors, and to the critical eyes of their readers. Review articles then provide crude sketch plans, elementary guides through the forests of the literature. Then come the monographs, exact surveys, mapping out the ground that has been won, adjusting claims for priority, putting each fact or theory into its place" (J. M. Ziman, *Principles of the Theory of Solids* (Cambridge University Press, 1972) p.v).

The main purpose of the book is to present the mechanism of high-T_c superconductivity discovered in 1986 by J. G. Bednorz and K. A. Müller, and to discuss the physics of high-T_c superconductors. The last chapter of the book presents analysis of tunneling measurements in cuprates. The book is addressed to researchers and graduate students in all branches of exact sciences.

A few words about the history of the book: I began to work in the field of high-T_c superconductivity in 1992; however, I remember quite well, when in June of 1987, a collaborator of our laboratory, where I was working at that time, brought from a neighboring Institute a small piece of pressed black ceramics and said: "Here is a new superconductor." Obviously, everybody, who was in the room, took it by *bare* hands and looked at it carefully trying to understand what is unique about this new superconductor. At that moment, I did not know that five years later high-T_c superconductivity was to be my topic of research. In 1992, I started running microwave measurements in cuprates, and four years later when I "felt a taste" of high-T_c superconductivity and got it under my skin, I decided that I would not be satisfied until I managed to unravel the mechanism of superconductivity in cuprates. So, this is how it started and how it became personal. Fortunately, in 1997, I began doing tunneling measurements in cuprates. Working nights and weekends, measuring and mainly, analyzing, after four years, I found a clue leading to what I was looking for. I was happy that I have satisfied my childish curiosity. To write a book was not foreseen, but I decided to spend some time to put down on paper my findings in order "to hammer in the first nail in the coffin" of high-T_c superconductivity, which remained a mystery for 15 years.

A final note: the book will appear in a book series "Fundamental Theories of Physics." However, one has to realize that the book does not introduce a theory of high-T_c superconductivity, but presents fundamentals of the mechanism of this remarkable phenomenon.

I am grateful to many people who have directly and indirectly contributed to the book. In particular I wish to thank F. Masin, G. Gusman, A. R. F. Barel, A. M. Gabovich, J. Delhiere, V. V. Moshchalkov, A. Volodin, A. V. Buryak, L. S. Brizhik, M. Remoissenet, P. Pirotte, D. N. Davydov, H. Hancotte, R. Deltour, J. Jeener, Y. De Wilde and N. Miyakawa.

Finally I thank W. Stone, J. Gegenberg, J. W. Turner, D. Johnson and J. Wickens for correcting English.

ANDREI MOURACHKINE
Bruxelles, April 2002

Chapter 1

INTRODUCTION

The whole of science is nothing more than a refinement of everyday thinking.

—Albert Einstein

1. Superconductivity: a brief sketch

Superconductivity was regarded as a major scientific mystery for a large part of last century: discovered in 1911, it was completely understood only in 1957. The discovery was followed by a large amount of experimental and theoretical studies. Many outstanding scientists such as Einstein, Landau and Heisenberg tried their hand at explaining the phenomenon. In 1957 the microscopic theory of superconductivity was formulated by Bardeen, Cooper and Schrieffer, which is now known as the BCS theory. The basis of the BCS theory is the interaction of a "gas" of conducting electrons with elastic waves of the crystal lattice (phonons). Two electrons in a vacuum repel each other by the Coulomb force, but in a superconductor below the critical temperature T_c there is a net attraction between two electrons that form the so-called Cooper pair. Each Cooper pair consists of two electrons of opposite momenta and spins.

In addition to electron pairing, superconductivity also requires long-range phase coherence among the Cooper pairs. In metallic superconductors, the pairing occurs due to the phonon-electron interaction, whereas the phase coherence among the Cooper pairs is established due to the overlap of their wave functions (the wave-function coupling). In conventional superconductors, the wave-function coupling can mediate the phase coherence because the average distance between Cooper pairs is much smaller than the coherence length (the size of a Cooper pair). In superconductors described by the BCS theory, the pairing and the long-range phase coherence occur simultaneously at T_c since the phase stiffness, which measures the ability of the superconducting state to carry supercurrent, is much larger than the energy gap Δ which reflects the strength of the binding of electrons into Cooper pairs (see Chapter 2). In unconventional superconductors, the situation is different: the pairing may occur above T_c without the phase coherence which appears at T_c.

The one effect that played the decisive role in showing the way to the correct theory of superconductivity in metals was the isotope effect. A study of different superconducting isotopes of mercury established a relationship between the critical temperature and the isotope mass M: $T_c M^{1/2}$ = constant, which turned out to be valid for most of conventional superconductors. Even nowadays the isotope effect serves as an indicator of the BCS mechanism of superconductivity.

After the appearance of the BCS theory, scientists who were working in the field of superconductivity in the 1960s and 1970s could relax and enjoy the BCS theory, but new superconducting materials were discovered in 1979: organic superconductors and heavy fermions. The experimental data obtained in organic superconductors and heavy fermions indicate that these compounds display an unconventional type of superconductivity. The year 1986 brought more "bad" news: the scientific world was astonished by the discovery of high-T_c superconductivity in copper oxides (cuprates). The most surprising fact is that high-T_c superconductivity occurs in oxides which are very bad conductors. The first reaction of most scientists working in the field of superconductivity was to think that there must be a new mechanism, because phonon-mediated superconductivity is impossible at so high temperature. The discovery of superconducting cuprates was followed by research growth at a rate unprecedented in the history of science: during 1987 the number of scientists working in the field of superconductivity increased, at least, by one order of magnitude. Data obtained in cuprates within year after the discovery of high-T_c superconductivity indeed showed that the characteristics of high-T_c superconductors deviate from the predictions of the BCS theory as those of organic superconductors and heavy fermions. For example, the BCS isotope effect is almost absent in cuprates. As a consequence, this has prompted the exploration of non-phonon electronic coupling mechanisms. Ph. Anderson was probably the first to suggest a theoretical model which did not incorporate phonon-electron interactions. Between 1987 and 2002, more than 100 theoretical models of high-T_c superconductivity were proposed. Most of these models consider phonons irrelevant. Looking ahead, it is worth noting that, as established by now, none of them can be fully applied to high-T_c superconductors; however, the combination of two proposed models, namely, the bisoliton theory and the theory based on spin-fluctuations, can in the first approximation describe the phenomenon of high-T_c superconductivity. In Chapter 10, we shall see that the electron pairing in cuprates occurs due to phonons and, moreover, the phonon-electron interaction in cuprates is moderately strong and nonlinear.

It is amazing that the isotope effect is manifest itself when the phonon-electron interaction is weak, and the isotope effect vanishes when the phonon-electron interaction becomes stronger! It is a paradox which is difficult to comprehend from common sense reasoning. As a matter of fact, this paradox played a negative role in understanding the mechanism of high-T_c superconductivity. *Nature has created many surprises for us, and we have to learn from it!*

However, this is not the full story. In the case of moderately strong phonon-electron interaction, the size of a Cooper pair becomes very small. If the density of small-size Cooper pairs is low, they have difficulties communicating with each other: The range in which the wave-function coupling is effective is insufficient to cover the distance between Cooper pairs. The Cooper pairs have to use another mechanism in order to establish the long-range phase coherence. In Chapter 9, we shall see that magnetic (spin) fluctuations mediate the phase coherence in high-T_c superconductors as well as in organic superconductors and heavy fermions.

To summarize, the *only* mechanism of electron pairing known *today* is the phonon-electron interactions. The strength of phonon-electron interactions is different in different superconducting materials: in metals, the phonon-electron interaction is weak, while in cuprates moderately strong and nonlinear. Thus, by convention, there are two types of electron pairing due to phonons: "linear" (the BCS type) and "non-linear". In addition, in the two cases, the mechanisms of the establishment of phase coherence are different: the wave-function coupling mediates the long-range phase coherence in metallic superconductors and, in high-T_c superconductors, organic compounds and heavy fermions, magnetic fluctuations establish the phase coherence.

The general belief which was predominant during the 15 years following the discovery of high-T_c superconductivity, namely that the superconductivity in cuprates has nothing to do with electron-phonon interactions, is wrong. On the contrary, the superconductivity in cuprates occurs due to the electron-phonon interaction, with the assistance of spin fluctuations.

2. High-T_c superconductivity: a brief historical introduction

This section is a *brief* reminder of key events which led to our understanding of the mechanism of high-T_c superconductivity. Generally speaking, the reality and the perception of the reality slightly differ from each other. So, the list of the events presented here is definitely subjective and not complete.

The first observation of what is now called the *soliton* was made by John Scott Russell near Edinburgh (Scotland) in 1834. He was observing a boat moving on a shallow channel and noticed that, when the boat suddenly stopped, the wave that it was pushing at its prow "rolled forward with great velocity, assuming the form of a large solitary elevation, a rounded, smooth and well defined heap of water which continued its course along the channel apparently without change of form or diminution of speed" [1]. He followed the wave along the channel for more than a mile.

The phenomenon of superconductivity was discovered by the Dutch physicist H. Kamerlingh Onnes in 1911 [2]. He found that *dc* resistivity of mercury suddenly drops to zero below 4.2 K. A year later, Kamerlingh Onnes discovered that a sufficiently strong magnetic field restores the resistivity in the sample as does a sufficiently strong electric current.

In 1933, W. Meissner and R. Ochsenfeld discovered one of the most fundamental properties of superconductors: perfect diamagnetism [3]. They found that the magnetic flux is expelled from the interior of the sample that is cooled below the critical temperature in weak external magnetic fields.

C. J. Gorter and H. B. G. Casimir introduced in 1934 a phenomenological theory of superconductivity based on the assumption that, in the superconducting state, there are two components of the conducting electron "fluid": "normal" and "superconducting" (hence the name given this theory, " the two-fluid model") [4]. The properties of the "normal" component are identical to those of the electron system in a normal metal, and the "superconducting" component is responsible for the anomalous properties. The two-fluid model proved to be a useful concept for analyzing the thermal properties of superconductors.

Vortices in superconductors were discovered by L. V. Shubnikov and coworkers in 1937 [5]. They found an unusual behavior for some superconductors in external magnetic fields. Actually, they discovered the existence of two critical magnetic fields for type-II superconductors and the new state of superconductors, known as the mixed state or Shubnikov's phase.

The isotope effect was found in 1950. A study of different superconducting isotopes of mercury established a relationship between the critical temperature and the isotope mass M: $T_c M^{1/2}$ = constant [6, 7].

In 1950, V. Ginzburg and L. Landau proposed an intuitive, phenomenological theory of superconductivity [8]. The theory uses the general theory of the second-order phase transitions, developed by L. Landau. The equations derived from the theory are highly non-trivial, and their validity was proven later on the basis of the microscopic theory. The Ginzburg-Landau theory played an important role in understanding the physics of the superconducting state.

By using the Ginzburg-Landau theory, A. A. Abrikosov theoretically found vortices and thus explained Shubnikov's experiments, suggesting that Shubnikov's phase is a state with vortices that actually form a periodic lattice [9]. This result seemed so strange that he could not publish his work during five years; and even after 1957, when it was published, this idea was only accepted after experimental proof of several predicted effects.

The microscopic theory of superconductivity in metals was proposed by J. Bardeen, L. Cooper and R. Schrieffer in 1957. This is usually referred to as the BCS theory [10]. The central concept of the BCS theory is a weak electron-phonon interaction which leads to the appearance of an attractive potential between two electrons. We shall discuss the BCS model in Chapter 2.

Quantum-mechanical tunneling of Cooper pairs through a thin insulating barrier (on the order of a few nanometers thick) between two superconductors was theoretically predicted by B. D. Josephson in 1962 [11]. After reading his paper, Bardeen publicly dismissed young Josephson's tunneling-supercurrent assertion [12]: "... pairing does not extend into the barrier, so that there can be no such superfluid flow." Josephson's predictions were confirmed within a year and the effects are now known as the Josephson effects. They play a special role in superconducting applications.

To the best of my knowledge, the soliton (or bisoliton) model of superconductivity was considered for the first time by L. S. Brizhik and A. S. Davydov in 1984 [13] in order to explain the superconductivity in organic quasi-one-dimensional conductors, the latter having been discovered by D. Jérome and co-workers [14].

In 1986, trying to explain the superconductivity in heavy fermions, discovered in 1979, K. Miyake, S. Schmitt-Rink and C. M. Varma considered the mechanism of superconductivity based on the exchange of antiferromagnetic spin fluctuations [15]. The calculations showed that the anisotropic even-parity pairings are assisted and the odd-parity as well as the isotropic even-parity are impeded by antiferromagnetic spin fluctuations.

Interest in the research of superconductivity was renewed in 1986 with the discovery of high-T_c superconductivity in copper oxides, made by J. G. Bednorz and K. A. Müller [16]. This book is devoted to a description of the mechanism of this remarkable phenomenon.

By analyzing the layered structure of cuprates, V. Z. Krezin and S. A. Wolf proposed in 1987, a model of high-T_c superconductivity based on the existence of two gaps: superconducting and induced [17]. Indeed, different experiments performed after 1987 have demonstrated the existence of two gaps, although they both have the superconducting origin.

Furthermore in 1987, L. P. Gor'kov and A. V. Sokol proposed the presence of two components of itinerant and more localized features in cuprates [18]. This kind of microscopic and dynamical phase separation was later rediscovered in other theoretical models.

Also in 1987, Ph. Anderson proposed a model of superconductivity in cuprates, in which the pairing mechanism and the mechanism for the establishment of phase coherence are different [19].

In 1988, A. S. Davydov suggested that high-T_c superconductivity occurs due to the formation of bisolitons, as it does in organic superconductors [20].

The pseudogap above T_c [21] was observed in 1989 in nuclear magnetic resonance (NMR) measurements [22]. The pseudogap is a partial energy gap, a depletion of the density of states above the critical temperature.

In 1990, A. S. Davydov presented a theory of high-T_c superconductivity, based on the concept of a moderately strong electron-phonon coupling which results in perturbation theory being invalid [23, 24]. The theory utilizes the concept of *bisolitons*, or electron (or hole) pairs coupled in a singlet state due to local deformation of the -O-Cu-O-Cu- chain in CuO_2 planes. We shall discuss the bisoliton model in Chapter 7.

In the early 1990s, proceeding from Anderson's suggestion, namely, that in cuprates the pairing mechanism and the mechanism for the establishment of phase coherence are different, a few theorists autonomously proposed that, *independently* of the origin of pairing mechanism, spin fluctuations mediate the long-range phase coherence in cuprates. It turned out that this suggestion was correct.

In 1994, A. S. Alexandrov and N. F. Mott pointed out that, in cuprates, it is necessary to distinguish the "internal" wave function of a Cooper pair and the order parameter of the Bose-Einstein condensate, which may have different symmetries [25].

In 1995, V. J. Emery and S. A. Kivelson emphasized that superconductivity requires pairing and long-range phase coherence [26]. They demonstrated that, in cuprates, the pairing may occur above T_c without the phase coherence.

In the same year 1995, J. M. Tranquada and co-workers found the presence of coupled, dynamical modulations of charges (holes) and spins in Nd-doped $La_{2-x}Sr_xCuO_4$ (LSCO) from neutron diffraction [27]. In LSCO, antiferromagnetic stripes of copper spins are separated by periodically spaced quasi-one-dimensional domain walls to which the holes segregate. The spin direction in antiferromagnetic domains rotates by 180° on crossing a domain wall.

In 1997, V. J. Emery, S. A. Kivelson and O. Zachar presented a theoretical model of high-T_c superconductivity [28]. It turned out that the model is incorrect (there is no charge-spin separation in cuprates); however, it is the first model of high-T_c superconductivity based on the presence of charge stripes in CuO_2 planes.

In 1999, analysis of tunneling and neutron scattering measurements showed that, in $Bi_2Sr_2CaCu_2O_{8+x}$ (Bi2212) and $YBa_2Cu_3O_{6+x}$ (YBCO), the phase coherence is established due to spin excitations [29, 30] which cause the appearance of the so-called magnetic resonance peak in inelastic neutron scattering spectra [31]. We shall consider the mechanism for establishing phase coherence in cuprates in Chapter 9.

In 2001, tunneling measurements provided evidence that the quasiparticle peaks in tunneling spectra obtained in Bi2212 are caused by condensed solitonlike excitations which form the Cooper pairs [32–34]. We shall discuss these data in Chapters 4 and 6.

3. Superconducting materials

The number of superconducting materials is slowly approaching tens of thousands. They can be classified into several groups according to their properties:

- metals;
- binary alloys and compounds;
- intermetallic compounds (A15 materials);
- Chevrel phases;
- semiconductors;
- organic quasi-one-dimensional superconductors;
- heavy fermions;
- oxides;
- high-T_c superconductors (copper oxides);
- and others.

The book presents a description of the mechanism of high-T_c superconductivity; however, in Chapters 9, 10 and 11, we shall discuss superconducting properties of organic superconductors, heavy fermions and some other unconventional superconductors. The first representatives of organic superconductors and heavy fermions were discovered in 1979.

The history of high-T_c superconductivity began in 1986 when Bednorz and Müller found evidence for superconductivity at $\sim$ 30 K in La-Ba-Cu-O ceramics. The maximum critical temperature of this family of cuprates is about 38 K. In 1987, groups at the Universities of Alabama and Houston under the direction of M. K. Wu and P. W. Chu, respectively, jointly announced the discovery of a 93 K superconductor YBCO. Just a year later, early in 1988, Bi- and Tl-based superconducting cuprates were discovered with T_c = 110 and 125 K, respectively. Finally, Hg-based cuprates with the highest T_c of 135 K were discovered in 1993 (at high pressure T_c increases up to 164 K). Figure 1.1 shows the superconducting critical temperature of different cuprates as a function of the year of discovery, as well as T_c of conventional superconductors.

All these cuprates are hole-doped. One family of cuprates which was discovered in 1989 is electron-doped: (Nd, Pr, Sm)CeCuO. Their maximum critical temperature is comparatively low, $T_c = 24$ K.

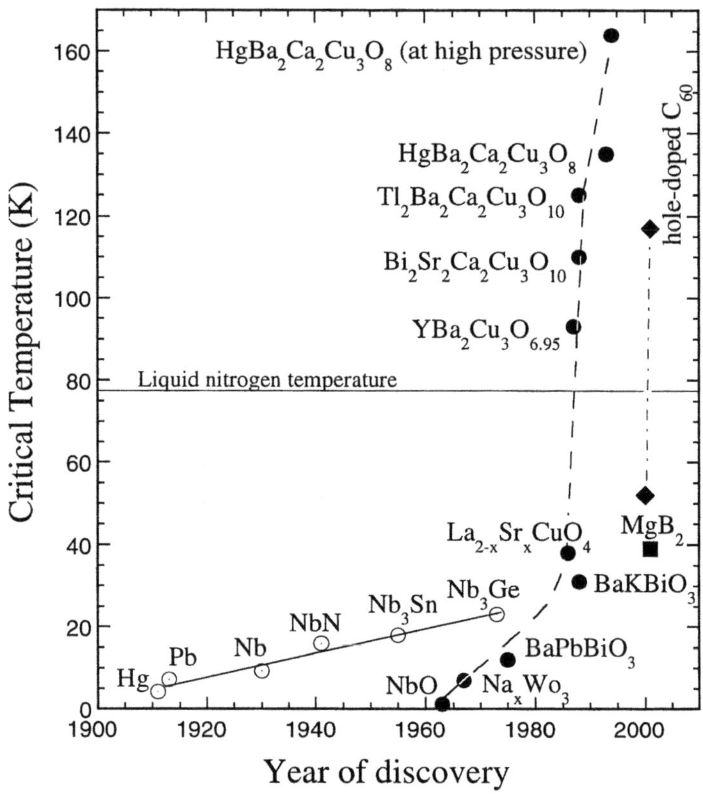

Figure 1.1. The time evolution of the superconducting critical temperature since the discovery of superconductivity in 1911. The solid line shows the evolution of conventional superconductors, and the dashed line marks the evolution of superconducting oxides. Quasi-one-dimensional organic superconductors and heavy fermions, having $T_{c,max} \leq 12$ K, are not shown.

All cuprates have a layered crystal structure, and they are very anisotropic. Copper oxide layers in these compounds are the crucial element for the occurrence of superconductivity. We shall discuss superconducting and normal-state properties of cuprates in Chapter 3.

Alternatively, all superconducting compounds can be classified into two groups: conventional and unconventional. Superconductivity in conventional superconductors is described by the equations of the BCS theory. Superconducting properties of unconventional superconductors are definitely different from those of conventional superconductors. We shall see that the mechanism of superconductivity in unconventional superconductors seems to be similar.

Moreover, some superconductors which were considered earlier as conventional are in fact unconventional.

Chapter 2

THE BCS MODEL OF SUPERCONDUCTIVITY IN METALS

What is now proved was once only imagined.

—William Blake (1751 - 1827)

This Chapter does not set out to cover all aspects of the BCS model of superconductivity in metals, but concentrates on a general description of the BCS mechanism and basic characteristics of conventional superconductors. There are many excellent books devoted to the BCS mechanism of superconductivity, and the reader who is interested to know all calculations in the framework of the BCS model is referred to the books (see Appendix).

Superconductivity was discovered by Kamerlingh Onnes in 1911: on measuring the electrical resistance of mercury at low temperatures, he found that, at 4.2 K, it dropped abruptly to zero. Subsequent investigations have shown that this sudden transition to perfect conductivity is characteristic of a number of metals and alloys. However, some metals never become superconducting. Bardeen, Cooper and Schrieffer [1] reported in 1957 the first successful microscopic theory of superconductivity (the BCS theory). Despite the existence of the BCS theory, there are no completely reliable rules for predicting whether a metal will be a superconductor at low temperature.

In *conventional* superconductors below the critical temperature T_c, the superconducting state is characterized by two distinctive properties: perfect electrical conductivity ($\rho = 0$) and perfect diamagnetism ($B = 0$), as shown in Figs 2.1 and 2.2, respectively. By measuring the damping of superconducting current, it was shown that the current lifetime in a superconducting ring is about 10^5 years. What is the mechanism which leads to the appearance of the superconducting state in metals?

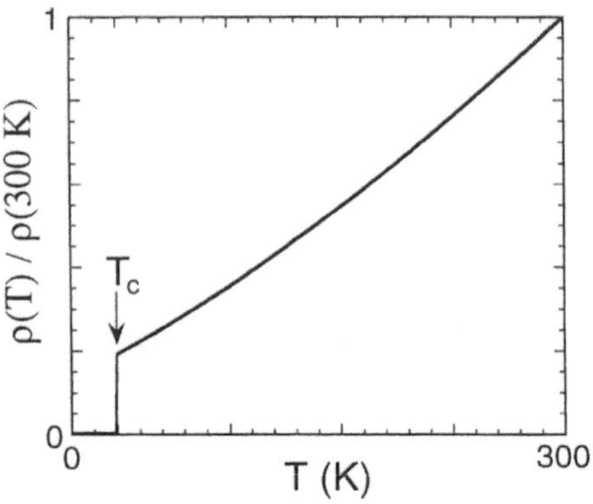

Figure 2.1. Temperature dependence of electrical resistivity of a superconductor. T_c marks the transition to the superconducting state.

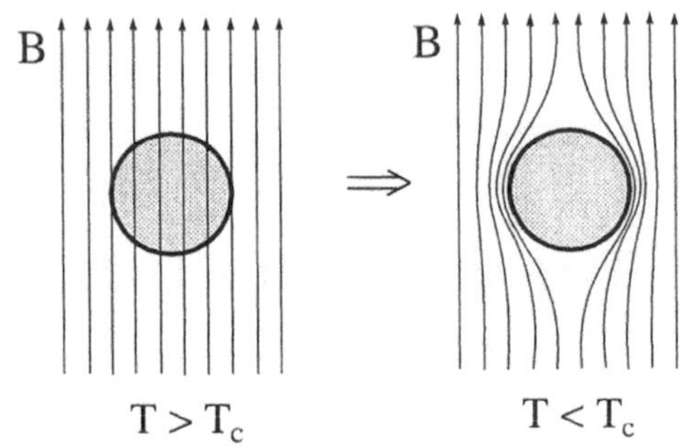

Figure 2.2. The Meissner effect: the expulsion of a weak, external magnetic field from the interior of a superconductor.

1. The BCS mechanism

In 1957, Bardeen, Cooper and Schrieffer showed how to construct a wave function in which the electrons are paired. The wave function, which is adjusted to minimize the free energy, is further used as the basis for a complete microscopic theory of superconductivity in metals. Thus, they showed that the superconducting state is a peculiar correlated state of matter—a quantum state on the macroscopic scale, in which all the electron pairs move in a single coherent motion. The success of the BCS theory and its subsequent elaborations are manifold. One of its key features is the prediction of an energy gap.

In Landau's concept of the Fermi liquid, excitations called *quasiparticles* are bare electrons dressed by the medium in which they move. Quasiparticles can be created out of the superconducting ground state by breaking up pairs, but only at the expense of a minimum energy of Δ per excitation. This minimum energy, Δ, is called the energy gap. The BCS theory predicts that, at T = 0, Δ is related to the critical temperature by $\Delta = 1.76 k_B T_c$ for any superconductor (k_B is the Boltzmann constant). This turns out to be nearly true, and where deviations occur they can be understood in terms of modifications of the BCS theory. The manifestation of the energy gap in tunneling provided strong conformation of the theory.

The key to the basic interaction between electrons which gives rise to superconductivity was provided by the *isotope effect*. The interaction between electrons and crystal lattice is one of the basic mechanisms of electrical resistance in an ordinary metal. It turns out that it is precisely the electron-lattice interaction that, under certain conditions, leads to an absence of resistance, i.e. to superconductivity. This is why, in excellent conductors such as copper, silver and gold, a rather weak electron-lattice interaction does not lead to superconductivity; however, it is completely responsible for their nonvanishing resistance near absolute zero.

The interaction mediated by the background crystal lattice can crudely be pictured as follows. An electron tends to create a slight distortion of the elastic lattice as it moves, because of the Coulomb attraction between the negatively charged electron and the positively charged lattice. If the distortion persists for a brief time, a second passing electron will feel the distortion and be affected by it. Under certain circumstances, this can give rise to a weak indirect attractive interaction between the two electrons which may more than compensate their Coulomb repulsion. Cooper first showed in 1956 that two electrons with an attractive interaction can bind together (in the momentum space, **not** in a real space) to form a bound pair, if they are in the presence of a high-density fluid of other electrons, no matter how weak the interaction is. This bound state of two electrons is today known as a Cooper pair. The two electrons in a Cooper pair have opposite momenta and spins.

1.1 Electron-electron attraction

The interaction of two electrons via phonons can be visualized as the emission of a "virtual" phonon by one electron, and its absorption by the other, as shown in Fig. 2.3. An electron in the state $\vec{k}_1$ emits a phonon, and is scattered into state $\vec{k}_1 - \vec{q}$. The electron in state $\vec{k}_2$ absorbs this phonon, and is scattered into $\vec{k}_2 + \vec{q}$. The diagram shown in Fig. 2.3 is the direct way of calculating the force which attracts the two electrons. However, it is necessary to know the spectrum of lattice vibrations in a solid because, in different solids, phonons propagate at different frequencies (energies).

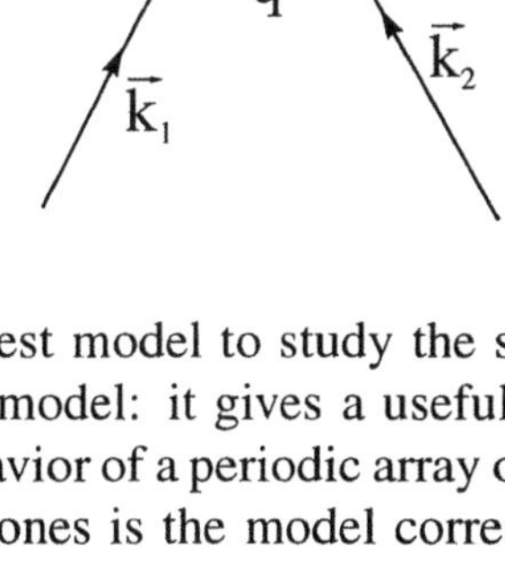

Figure 2.3 Diagram illustrating electron-electron interaction via exchange of a virtual phonon of momentum $\hbar\vec{q}$.

The simplest model to study the spectrum of lattice vibrations is the one-dimensional model: it gives a useful picture of the main features of the mechanical behavior of a periodic array of atoms. The simplest model among one-dimensional ones is the model corresponding to a monatomic crystal, which can be visualized as a linear chain of masses M with the same spacing, a, and connected to each other by massless springs. However, it is more practical to consider the one-dimensional model for a diatomic crystal in which a unit cell contains two different atoms. In this model, the linear chain consists of two different masses, M and m, placed alternatively. Figure 2.4 shows schematically the energy-momentum relation $E(k)$, obtained in the framework of the one-dimensional model for a diatomic crystal. The $E(k)$ relation is generally known as a *dispersion relation.* The momentum space in the range $\pm\pi/2a$, where $2a$ is the periodicity of the lattice, is known as the *Brillouin zone.* In Fig. 2.4, the higher-energy oscillations are conventionally called an *optical mode*, and the lower-energy oscillations an *acoustic mode.* The situation in three dimensions becomes more complicated and, in general, there are different dispersion relations for waves propagating in different directions in a crystal as a result of anisotropy of the force constants.

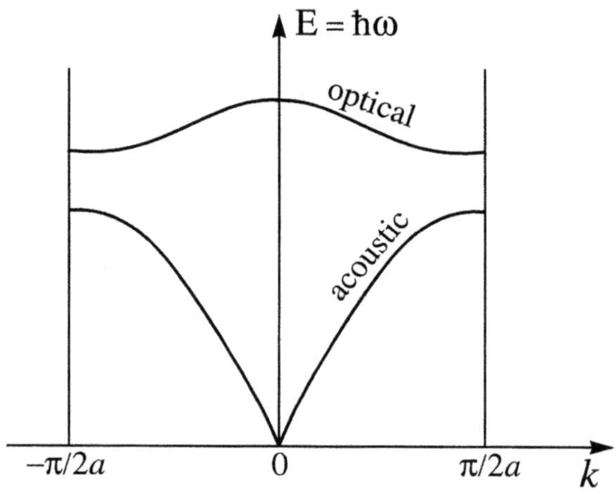

Figure 2.4. Vibration frequencies of diatomic chain.

In the BCS theory, the Debye spectrum of phonon frequencies is used to determine a critical temperature T_c. The Debye model assumes that the energies available are insufficient to excite the optical modes, so the BCS theory considers only low-energy (acoustic) phonons. In the Debye model, the Brillouin zone, which bounds the allowed values of $\vec{k}$, is replaced by a sphere of the same volume in $\vec{k}$-space. The Debye temperature Θ is defined by

$$k_B\Theta = \hbar w_D, \tag{2.1}$$

where w_D is the phonon frequency at the edge of the Debye sphere. So, $k_B\Theta$ (or $\hbar w_D$) is the energy of the highest-energy phonon in the Debye sphere.

Let us go back to our electrons shown in Fig. 2.3. To enable an electron to scatter from the state $\vec{k}_1$ to to the state $\vec{k}_1 + \vec{q}$, the latter must be free (due to the Pauli exclusion principle). This is possible only in the vicinity of the Fermi surface which is represented in momentum space by a sphere of radius $\vec{k}_F$, as shown in Fig. 2.5. Now we are ready to formulate the law of phonon-mediated interaction between electrons which forms the foundation of the BCS theory: *Electrons with energies that differ from the Fermi energy by no more than $\hbar w_D$ are attracted to each other.* Thus, in the BCS model, only those electrons that occupy the states within a narrow spherical layer near the Fermi surface experience mutual attraction. The thickness of the layer $2\Delta k$ is determined by the Debye energy:

$$\frac{\Delta k}{k_F} \sim \frac{\hbar w_D}{E_F}, \quad \text{where} \quad E_F = \frac{\hbar^2 k_F^2}{2m}, \tag{2.2}$$

and m is the electron mass. In the framework of the BCS model, two electrons in a Cooper pair have opposite momenta, $\vec{k}_1 + \vec{k}_2 = 0$, and spins, $\vec{S}_1 + \vec{S}_2 = 0$.

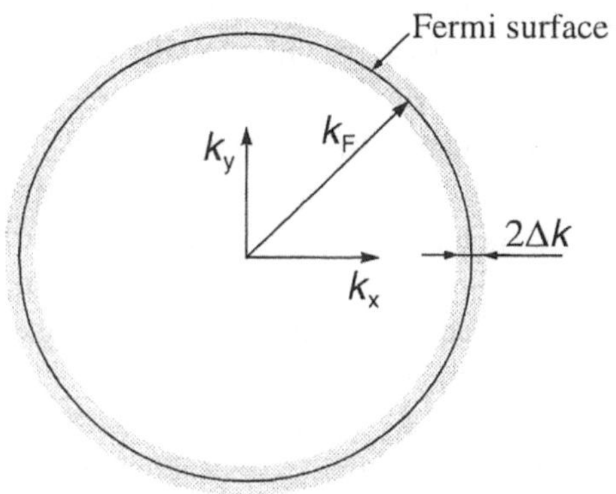

Figure 2.5. In the BCS picture, only the electrons within the $2\Delta k$ layer near the Fermi surface interact via phonons.

1.2 Critical temperature

In the framework of the BCS theory, the transition temperature T_c depends upon the product $\lambda = N(0)V$ of the single spin density of states at the Fermi surface $N(0)$ with the pairing potential V, and a cutoff frequency ω_c of the order of the Debye frequency. When λ is small (in practice less than 1/3) the BCS theory predicts that

$$k_B T_c = 1.14\omega_c \exp(-1/\lambda). \tag{2.3}$$

In this weak-coupling limit, the gap at zero temperature is given by

$$\Delta(0) = 2\omega_c \exp(-1/\lambda), \tag{2.4}$$

and, as a consequence,

$$2\Delta(0)/k_B T_c = 3.52. \tag{2.5}$$

If the Coulomb interaction is included, in weak-coupling limit $\lambda \ll 1$, the transition temperature changes to

$$T_c = \omega_0 \exp(-\frac{1}{\lambda - \mu^*}), \tag{2.6}$$

where μ^* is $N(0)$ times the Coulomb pseudo-potential, and ω_0 is the typical phonon frequency for a given metal. McMillan [2] extended this result to the case of strong-coupling superconductors, and obtained

$$T_c = \frac{\Theta}{1.45} \exp\left[-\frac{1.04(1+\lambda)}{\lambda - \mu^*(1+0.62\lambda)}\right]. \tag{2.7}$$

Here the Debye temperature Θ is used as the typical phonon frequency. When the Coulomb interaction is taken into account, the M dependence of the cutoff frequency appearing in μ^* modifies the isotope effect, and thus explains the deviation of the isotope effect from the ideal value of 1/2 (see Section 1.4).

In BCS-McMillan's expression for T_c, the Debye temperature Θ occurs not only in the pre-exponential factor in the expression $T_c \propto \Theta \exp(-1/\lambda)$, but also in the electron-phonon coupling constant λ which can be presented as $\lambda \approx C/M\langle\omega^2\rangle$, where C is a constant for a given class of materials, M is the mass, and $\langle\omega^2\rangle$ is the mean-square average phonon frequency, and $\langle\omega^2\rangle \propto \Theta$. Consequently, in BCS-type superconductors, T_c increases as Θ decreases. In other words, T_c increases with lattice softening.

In high-T_c superconductors, T_c increases with lattice stiffening, thus, contrary to conventional superconductors. This fact points out that *the superconducting state of high-T_c superconductors cannot be described by the BCS theory for conventional superconductors.*

1.3 Strength of the electron-phonon interaction

The coupling constant λ can be determined experimentally. In a given material, the strength of the electron-phonon interaction depends on the function $\alpha^2(\omega)F(\omega)$, where $F(\omega)$ is the density of states of lattice vibrations (the phonon spectrum); $\alpha^2(\omega)$ describes the interaction between the electrons and the lattice, and ω is the phonon frequency. The spectral function $\alpha^2(\omega)F(\omega)$ is the parameter of the electron-phonon interaction in the Eliashberg equations, and if this function is known explicitly, one can calculate the coupling constant λ with the help of the following relation $\lambda = 2\int \alpha^2(\omega)F(\omega)\,\frac{d\omega}{\omega}$.

The phonon spectrum $F(\omega)$ can be determined by inelastic neutron scattering. The product $\alpha^2(\omega)F(\omega)$ can be obtained in tunneling measurements. A comparison of the data obtained by these two different techniques reveals in many superconducting materials a remarkable agreement of the spectral features. The function $\alpha(\omega)$ can be explicitly determined from these data, which is usually smooth relative to $F(\omega)$. As an example, Figure 2.6 shows the phonon spectrum $F(\omega)$ and the function $\alpha^2(\omega)F(\omega)$ for Nb, obtained by neutron and tunneling spectroscopies, respectively. In Fig. 2.6, one can see that there is good agreement between the two spectra. In different materials, the phonon spectrum often has two peaks, as shown in Fig. 2.6. These peaks originate from the longitudinal and transverse phonons.

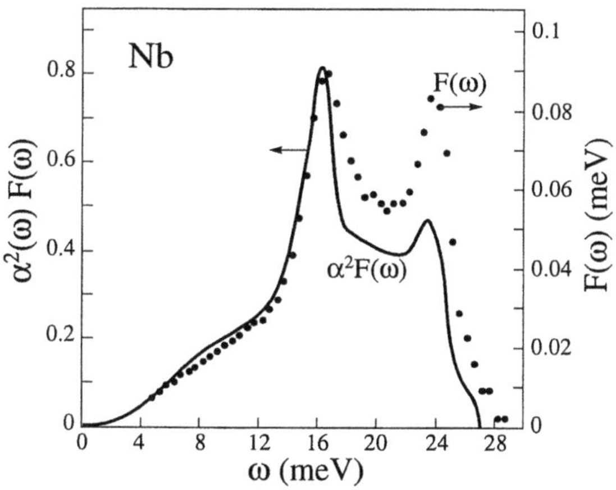

Figure 2.6. Tunneling and neutron spectroscopic data for Nb [3].

1.4 The isotope effect

The isotope effect provided a crucial key to the development of the BCS microscopic theory of superconductivity. It was found that, for a given element, T_c was proportional to $M^{-1/2}$, where M is the isotope mass. The vibrational frequency of a mass M on a spring is proportional to $M^{-1/2}$, and the same relation holds for the characteristic vibrational frequencies of the atoms in a crystal lattice. Thus, the existence of the isotope effect indicated that, although superconductivity is an electronic phenomenon, it nevertheless related in an important way to the vibrations of the crystal lattice in which the electrons move. Luckily, not until after the development of the BCS theory was it discovered that the situation is more complicated than it had appeared. For some conventional superconductors, the exponent of M is not -1/2, but near zero, as listed in Table 2.1 [3].

In high-T_c superconductors, the exponent is also near zero. As a consequence, this fact has prompted the exploration of non-phonon electronic coupling mechanisms of high-T_c superconductivity.

Table 2.1. Isotope effect ($T_c \propto M^{-\alpha}$)

Element	α
Mg	0.5
Sn	0.46
Re	0.4
Mo	0.33
Os	0.21
Ru	0 ($\pm$0.05)
Zr	0 ($\pm$0.05)

1.5 Energy gap

The energy gap in a superconductor is quite different in its origin from that in a semiconductor. From the band theory, energy bands are a consequence of the static lattice structure. In a superconductor, the energy gap is far smaller, and results from an attractive force between electrons in the lattice which plays only an indirect role. In conventional superconductors, the energy gap arises only at temperatures below T_c, and varies with temperature. The gap occurs on either side of the Fermi level, as shown in Fig. 2.7. Thus, if, in a semi-

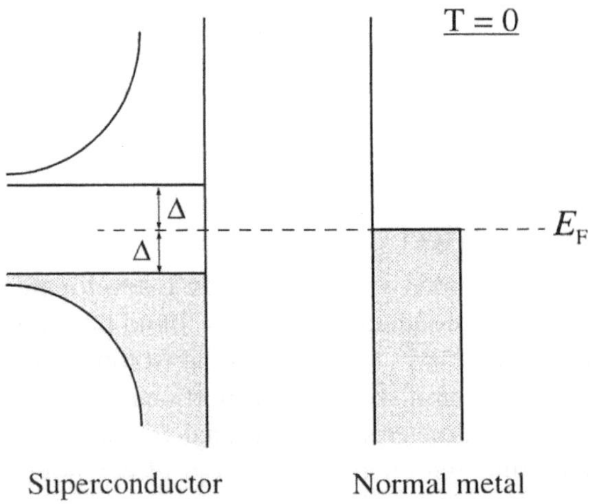

Figure 2.7. The density of states near the Fermi level E_F in a superconductor, showing the energy gap 2Δ at $T = 0$, and in a normal metal. All the states above the gap are assumed empty and those below, full.

conductor, the energy gap is tied to the Brillouin zone, in a superconductor, the energy gap is carried by the Fermi surface. At $T = 0$ all electrons are accommodated in states below the energy gap, and a minimum energy 2Δ must be supplied to produce an excitation across the gap. The BCS theory predicts that $2\Delta(T = 0) = 3.52k_BT_c$. In real superconductors, the ratio $\frac{2\Delta}{k_BT_c}$ varies between 3.5 and 4.9. Figure 2.8 shows the temperature dependence of the magnitude of energy gap, predicted by the BCS theory.

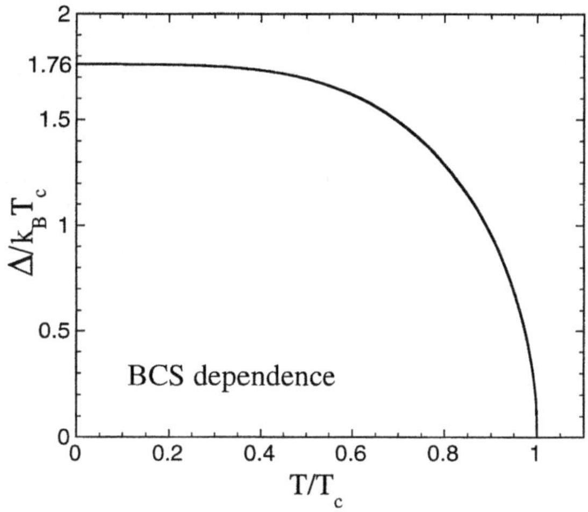

Figure 2.8. The BCS temperature dependence of the energy gap $\Delta(T)$.

1.6 Coherence length

The "distance" between the two electrons of a Cooper pair is called the *coherence length*, ξ. In the framework of the BCS theory, the coherence length and the energy gap relate to each other at $T = 0$ as

$$\xi_0 = \frac{\hbar v_F}{\pi \Delta(0)}, \tag{2.8}$$

where v_F is the Fermi velocity (on the Fermi surface). The quantity ξ_0 is the intrinsic coherence length which is temperature-independent. In the framework of the Ginzburg-Landau theory, the relation between the temperature-dependent coherence length and the intrinsic coherence length is given by $1/\xi = 1/\xi_0 + 1/\ell$, where ℓ is the mean free path of an electron. Since at low temperature ℓ can be centimeters long, then $\xi \approx \xi_0$. The coherence length is large in metal superconductors: ξ = 16000 Å in Al, and ξ = 380 Å in Nb. In spite of the fact that two electrons in a Cooper pair are far apart from each other, the other Cooper pairs are only a few nanometers away (the period of a crystal lattice is less than a nanometer, a few Å). In most unconventional superconductors, the "size" of the Cooper pairs is very small: in cuprates, for example, ξ is only a few periods of the crystal lattice, $\xi_{ab} \approx 15$ Å and $\xi_c \approx$ 1–3 Å.

1.7 Penetration depth

The way in which a superconductor expels from its interior an applied magnetic field with the small magnitude (the Meissner effect) is by establishing a persistent supercurrent on its surface which exactly cancels the applied field inside the superconductor. This surface current flows in a very thin layer of thickness λ, which is called the *penetration depth* (not to be confused with the electron-phonon coupling constant $\lambda = N(0)V$). The external field also actually penetrates the superconductor within λ. The magnitude of the penetration depth depends on the material and temperature, and decreases exponentially towards the core of a superconductor, as shown in Fig. 2.9. In the clean limit (the mean free path ℓ is much greater than ξ), the magnitude of the penetration depth is directly related to the superfluid density n_s as

$$\lambda^2(T) = \frac{m^* c^2}{4\pi e^2 n_s(T)}, \tag{2.9}$$

where m^* is the effective mass of the charge carries; e is the electron charge, and c is the speed of light in vacuum.

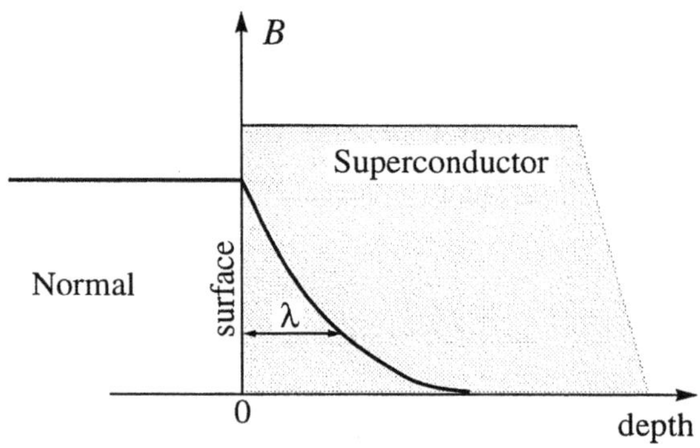

Figure 2.9. The penetration of the magnetic field into the superconducting sample.

1.8 Symmetry of the order parameter

In metals, below T_c, the layer of 2Δ literally covers all Fermi surface (above and below), so there are no nodes. In order to produce an excitation across the gap in any direction $\vec{k}$ (or $\vec{r}$), the same minimum energy of 2Δ must be supplied. As shown by Gor'kov, the amplitude of the wave function $\Psi(\vec{r})$ describing the superconducting state, also called the order parameter, is proportional to $\Delta(\vec{r})$. Thus, in conventional superconductors, the symmetry of the order parameter is isotropic, $\Delta(\vec{r}) = \Delta_0$, or an s-wave (in some cases, a slightly anisotropic s-wave). It is experimentally established that in unconventional superconductors such as organic superconductors, heavy fermions, high-T_c superconductors etc., the energy gap is highly anisotropic and, even, vanishes in some directions on the Fermi surface—at nodes (points, lines etc.). Hence, the symmetry of the order parameter in these unconventional superconductors is not an s-wave but rather a d-wave (or a p-wave). We shall discuss this issue in the following Chapter.

2. Characteristics of the superconducting state

Here we consider basic characteristics of the superconducting state.

2.1 Type-I and type-II superconductors

In 1937, Shubnikov found an unusual behavior of some superconductors in external magnetic fields. Actually, he discovered the new state of superconductors known as the *mixed state*. Twenty years later, by using the Ginzburg-Landau theory, Abrikosov theoretically found vortices in 1957 and thus explained Shubnikov's experiments, suggesting that Shubnikov's phase is a state with vortices that actually form a periodic lattice. The Abrikosov results showed that there are two types of superconducting materials: type I and type II. While the former expel magnetic flux completely from their interior, the latter do it completely only at small magnetic field magnitudes, but partially in higher external fields.

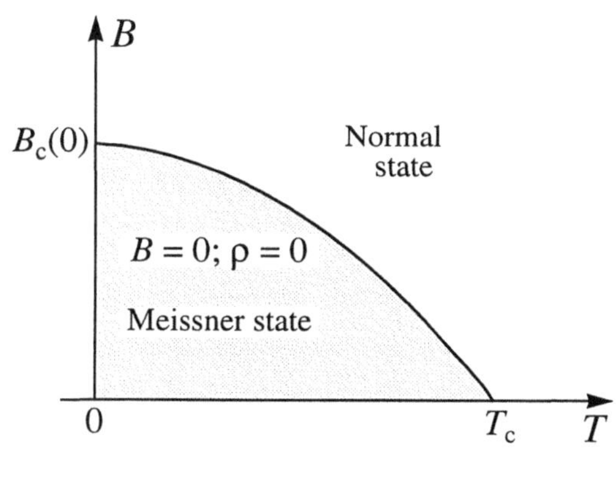

Figure 2.10. $B_c(T)$ dependence for a type-I superconductor, shown schematically.

With the exception of Nb and V, all superconducting elements and most of their alloys are type-I superconductors. As schematically shown in Fig. 2.10, the variation of the critical field B_c with temperature for a type-I superconductor is approximately parabolic:

$$B_c(T) = B_c(0)[1 - (T/T_c)^2], \tag{2.10}$$

where $B_c(0)$ is the value of the critical field at absolute zero. For a type-II superconductor, there are two critical fields, the lower critical fields B_{c1} and the upper critical fields B_{c2}, as shown in Fig. 2.11. In applied fields less than B_{c1}, the superconductor completely expels the field, just as a type-I superconductor does below B_c. At fields just above B_{c1}, however, flux begins to penetrate the superconductor in microscopic filaments called *vortices* which form a regular (hexagonal) lattice. Each vortex consists of a normal core in which the magnetic field is large, surrounded by a superconducting region, and can be approximated by a long cylinder with its axis parallel to the external magnetic field. Inside the cylinder, the superconducting order parameter Ψ is zero. The radius of the cylinder is of the order of the coherence length ξ.

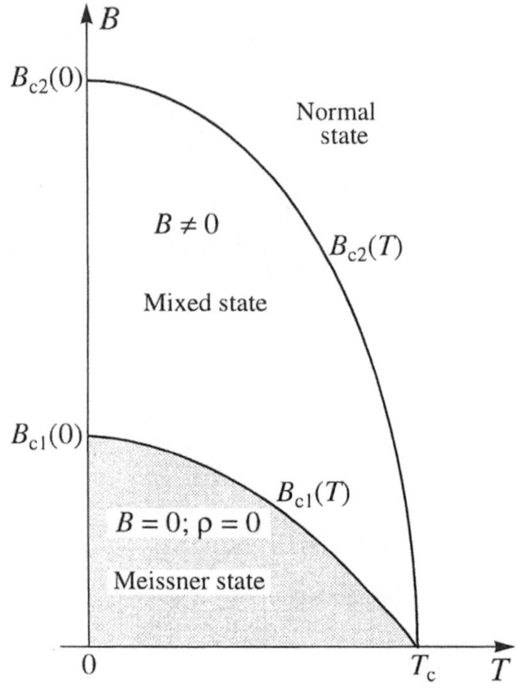

Figure 2.11. $B_{c1}(T)$ and $B_{c2}(T)$ dependences for a type-II superconductor, shown schematically.

The supercurrent circulates around the vortex within an area of radius $\sim \lambda$, the penetration depth. The spatial variations of the magnetic field and the order parameter inside and outside an isolated vortex are illustrated in Fig. 2.12. The vortex state of a superconductor, discovered experimentally by Shubnikov and theoretically by Abrikosov, is known as the mixed state. It exists for applied fields between B_{c1} and B_{c2}. At B_{c2}, the superconductor becomes normal, and the field penetrates completely. Depending on the geometry of a supercon-

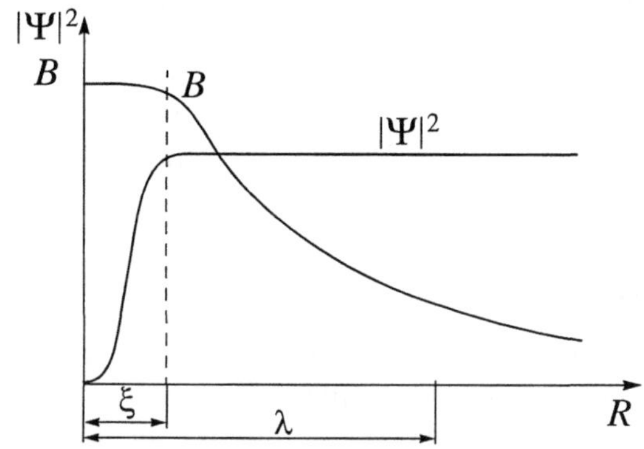

Figure 2.12. The spatial variations of the magnetic field and the order parameter inside and outside an isolated vortex in an infinite superconductor. R is the distance from the center of the vortex, and ξ and λ are the coherence length and the penetration depth of the superconductor, respectively (in type-II superconductors, $\xi < \lambda$).

ducting sample and the direction of an applied field, the surface sheath of the superconductor may persist to even higher critical field B_{c3}, which is approximately $1.7B_{c2}$. In type-I superconductors, the coherence length is larger than the penetration depth, $\xi > \lambda$, and in type-II superconductors, it is vice versa, $\xi < \lambda$.

The Ginzburg-Landau theory predicts that

$$B_c(T)\lambda(T)\xi(T) = constant. \tag{2.11}$$

(For type-II superconductors, $B_c = \sqrt{B_{c1}B_{c2}}$). This important relation, verified experimentally in conventional superconductors, is often used to obtain the value of the coherence length.

The magnitude of the critical magnetic fields of most conventional type-II superconductors is very small, less than 1 Tesla; however, in $PbMo_6S_8$ Chevrel phase, B_{c2} = 60 Tesla. All high-T_c superconductors are type-II superconductors, and they have very high values of the critical magnetic field. For example, in three- and four-layer compounds, the critical magnetic field parallel to the c-axis can reach the value of $B_{c2}(0) \sim 250$ Tesla, and with the field parallel to the ab-planes $B_{c2} \sim 650$ Tesla.

2.2 Critical current

Characterized by a critical temperature and a critical magnetic field, a superconductor also has a critical *dc* current density J_c (current divided by the cross-sectional area through which it flows). Above this limit, the *dc* current destroys the superconducting state. At $T = 0$, the critical current density can be estimated by using the electron velocity on the Fermi surface, $v_F = \pi\Delta\xi_0/\hbar$, the well-known expression for n_s, and the critical (maximum) velocity of a Cooper pair, $v_c \simeq \Delta/mv_F$, as

$$J_c = n_s e v_c \simeq n_s e \frac{\Delta}{mv_F} = \frac{\hbar c^2}{16\pi e} \frac{1}{\lambda^2 \xi_0}, \tag{2.12}$$

where m is the electron mass.

2.3 Phase stiffness

The superconducting state requires the electron pairing *and* the onset of long-range phase coherence. The superfluid density in conventional superconductors is relatively high. As a consequence, phase fluctuations in metal superconductors are practically absent. In other words, the phase "stiffness" is extraordinary high. Let us determine T_{PO} as the temperature at which the phase order would disappear if the disordering effects of all degrees of freedom were ignored [4]. If, $T_c \ll T_{PO}$, phase fluctuations are relatively unimportant, and T_c will be close to the mean-field transition temperature T_{MF}, predicted by BCS theory. On the other hand, if $T_c \approx T_{PO}$, then the value of T_c is determined primarily by phase ordering, and T_{MF} is simply the characteristic temperature below which pairing becomes significant locally. The ratio T_{PO}/T_c in conventional superconductors lies between 2×10^2 and 2×10^5 [4]. So, in low-T_c superconductors, the pairing and the onset of the long-range phase coherence occur practically simultaneously, and phase fluctuations are indeed unimportant. However, phase fluctuations play an important role in unconventional superconductors. The ratio T_{PO}/T_c in superconductors with low superfluid density and small coherence length, such as organic and high-T_c superconductors, is small and lies between 0.7 and 16 [4]. Thus, the pairing may occur well above T_c which is controlled by the onset of long-range phase order.

It is important emphasizing that, in conventional superconductors and, as we shall see latter, in unconventional superconductors as well, the pairing mechanism and the mechanism for the establishment of long-range phase coherence are different. We shall discuss this issue in Chapters 3, 10 and 11.

2.4 Josephson effects

In 1962, Josephson calculated the currents that could be expected to flow during tunneling of Cooper pairs through a thin insulating barrier (on the order of a few nanometers thick), and found that a current of paired electrons (supercurrent) would flow in addition to the usual current that results from the tunneling of single electrons (single or unpaired electrons are present in a superconductor along with bound pairs). Tunneling between two superconductors and between a superconductor and a normal metal is considered in detail below. The Josephson effect which is a manifestation of quantum mechanics on a macroscopic scale was experimentally confirmed in 1963. The zero-voltage current flow resulting from the tunneling of Cooper pairs is known as the *dc Josephson effect.* Josephson also predicted that if a constant nonzero voltage V is maintained across the tunnel barrier, an alternating supercurrent will flow through the barrier in addition to the *dc* current produced by the tunneling of single electrons. The frequency ν of the *ac* supercurrent is given by

$$\nu = \frac{2eV}{h}, \tag{2.13}$$

where e is the electron charge, and h is the Planck constant. The oscillating current of Cooper pairs that flows when a steady voltage is maintained across a tunnel barrier is known as the *ac Josephson effect.* The Josephson effects play a special role in superconducting applications.

2.5 Effect of impurities

How do magnetic and non-magnetic impurities affect the critical temperature in conventional superconductors? It appears that, in superconductors described by the BCS theory, non-magnetic impurities play an unimportant role in the behavior of T_c, whereas magnetic impurities drastically suppress the superconducting transition temperature. This effect is often used to determine the nature of electron-electron attraction in new superconducting materials. For example, the effect of impurities on the superconducting transition temperature in superconductors with the magnetic electron-electron attraction is anticipated to be opposite to that in conventional superconductors.

2.6 High-frequency residual losses

The vanishing of the *dc* resistance is the most striking feature of the superconducting state. However, the losses in a superconductor are non-zero for *ac* current which flows in a thin surface layer having the thickness of λ. The surface resistance depends on the frequency ω of an *ac* current (or an electromagnetic field), as well as on the temperature and the energy gap of the superconductor. At microwave frequencies and at temperatures less than half the transition temperature, the surface resistance has the following approximate BCS form:

$$R_s(T, \omega) = \frac{C\omega^2}{T} e^{-\Delta/k_B T} + R_0(\omega), \tag{2.14}$$

where C is a constant that depends on the penetration depth in the material, and R_0 is the residual resistance. Although it is not entirely clear where the residual losses originate, it has been determined experimentally that trapped flux and impurities are principal causes. An additional source of the residual losses can be imperfection of the surface of the superconductor, which may not be perfectly flat or contains nonsuperconducting regions (for example, oxides etc.). Nevertheless, in conventional superconductors, the residual losses which are determined by R_0 are small in comparison, for example, with those in high-T_c superconductors.

2.7 Acoustic properties

Let us consider the case in which pulsed ultrasound travels in a sample and is reflected at the surface. If the sample of length x has two parallel surfaces on both sides, a train of echo pulses will be observed with a time interval of $2\Delta t$. The height of successive echo pulses decreases exponentially with time t: $I = I_0 \exp(-\beta t)$. From this information, one can obtain important physical quantities such as the sound velocity $v = x/\Delta t$, the attenuation coefficient $\alpha = \beta/v$ and the elastic coefficient $c = \rho v^2$, where ρ is the mass density. These coefficients provide information not only about lattice properties but also about the electron-phonon interaction and electronic properties. Transverse ultrasound measurements are even able to probe the presence of electronic gap nodes in a superconductor.

Ultrasound waves are scattered in a superconductor by normal electrons, not by Cooper pairs, so that their attenuation is a measure of the fraction of normal electrons. As a consequence, superconductors absorb sound waves more weakly than normal metals. The ratio of the attenuation measured in conventional superconductors indeed follows the prediction of the BCS theory, confirming the validity of the principal ideas of the theory.

In conventional superconductors, when the sample is cooled to the superconducting state, the elastic coefficients *always* decrease. However, the decrease is small, less than 0.01%. The difference between the sound velocity in the superconducting state and in the normal state is also small, usually on the order of 10^{-6} (in lead, $\Delta v/v \approx 10^{-4}$). This means that the crystal lattice is not altered significantly by the superconducting transition. This is because in conventional superconductors $\Delta(0) \ll E_F$, where E_F is the Fermi energy (see Fig. 2.7). However, the situation is different in high-T_c superconductors: acoustic properties of cuprates will be discussed in Chapters 3, 6 and 8.

2.8 Thermal properties

The appearance of the superconducting state is accompanied by quite drastic changes in both the thermodynamic equilibrium and thermal properties of a superconductor. The normal–superconducting transition is a second-order phase transition accompanied by a jump in the heat capacity. On cooling, the heat capacity of a superconductor has a discontinuous jump at T_c and then the exponential falls to zero. The thermal conductivity of a pure superconductor is less in the superconducting state than in the normal state, and at very low temperatures approaches zero.

are determined by R_0 are small in comparison, for example, with those in high-T_c superconductors.

2.7 Acoustic properties

Let us consider the case in which pulsed ultrasound travels in a sample and is reflected at the surface. If the sample of length x has two parallel surfaces on both sides, a train of echo pulses will be observed with a time interval of $2\Delta t$. The height of successive echo pulses decreases exponentially with time t: $I = I_0 \exp(-\beta t)$. From this information, one can obtain important physical quantities such as the sound velocity $v = x/\Delta t$, the attenuation coefficient $\alpha = \beta/v$ and the elastic coefficient $c = \rho v^2$, where ρ is the mass density. These coefficients provide information not only about lattice properties but also about the electron-phonon interaction and electronic properties. Transverse ultrasound measurements are even able to probe the presence of electronic gap nodes in a superconductor.

Ultrasound waves are scattered in a superconductor by normal electrons, not by Cooper pairs, so that their attenuation is a measure of the fraction of normal electrons. As a consequence, superconductors absorb sound waves more weakly than normal metals. The ratio of the attenuation measured in conventional superconductors indeed follows the prediction of the BCS theory, confirming the validity of the principal ideas of the theory.

In conventional superconductors, when the sample is cooled to the superconducting state, the elastic coefficients *always* decrease. However, the decrease is small, less than 0.01%. The difference between the sound velocity in the superconducting state and in the normal state is also small, usually on the order of 10^{-6} (in lead, $\Delta v/v \approx 10^{-4}$). This means that the crystal lattice is not altered significantly by the superconducting transition. This is because in conventional superconductors $\Delta(0) \ll E_F$, where E_F is the Fermi energy (see Fig. 2.7). However, the situation is different in high-T_c superconductors: acoustic properties of cuprates will be discussed in Chapters 3, 6 and 8.

2.8 Thermal properties

The appearance of the superconducting state is accompanied by quite drastic changes in both the thermodynamic equilibrium and thermal properties of a superconductor. The normal–superconducting transition is a second-order phase transition accompanied by a jump in the heat capacity. On cooling, the heat capacity of a superconductor has a discontinuous jump at T_c and then the exponential falls to zero. The thermal conductivity of a pure superconductor is less in the superconducting state than in the normal state, and at very low temperatures approaches zero.

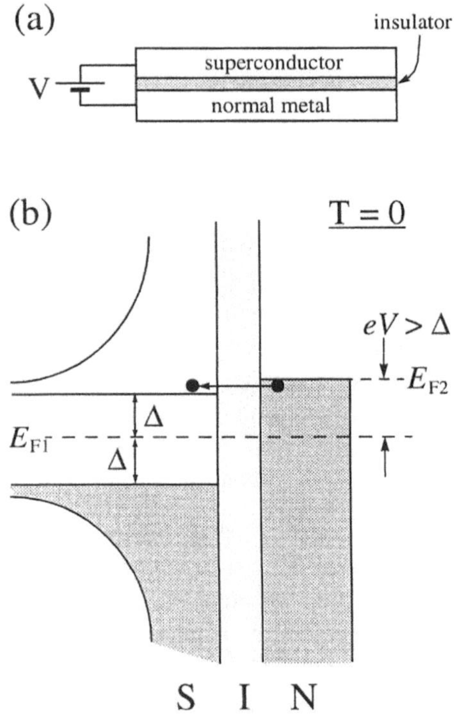

Figure 2.13. (a) SIN tunneling junction, and (b) corresponding energy diagram at $T = 0$ in the presence of an applied voltage: quasiparticles can tunnel when $|V| \geq \Delta/e$.

3.2 Density of states

In the framework of the BCS theory, the density of states of quasiparticle excitations in the superconducting state, $N_s(E)$, and the density of states in the normal state $N_n(E)$ relate to each other at $T = 0$ as

$$N_s/N_n = \begin{cases} \frac{E}{\sqrt{E^2-\Delta^2}} & \text{for } |E| \geq \Delta \\ 0 & \text{if } |E| < \Delta. \end{cases} \tag{2.15}$$

The density of states in the normal state N_n can be obtained by applying a magnetic field with magnitude $B > B_c$ (B_{c2}). In most low-T_c superconductors, N_n is normally constant over the energy range of interest.

A $dI(V)/dV$ tunneling characteristic measured in an SIN junction corresponds directly to the density of states of quasiparticle excitations in the superconductor. In first approximation, assuming that the normal metal has a constant density of states near the Fermi level and the transmission of the barrier (insulator) is independent of energy, the tunneling conductance $dI(V)/dV$

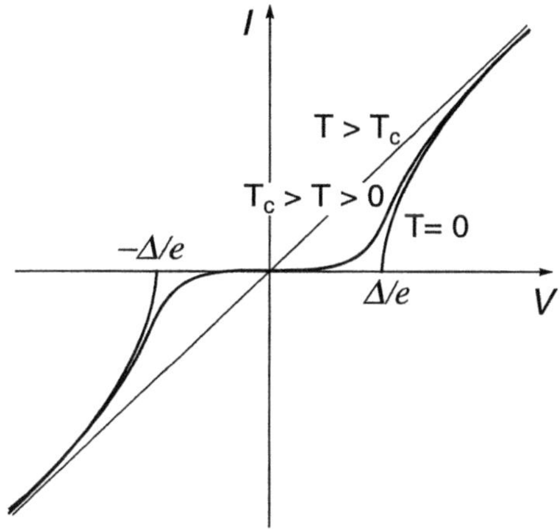

Figure 2.14. Tunneling $I(V)$ characteristics obtained in an SIN junction at different temperatures: $T = 0$; $0 < T < T_c$, and $T > T_c$ (the latter case corresponds to a NIN junction). At $0 < T < T_c$, quasiparticle excitations exist at any applied voltage.

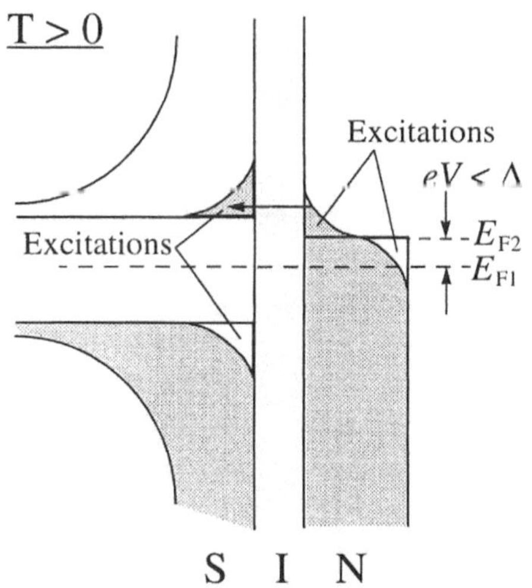

Figure 2.15. The density of states near the Fermi level E_F in a superconductor and a normal metal in an SIN junction at $0 < T < T_c$. Due to thermal excitations, there are states above the gap in the superconductor and above the Fermi level in the metal. Quasiparticles can tunnel at any applied voltage, as shown in Fig. 2.14.

is proportional to the density of states of the superconductor, broadened by the Fermi function $f(E,T) = [\exp(E/k_BT) + 1]^{-1}$. Thus, at low temperature

$$\frac{dI(V)}{dV} \propto \int_{-\infty}^{+\infty} N_s(E) \left[-\frac{\partial}{\partial(eV)} f(E+eV,T) \right] dE \cong N_s(eV), \qquad (2.16)$$

where e is the electron charge, and the origin of the energy scale E in the tunneling spectra corresponds to the Fermi level of the superconductor. Consequently, the differential conductance at negative (positive) voltage reflects the density of states below (above) the Fermi level E_F.

In order to smooth the gap-related structures in the density of states $N_s \propto E/\sqrt{E^2 - \Delta^2}$, a phenomenological smearing parameter Γ was introduced, which accounts for a lifetime broadening of quasiparticles ($\Gamma = \hbar/\tau$, where τ is the lifetime of quasiparticle excitations) [5]. The energy E in the density-of-state function is replaced by $E - i\Gamma$ as

$$N_s(E,\Gamma) \propto Re\left\{ \int \frac{E - i\Gamma}{\sqrt{(E - i\Gamma)^2 - (\Delta(\vec{k}))^2}} d\vec{k} \right\}, \qquad (2.17)$$

where we introduced $\Delta(\vec{k})$ which is the $\vec{k}$-dependent energy gap for the general case of an anisotropic gap. In the two-dimensional case, the integration is reduced to integrating over the in-plane angle $0 \leq \theta < 2\pi$.

In SIN tunneling junctions of conventional superconductors, there is good agreement between the theory and experiment. As an example, Figure 2.16 shows the correspondence between experimental data obtained in Nb and the theoretical curve.

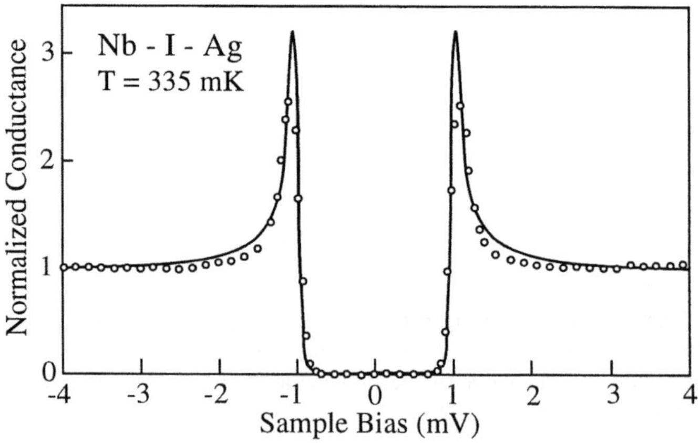

Figure 2.16. Tunneling conductance (circles) versus sample bias for a Nb/I/Ag junction, measured at 335 mK [6]. The solid line is the theoretical curve calculated from Eqs. (2.16) and (2.17).

3.3 SIS tunneling

In a superconductor-insulator-superconductor (SIS) junction, the tunneling conductance is proportional to the convolution of the density-of-states function of the superconductor with itself. In the case of symmetrical SIS contacts, the expression for the tunneling current through contacts at finite temperature is

$$I(V) = K \int_0^{\infty} N(E,\Gamma)N(E-eV,\Gamma)[f(E,T)-f(E-eV,T)]dE, \quad (2.18)$$

where K is the constant which contains the tunneling probabilities. The expression for tunneling conductance at finite temperature can be presented as

$$\frac{(dI/dV)_s}{(dI/dV)_n} = \frac{d}{d(eV)} \int_0^{eV} N(E,\Gamma)N(E-eV,\Gamma)[f(E,T)-f(E-eV,T)]dE, \quad (2.19)$$

where $(dI/dV)_n$ is the conductance in the normal state.

If, in SIN tunneling junctions of s-wave superconductors, there is good agreement between the theory and experiment, as shown in Fig. 2.16; in SIS junctions, the correspondence between the BCS density of states and experimental data is poor. Figure 2.17a shows the density of states which is expected to be observed in an SIS junction between two identical s-wave superconductors. As one can see in Fig. 2.17a, the calculated density of states has two specific features which are present for any $\Gamma \neq 0$: it is not zero at zero bias, and displays the presence of a "knee" at low bias. Figure 2.17b shows the conductance curve dI/dV measured in a Pb-insulator-Pb junction. In Fig. 2.17, one can see that the correspondence between the two curves is poor (see also Fig. 4(a) in [6]). This issue has not been discussed in the literature to date. We shall return to this question in Chapter 12.

3.4 The Josephson $I_c R_n$ product

The magnitude of the zero-voltage current resulting from the tunneling of Cooper pairs, known as the *dc* Josephson effect, depends on the phase difference between two superconductors as

$$I = I_c \sin(\varphi_2 - \varphi_1), \quad (2.20)$$

where $(\varphi_2 - \varphi_1)$ is the phase difference, and I_c is the critical Josephson current. The Ambegaokar-Baratoff theory [8] for BCS superconductors predicts that the product $I_c R_n$, where R_n is the normal-state zero-bias resistance of the

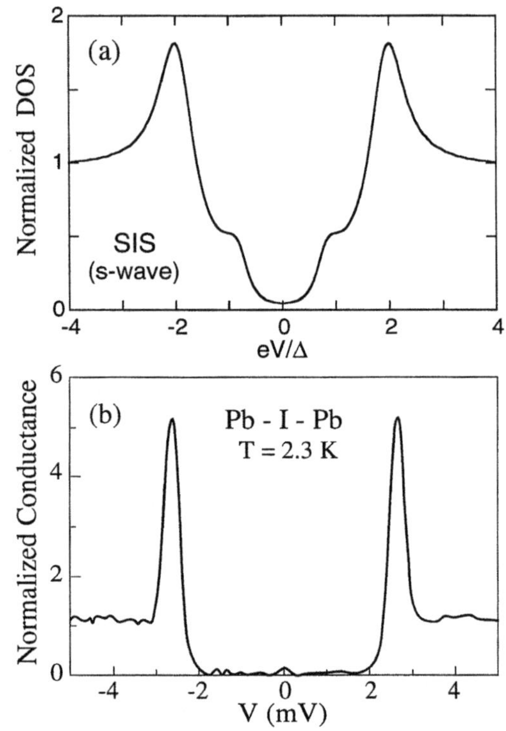

Figure 2.17. (a) Normalized density of states expected to be observed in an SIS junction, obtained from Eqs. (2.19) and (2.17) and by using $\Gamma = 0.2\Delta$. (b) Normalized dI/dV conductance measured in a Pb-insulator-Pb junction at 2.3 K (T_c = 6.8 K). The zero-bias peak due to the Josephson current is absent because of the large value of the normal resistance of the junction, $R_n = 100M\Omega$ [7].

junction, is constant, and at $T = 0$ proportional to the gap magnitude as

$$I_c R_n = \frac{\pi\Delta(0)}{2e}, \tag{2.21}$$

where e is the electron charge, and $\Delta(0)$ is the energy gap at $T = 0$. The product $I_c R_n$ is known as the *Josephson product.* The temperature dependence of the Josephson product is similar to the temperature dependence of the energy gap.

3.5 Andreev reflections

In an SIN junction, in tunneling regime, the normal resistance of the junction R_n is usually high ($\sim$ MΩ). Tunneling can still be observed in junctions with $R_n \sim 100\ \Omega$. We are interested in what happens in a junction with $R_n \sim 1\ \Omega$ to an electron in the normal metal moving towards the superconductor, when it encounters the NS interface. If the electron energy is less than the energy gap of the superconductor, the electron is reflected back from the interface. The propagation of a negative charge in the normal metal from the interface is equivalent to propagation of a positive charge in the superconductor in the opposite direction. Therefore, the process of the electron reflection gives rise to a charge transfer from the normal metal to the superconductor, i.e. to an electrical current.

This process was first proposed theoretically by Andreev [9] and is called now the *Andreev reflection.*

In a sense, the tunneling and the Andreev reflection are two opposite processes. In an SIN junction at $T = 0$, in the tunneling regime (high R_n), the current is absent at $|V| < \Delta/e$, whereas in Andreev-reflection regime (small R_n), the current at $|V| < \Delta/e$ is twice as large as the current at high bias.

It is important noting that tunneling spectroscopy is a phase-insensitive probe (*at least*, for a s-wave superconductor), whereas Andreev reflections are exclusively sensitive to coherence properties of the condensate.

3.6 Tunneling techniques

The tunneling junction, shown schematically in Fig. 2.13a, can be realized, in practice, in a few ways. A thin insulating layer between a superconductor and a normal metal or between two superconductors is the necessary feature of any tunneling junction. The insulator can be a thin insulating film, a natural oxide layer on the surface, a metal, vacuum, an He ambient etc. Basically, there are a few techniques which are usually used in tunneling measurements, such as point contacts (SIN junctions), STM (surface tunneling microscope) (SIN), overlapping bridges (SIN), thin-film microbridges (SIS), intrinsic interlayer tunneling junctions in high-T_c superconductors (SIS), and break-junctions (SIS).

In Andreev-reflection measurements, the necessary condition for a normal contact is to be within the Sharvin limit—that is, its size has to be smaller than the mean free path in the electrodes. Hence, point contacts sharpened to a point of radius usually less than 1 μm are used in Andreev-reflection measurements.

APPENDIX: Books recommended for further reading

- V. V. Schmidt, *The Physics of superconductors: Introduction to Fundamentals and Applications* (Springer, Berlin, 1997).
- *Superconductivity*, Vols. 1 and 2, edited by R. D. Parks (Marcel Deker, New York, 1969).
- J. R. Schrieffer, *Theory of Superconductivity* (Benjamin/Cummings, New York, 1983).
- T. Van Duzer and C. Turner, *Principles of Superconducting Devices and Circuits* (Elsevier, New York, 1985).

- M. Tinkham, *Introduction to Superconductivity* (McGraw-Hill, New York, 1975).
- V. Z. Kresin and S. A. Wolf, *Fundamentals of Superconductivity* (Plenum Press, New York, 1990).
- M. Cyrot and D. Pavuna, *Introduction to Superconductivity and High-T_c Materials* (World Scientific, Singapore, 1992).

Chapter 3

CUPRATES AND THEIR BASIC PROPERTIES

How much better is it to get understanding than gold.

—Proverb

A superconducting compound is said to belong to the family of high-T_c superconductors (cuprates) if it has CuO_2 planes. The parent compounds of superconducting cuprates are antiferromagnetic Mott insulators.

A Mott insulator is a material in which the conductivity vanishes as temperature tends to zero, even though band theory would predict it to be metallic. A Mott insulator is fundamentally different from a conventional (band) insulator. If, in a band insulator, conductivity is blocked by the Pauli exclusion principle, in a Mott insulator charge conduction is blocked by electron-electron repulsion. Quantum charge fluctuations in a Mott insulator generate the so-called *superexchange interaction*, which favors antiparallel alignment of neighboring spins. Thus, a Mott insulator has a charge gap of $\sim$ 2 eV, whereas the spin wave spectrum extends to zero energy. When cuprates are slightly doped by holes or electrons (the hole/electron concentration is changed from one per cell), on cooling they become superconducting. The cuprates are the only Mott insulators known to become superconducting.

The first high-T_c superconductor was discovered in 1986 by Bednorz and Müller at IBM Zurich Research Laboratory [1]. Without any doubts, the discovery was revolutionary because it showed that, contrary to general belief, superconductivity can exist above 30 K, and it can occur in very bad conductors—in ceramics. Fortunately, Bednorz and Müller didn't completely rely on the general belief.

Before we discuss the mechanism of high-T_c superconductivity, it is necessary to consider the structure and basic properties of cuprates. There are many cuprates which become superconducting at low temperature. They can be classified in a few groups according to their chemical formulas which are sufficiently complicated, therefore it is useful to use abbreviations. The abbreviations which will be used further are summarized in Table 3.1.

Table 3.1. Abbreviations for different cuprates

Cuprate	*CuO_2 planes*	*T_c (K)*	*abbreviation*
$La_{2-x}Sr_xCuO_4$	1	38	LSCO
$Nd_{2-x}Ce_xCuO_4$	1	24	NCCO
$YBa_2Cu_3O_{6+x}$	2	93	YBCO
$Bi_2Sr_2CuO_6$	1	~12	Bi2201
$Bi_2Sr_2CaCu_2O_8$	2	95	Bi2212
$Bi_2Sr_2Ca_2Cu_3O_{10}$	3	110	Bi2223
$Tl_2Ba_2CuO_6$	1	95	Tl2201
$Tl_2Ba_2CaCu_2O_8$	2	105	Tl2212
$Tl_2Ba_2Ca_2Cu_3O_{10}$	3	125	Tl2223
$TlBa_2Ca_2Cu_4O_{11}$	3	128	Tl1224
$HgBa_2CuO_4$	1	98	Hg1201
$HgBa_2CaCu_2O_8$	2	128	Hg1212
$HgBa_2Ca_2Cu_3O_{10}$	3	135	Hg1223

1. Structure

Cuprates are variations of the crystal type known as *perovskites*. Perovskites are minerals whose chemical formula is ABX_3 or AB_2X_3. Thus, perovskites contain three elements A, B, X in the proportion 1:1:3 or 1:2:3. Atoms A are metal cations, atoms B and X are nonmetal anions. Element X is often represented by oxygen.

Knowledge of structure (the location of atoms) is a prerequisite for understanding the physical properties of cuprates. The crystal structure of copper oxides is highly *anisotropic*. Such a structure defines most physical properties of cuprates. In conventional superconductors, there are no important structural effects since the coherence length is much longer than the penetration depth. This, however, is not the case for cuprates.

In general, the high-T_c materials are basically tetragonal, and all of them have one or more CuO_2 planes. Superconductivity in cuprates occur in the copper-oxide planes. The CuO_2 layers are always separated by layers of other atoms such as Bi, O, Y, Ba, La etc., which provide the charge carriers into CuO_2 planes. These layers are often called *charge reservoirs*. In the CuO_2 planes, each copper ion is strongly bonded to four oxygen ions separated by a distance approximately 1.9 Å. A large number of compounds with the characteristic CuO_2 planes have been synthesized (see Table 3.1). In different synthesized cuprates, the number of CuO_2 layers per unit cell, N_ℓ, is different. In general, the critical temperature T_c correlates with the number of CuO_2 layers: at fixed doping level, by increasing the number of CuO_2 layers, T_c first increases, reaching the maximum at $N_\ell = 3$, and then decreases. In most superconducting cuprates, by changing the doping level at fixed N_ℓ, the $T_c(p)$ dependence has the bell-like shape, where p is the hole concentration in CuO_2 planes, as shown in Fig. 3.1. The bell-like curve of $T_c(x)$ dependence is more or less universal in cuprates, where x can be p, N_ℓ, the lattice constants a, b or c, the buckling angle of CuO_2 planes etc. Thus, the maximum T_c value can only be achieved when all the necessary parameters have their optimal values. If one of the parameters is changed from its optimal value, the T_c value will, as a consequence, decrease.

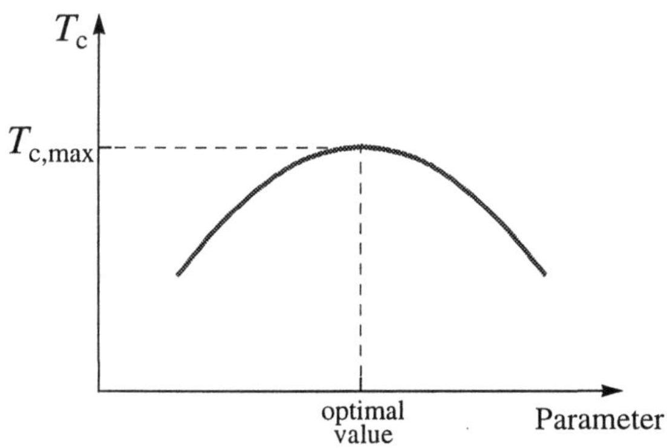

Figure 3.1. Critical temperature as a function of a parameter which can be the doping level p, the number of CuO_2 planes per unit cell, the unit-cell constants, the buckling angle of CuO_2 planes etc.

1.1 LSCO

This compound was the first high-T_c superconductor discovered. The maximum value of T_c is 38 K. The tetragonal unit cell of LSCO is shown in Fig. 3.2a. The lattice constants are $a \approx 5.35$ Å, $b \approx 5.40$ Å and $c \approx 13.15$ Å. This compound is often termed the *214 structure*, because it has two La (Sr), one Cu and four O atoms. Upon examining the unit cell, Fig. 3.2a reveals that the basic 214 structure is doubled to form a unit cell. Therefore a more proper label might be 428. The reason for this doubling is that every other CuO_2 plane is offset by one-half a lattice constant, so that the unit cell would not be truly repetitive if we stopped counting after one cycle of the atoms.

In LSCO, the conducting CuO_2 planes are ~ 6.6 Å apart, separated by two LaO planes which form the charge reservoir that captures electrons from the conducting planes upon doping. In the crystal, oxygen is in a O^{2-} valence state that completes the p shell. Lanthanum loses three electrons and becomes La^{3+},

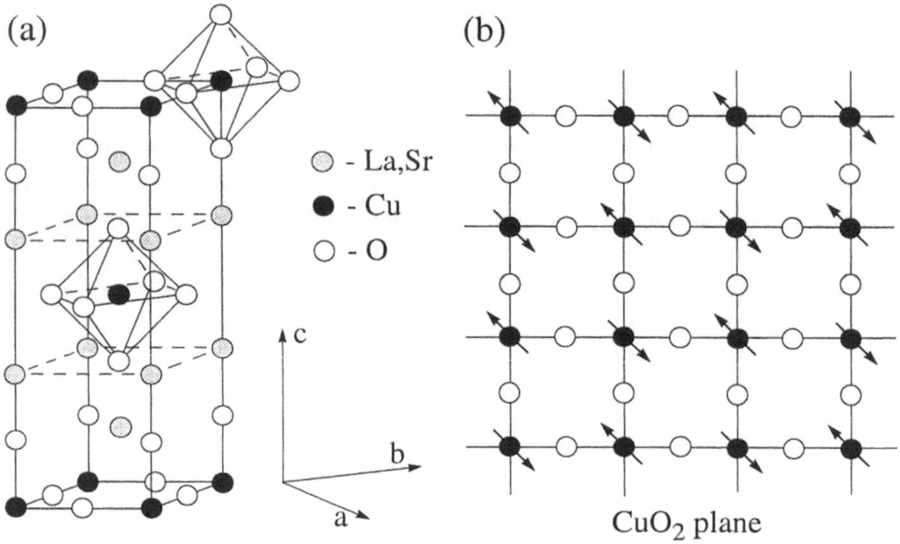

Figure 3.2. (a) Crystal tetragonal structure of LSCO compound. (b) Schematic of CuO_2 plane, the crucial subunit for high-T_c superconductivity. The arrows indicate a possible alignment of spins in the antiferromagnetic ground state.

which is in a stable closed-shell configuration. To conserve charge neutrality, the copper atoms must be in a Cu^{2+} state, which is obtained by losing the $(4s)$ electron and also one d electron. This creates a hole in the d shell, and thus Cu^{2+} has a net spin of $\frac{1}{2}$ in the crystal. Each copper atom in the conducting planes has an oxygen above and below in the c direction. These oxygen atoms are called *apical*. Thus, in LSCO, the copper ions are surrounded by octahedra of oxygens, as shown in Fig. 3.2a. However, the distance between a Cu atom and an apical O is $\sim$ 2.4 Å, which is considerably larger than the distance Cu–O in the planes (1.9 Å). Consequently, the dominant bonds are those on the plane, and the bonds with apical oxygens are much less important. Many high-T_c compounds have apical oxygens always separated from the conducting planes by a distance of about 2.4 Å.

Figure 3.2b schematically shows a CuO_2 plane, the crucial subunit for high-T_c superconductivity. In Fig. 3.2b, the arrows indicate a possible alignment of spins in the antiferromagnetic ground state of La_2CuO_4.

The phase diagram of this material is shown in Fig. 3.3. Near half-filling, the antiferromagnetic order is clearly observed. For higher Sr doping, 0.02 $\leq x \leq$ 0.08, there is no long-range antiferromagnetic order, however, at very low temperatures, there is a spin-glass-like phase. This phase is **not** a conventional spin glass, but rather an inhomogeneous phase consisting of, depending on hole concentration, either insulating or superconducting islands and spin-glass islands with domain size $L(x) \propto x^{-1/2}$. The insulator-metal transition occurs at about $x \sim 0.04$–0.05. For Sr dopings between $x \sim 0.05$ and ~ 0.27, a superconducting phase is found at low temperatures. The maximum T_c is observed at the "optimal" doping $x \sim 0.16$. As one can see in Fig. 3.3, the Sr substitution for La in LSCO induces a structural phase transition from the high-temperature tetragonal (HTT) to low-temperature orthorhombic (LTO) and, at low temperatures, from the LTO phase to the low-temperature tetragonal (LTT) phase (not shown). The HTT→LTO and LTO→LTT phase transitions will be discussed latter.

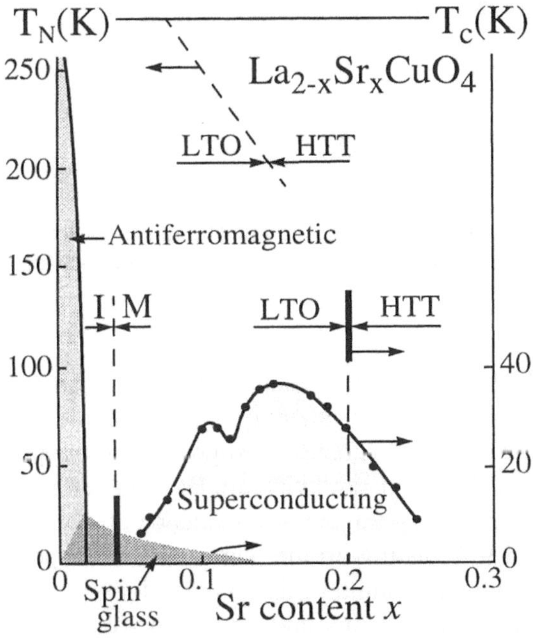

Figure 3.3. Phase diagram of LSCO.

In Fig. 3.3, the $T_c(x)$ dependence has a nearly bell-like shape. However, at doping $x = \frac{1}{8}$, the curve has a dip. This dip is the so-called $\frac{1}{8}$ *anomaly* and inherent exclusively to LSCO. Recently, it was shown that the local *static* magnetic order is responsible for the $\frac{1}{8}$ anomaly [2]. The dip at $\frac{1}{8}$ is not present in the $T_c(p)$ dependence of other high-T_c superconductors.

1.2 YBCO

YBCO is the first superconductor found to have $T_c > 77$ K, which is commonly termed "123". The orthorhombic unit cell of YBCO is shown in Fig. 3.4. The lattice constants are $a \approx 3.82$ Å, $b \approx 3.89$ Å, and $c \approx 11.7$ Å. The two CuO_2 layers are separated by a single yttrium atom. The role of yttrium is very

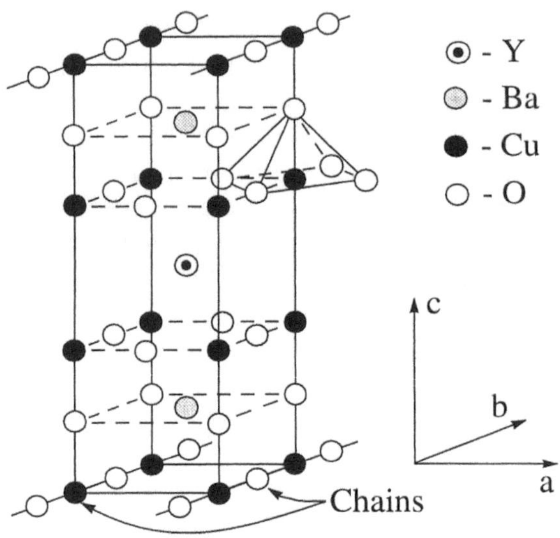

Figure 3.4. Crystal orthorhombic structure of YBCO compound. Note that the orthorhombic lattice vectors are at 45° relatively to the tetragonal lattice vectors (see Fig. 3.2a).

minor, it just holds the two CuO_2 layers apart. In the crystal, Y has a valence of +3. The replacement of Y by many of the lanthanide series of rare-earth elements causes no appreciable change in superconducting properties. Outside the layer CuO_2–Y–CuO_2 layer sandwich are BaO planes and Cu–O chains, as shown in Fig. 3.4. In the crystal, Ba has a valence of +2. The distance Cu–O in the chains is $\sim$ 1.9 Å, as it is in the planes. Each copper ion in the CuO_2 planes is surrounded by a pyramid of five oxygen ions.

YBCO is the only high-T_c compound having one-dimensional Cu–O chains. In $YBCO_6$, there are no Cu–O chains, and the compound is an antiferromagnetic insulator, as shown in Fig. 3.5. It has to be doped to gradually become a metallic conductor and a superconductor at low temperature. The doping is achieved by adding additional oxygen atoms which form the Cu–O chains. So, the oxygen content can be changed reversibly from 6.0 to 7.0 simply by pumping oxygen in and out of the parallel CuO chains running along the b axis. Thus, the CuO chains play the role of charge reservoirs. At an oxygen content of 6.0, the lattice parameters $a \neq b$, and the unit cell is orthorhombic. The increase of oxygen content causes the unit cell to have square symmetry, i.e. $a = b$. So, the crystal becomes tetragonal at low temperatures, as shown in Fig. 3.5. At an oxygen content of 6.4, antiferromagnetic long-range order disappears and the superconducting phase starts developing. The maximum value of T_c is achieved at doping level of about 6.95 (the optimal doping). Unfortunately, it is not possible to explore the phase diagram above the oxygen

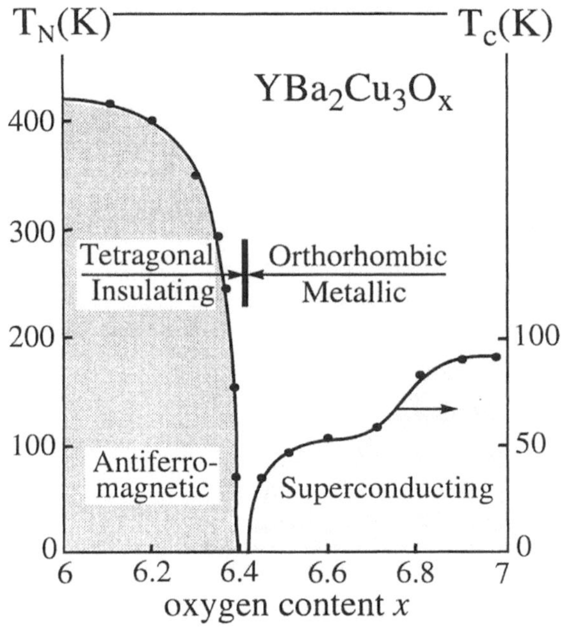

Figure 3.5. Phase diagram of YBCO.

content of 7.0, since the Cu–O chains are completed. $YBCO_7$ is a stoichiometric compound with the highest T_c.

In Fig. 3.5, one can see that, at $x \simeq 6.7$, there is a plateau at $T_c \simeq 60$ K. This is the so-called *60 K plateau.* There are two different explanations for the origin of the 60 K plateau: the hole concentration in CuO_2 planes at $x \simeq 6.7$ remains unchanged, and the change of hole concentration occurs exclusively in Cu–O chain layers. The second explanations is that this plateau relates directly to the $\frac{1}{8}$ anomaly in LSCO.

1.3 Bi2212

Figure 3.6 shows the unit cell of Bi-2212 which has the maximum T_c of 95 K. The dimensions of the tetragonal lattice constants of Bi2212 are $a \simeq b \approx 5.4$ Å and $c \approx 30.89$ Å. In addition to the CuO_2 double layer intercalated by Ca, the unit cell also contains two semiconducting BiO and two insulating SrO layers, as shown in Fig. 3.6. In the crystal, Bi and Sr have a valence of +3 and +2, respectively. In Bi2212 crystals, there is a lattice modulation along the b axis, with the period of 4.76 b. The family of the bismuth cuprates consists of three members: Bi2201, Bi2212 and Bi2223 with the unit having 1, 2 and 3 CuO_2 planes, respectively. The maximum T_c increases with increasing number of CuO_2 planes.

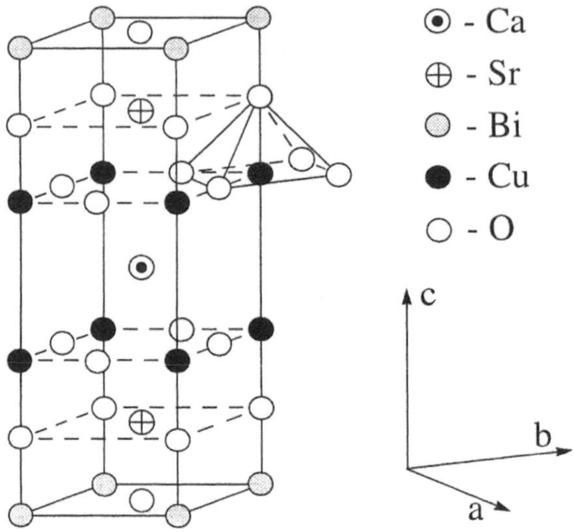

Figure 3.6. Crystal structure of Bi2212 compound.

The structure of bismuth cuprates is very similar to the structure of thallium cuprates such as Tl2201, Tl2212 and Tl2223, with bismuth replaced by thallium, and strontium replaced by barium. In spite of similar structural features of bismuth and thallium compounds, there are differences in superconducting and normal-state properties.

Bismuth cuprates are very suitable for tunneling and other measurements: the oxygen content in Bi2212 (Bi2201 or Bi2223) at room temperature is stable, unlike that in YBCO. Second, the bonds between BiO layers in the crystal are weak, so it is easy to cleave a Bi2212 crystal. After the cleavage, the Bi2212 crystal has a BiO layer on the surface. However, it is generally agreed that Bi2212 samples have not reached the degree of purity and structural perfection obtained in YBCO.

Since in bismuth, thallium and mercury cuprates, the lattice constants $a = b$, there is no twinning within a crystal.

1.4 NCCO

The structure of electron-doped NCCO cuprate, shown in Fig. 3.7, is body-centered tetragonal like that of LSCO. The difference between the two lies in the position of the oxygen atoms of the charge reservoirs. The tetragonal lattice constants of NCCO are $a \simeq b \approx 5.5$ Å and $c \approx 12.1$ Å. In the crystal, Nd and Ce have a valence of +3 and +4, respectively. When a Nd^{3+} ion is replaced by Ce^{4+}, the CuO_2 planes get an excess of electrons. It is believed that an

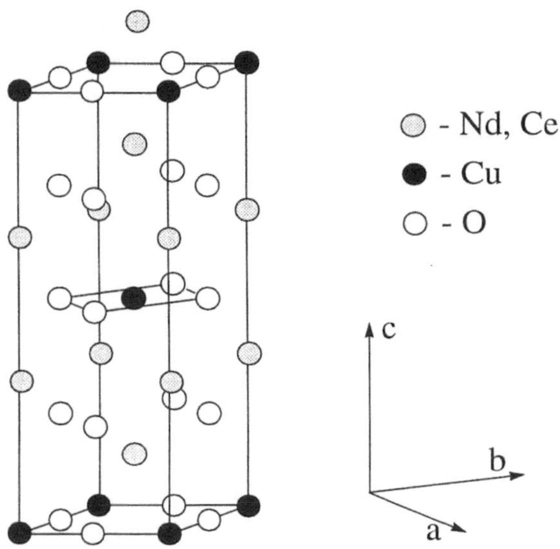

Figure 3.7. Crystal structure of NCCO compound.

added electron occupies a hole in the d shell of copper, producing an $S = 0$ closed-shell configuration.

Superconductivity in NCCO is observed when the Ce content varies between $x \simeq 0.14$ and 0.18. The phase diagram of NCCO is compared in Fig. 3.8 with a hole-doped LSCO cuprate. As one can see in Fig. 3.8, the two diagrams are very similar. Both present an antiferromagnetic phase with similar Néel temperature. When x is increased further, a superconducting phase appears close to antiferromagnetism, although the width of the superconducting phase in the two cases differs by a factor of 3. The maximum value of T_c in NCCO is 24 K, so it is almost twice as small as that of LCSO.

1.5 Structural phase transitions

In low-T_c superconductors, there are no important structural effects since the coherence length is much longer than the penetration depth. In cuprates, however, the structural effects are very important because $\xi \ll \lambda$. Moreover, in cuprates, *the lattice is very unstable.* As a consequence, in cuprates, there are many structural phase transitions which result in the appearance of the superconducting state.

Structural phase transitions are probably the best documented in LSCO and $La_{2-x}Ba_xCuO_4$ (LBCO) because of their relatively simple crystal structure and the availability of large high-quality single crystals. As shown in Figs 3.3 and 3.9, the Sr (Ba) substitution for La in LSCO (LBCO) induces a structural phase transition from the high-temperature tetragonal (HTT) to low-

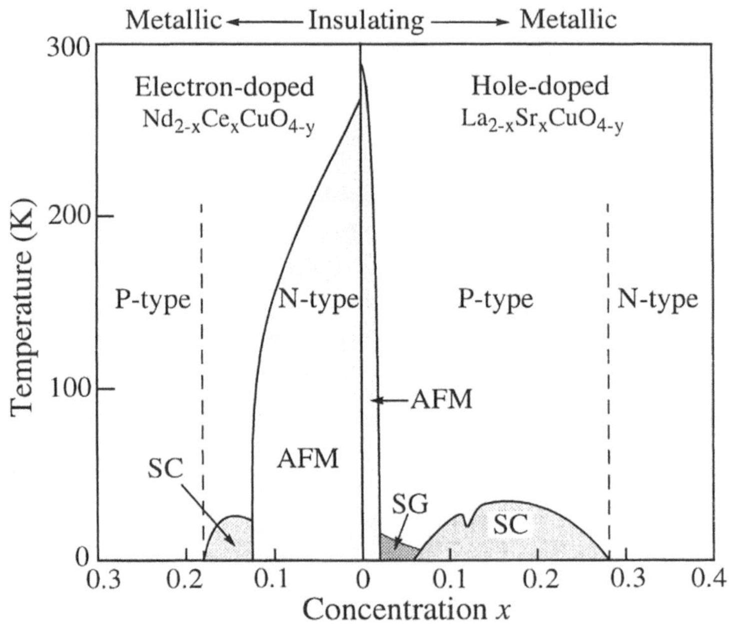

Figure 3.8. Schematic phase diagram of NCCO and LSCO shown together for better comparison.

temperature orthorhombic (LTO). The HTT→LTO transition, which had been studied a decade before the discovery of high-T_c superconductivity, is driven by soft phonons. At high temperatures, LSCO has the same body-centered tetragonal structure as K_2NiF_4. On cooling, the HTT phase transforms into the LTO phase, because of a staggered tilt of the CuO_6 octahedra. The tilt angle of the octahedra is about 3.6° and uniform, as shown in Fig. 3.10. In LBCO, on further cooling, the LTO phase transforms into new phases with doubled unit cells, one of which is called the low-temperature tetragonal (LTT) phase [3]. In the LTT phase, the CuO_6 octahedra rotate about alternate orthogonal axes in successive layers with no change in the magnitude of the tilt, as shown in Fig. 3.10. The underlying interaction that gives rise to the low-temperature structural phases, including the LTT phase, is the mismatch in preferred lattice constants of the CuO_2 layer and the intervening rare-earth oxide layers. At high Ba doping ($0.2 < x$), there is no well defined LTO→LTT phase transition below 300 K, so a crystal is a mixture of the LTO and LTT phases. Even at low Ba (Sr) doping level, neutron and X-ray scattering measurements performed on LBCO and Nd-doped LSCO show that the LTO and LTT phases coexist well below the onset of the LTO→LTT structural transition [2–4].

The small structural differences between the LTO and LTT phases have a drastic influence on electronic properties of LSCO and LBCO. Superconduc-

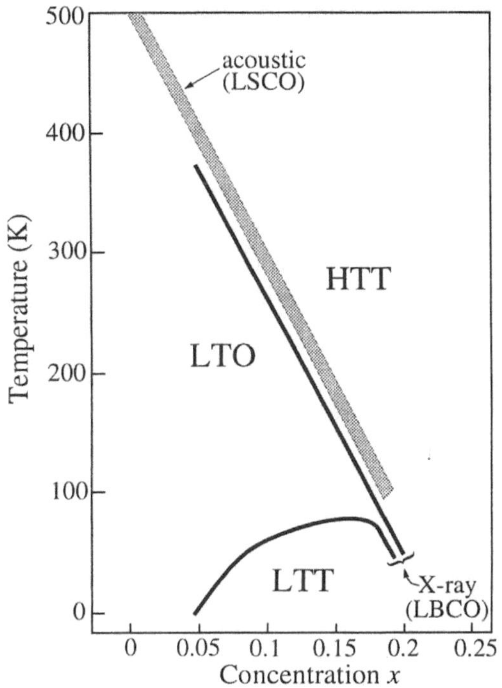

Figure 3.9 Structural phase diagram of LSCO and LBCO obtained by X-ray [3] and acoustic measurements [7].

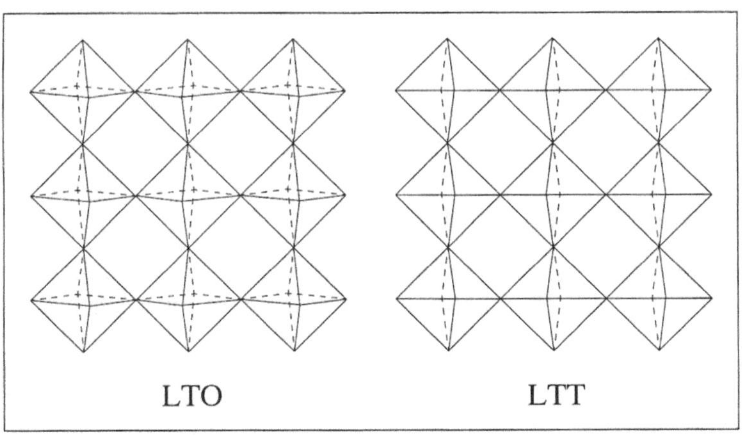

Figure 3.10. Tilt pattern of CuO_6 octahedra in the LTO and LTT phases.

tivity is suppressed in the LTT phase and static antiferromagnetism occurs in compounds with large hole content. In a sense, the LTO→LTT structural transition is intrinsically magnetic in nature [5]. Observations made on the basis of transport measurements suggest that the LTO→LTT transition induces intragap electronic states in the middle of the normal-state gap (the pseudogap)

[6]. Independently, and well in agreement with the latter finding, it has been argued that the LTT structure provides a pinning potential for the so-called *charge stripes* (see Section 3.2).

The structural transitions in LSCO are clearly observed in acoustic measurements: at different dopings, the HTT→LTO transition is manifest by a dip in the temperature dependence of the longitudinal sound velocity [7] and the longitudinal C_{11} elastic coefficient [8], as shown in Fig. 3.9. The LTO→LTT transition is observed in the $(C_{11} - C_{12})/2$ transverse elastic coefficient [8]. However, acoustic measurements alone cannot identify the type of the structural transition.

The HTT→LTO transition in YBCO is shown in Fig. 3.5. It is assumed that the superconductivity in YBCO appears in the pure LTO phase. In fact, it is not completely true: there are at least three structural transitions in YBCO, which occur at T_c, 140–150 K and 220–250 K [9]. The transition at 220–250 K is close to that shown in Fig. 3.5, therefore it can be associated with the HTT→LTO transition. Unfortunately, the other two structural transitions in YBCO are unknown: the measurements performed in YBCO with different dopings ($0.55 \leq x \leq 1$) by ion channeling spectrometry cannot identify the types of these structural transitions. Nevertheless, it is evident that the LTO phase in YBCO is transformed twice before superconductivity appears.

Three structural phase transitions are also observed in Bi2212 by acoustic measurements [10], which occur at temperatures similar to those in YBCO [9]. Again, acoustic measurements cannot determine the type of the structural transitions, even if they are able to observe them. Acoustic measurements in cuprates will be discussed at the end of this Chapter (see Section 3.19).

1.6 Crystal structure and T_c

As mentioned above, in conventional superconductors, there are no important structural effects. This, however, is not the case for cuprates. Therefore, it is useful to consider parameters which affect T_c. Each parameter has it's optimal value, as shown in Fig. 3.1. The maximum T_c corresponds to the case when all the parameters have their optimal values. If one of the parameters is changed from it's optimal value, then T_c will inevitably decrease. The main problem in identifying the parameters which have more influence on T_c than the others is that by changing one parameter, the others are changed too. So it is impossible to conclude whether the change in T_c is defined by the change of the first parameter or by the variations of the other parameters induced by the change of the first one. In other words, the system is nonlinear, and it is difficult to predict the response of the system to a disturbance without knowing the set of nonlinear equations. So, even at this stage, it is possible to argue that superconductivity in cuprates is “nonlinear”.

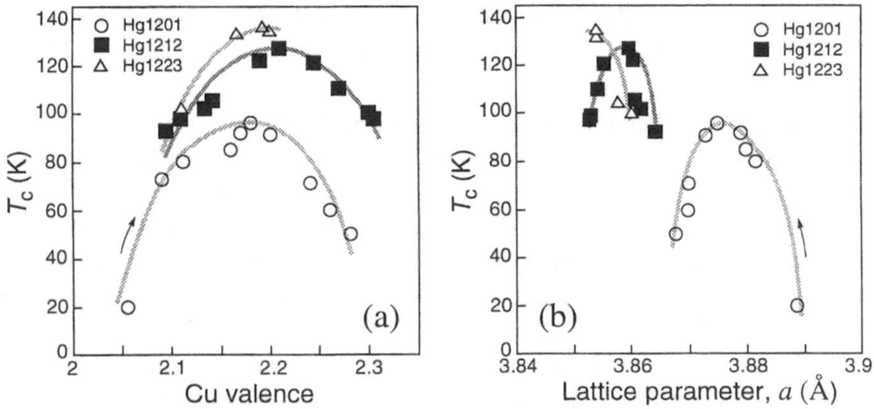

Figure 3.11. (a) Doping dependence of the critical temperature in Hg-based cuprates. (b) Critical temperature from frame (a) as a function of the a-axis lattice parameter [11].

There is a consensus that superconductivity in cuprates occurs in CuO_2 planes. Thus the structural parameters of CuO_2 planes affect T_c the most. The geometry of a CuO_2 plane is defined by the following factors: the length of the Cu–O bond; the degree of an orthorhombic distortion from square, and the degree of deviation from a flat plane (a buckling angle). The influence of the Cu–O length on T_c is well demonstrated in mercury compounds. Figure 3.11a shows the critical temperature in different mercury compounds as a function of hole concentration: all the $T_c(p)$ dependences have bell-like shapes. Figure 3.11b shows, the critical temperature of the same compounds as a function of length of the Cu–O bond. In Fig. 3.11b, there is a clear correlation between the Cu–O length and T_c: the highest T_c corresponds to the smallest Cu–O length. It is obvious that, if it were possible to decrease the Cu–O length further, then below a certain value, T_c would decrease. However, one has to realize that the dependence shown in Fig. 3.11b is not universal for other cuprates because, in different cuprates, the intervening layers between the CuO_2 planes are different, so that their influence on CuO_2 layers is different.

The buckling angle of a CuO_2 plane is defined as the angle at which the plane oxygen atoms are out of the plane of the copper atoms. The critical temperature of YBCO correlates with the magnitude of the buckling angle: the highest maximum T_c corresponds to the smallest maximum buckling angle [12]. So, at constant doping level, the appearance of the CuO_2 plane buckling lowers T_c. The buckling induces the change in electronic structure. The highest critical temperature T_c = 135 K is observed in mercury compounds which have perfectly flat CuO_2 planes. The orthorhombic distortion is defined by the parameter $\frac{b-a}{b+a}$, where a and b are the lattice constants. All cuprates with the high critical temperatures ($>$ 100K) have tetragonal crystal structure. Therefore, for increasing T_c the degree of orthorhombic distortion should be as small as possible. Thus, at fixed doping level, the highest T_c corresponds to flat and square CuO_2 planes.

Consider now other parameters of the crystal structure (outside the CuO_2 planes) which affect the critical temperature. First, consider the effect of the distance between the CuO_2 planes on T_c. In cuprates with two or more CuO_2 layers, there are two interlayer distances: the distance between CuO_2 layers in a bi-layer (three-layer, four-layer) block, d_{in}, and the distance between the bi-layer (three-layer, four-layer) blocks, d_{ex}. Usually, $d_{in} + d_{ex} \simeq 15$ Å, and $d_{in} \approx 3$–6 Å, $d_{ex} \approx 9$–12 Å. The intervening layers between the group of CuO_2 planes are semiconducting or insulating. Transport measurements in the c-axis direction show that the c-axis resistivity depends exponentially on d_{ex}; however, there is no strong dependence of T_c on d_{ex}. For example, in the infinite-layer cuprate (Sr,Ca)CuO_2 which has no charge reservoirs, the distances d_{in} and d_{ex} are equal and short, $d_{in} = d_{ex} \simeq 3.5$ Å; however, $T_c \simeq 110$ K. So, the "optimal" region of the d_{in} and d_{ex} parameters is rather wide (see Fig. 3.1). Considering three superconducting one-layer cuprates LSCO ($d_{ex} \simeq 6.6$ Å, $T_{c,max} = 38$ K), Hg1201 ($d_{ex} \simeq 4.75$ Å, $T_{c,max} = 98$ K) and Tl2201 ($d_{ex} \simeq 11.6$ Å, $T_{c,max} = 95$ K) it is clear that there is no correlation between T_c and d_{ex}. The large difference in T_c, for example, between LSCO and Tl2201, is not due to the difference between the c-axis distances in these cuprates, but due to the difference in the structural parameters of CuO_2 planes, which were discussed above.

Next is the question of the affect of the intervening layers on T_c. The intervening layers can be divided into two categories: "structural" layers and charge reservoirs. The structural layers, like Y in YBCO, play a minor role in the variation of T_c. At the same time, the charge reservoirs make a large impact on T_c. Different charge reservoirs have different polarized abilities and different abilities to polarize other ions: the higher are better. The distance between charge reservoirs and CuO_2 planes is also important: the shorter is better. In addition, the charge reservoirs also play the role of the structural layers. For example, in LSCO, the critical temperature is very sensitive to lattice strains induced by substituting Sr for different cations having different ionic radius, $\langle r_A \rangle$. The $T_c(\langle r_A \rangle)$ dependence has a bell-like shape with a maximum near 1.218–1.22 Å [13]. In Bi2201 which has the lowest T_c among all cuprates, by partially replacing Sr by La the maximum critical temperature can be tripled, as shown in Fig. 3.12. The partial replacement of La in LSCO by Nd makes a large impact on the critical temperature. These examples show that the intervening layers can affect the electronic structure of the CuO_2 planes drastically, especially in single-layer compounds.

It is important noting that an *isolated* CuO_2 layer will not superconduct. Even a CuO_2 layer situated on the surface of a crystal (and this does happen

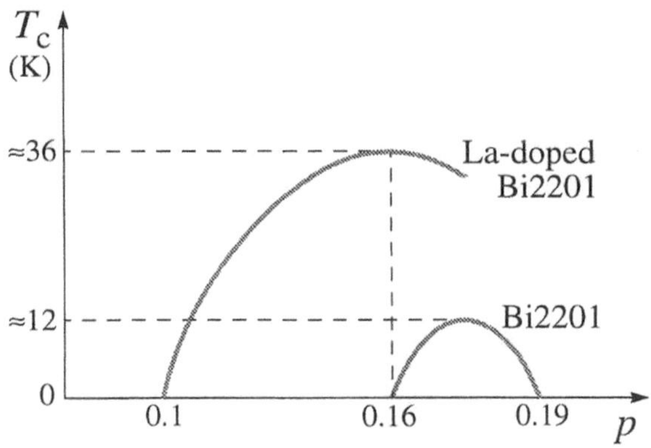

Figure 3.12. Critical temperature as a function of doping for Bi2201 and La-doped Bi2201.

occasionally) will be, most likely, not superconducting but semiconducting. This can be easily understood: in order to become superconducting, a CuO_2 layer has to be structurally stabilized from both sides, above and below. Of course, this is a necessary but not sufficient condition.

1.7 Structural defects

Superconductivity in all high-T_c superconductors depends delicately on the details of their layered structure; and that, in turn, depends on the distortions produced by lattice defects or by various dopant atoms. For this reason, it is difficult to grow perfect single crystals and thin films. As a consequence, when studying high-T_c superconductors one has to take into account the effect of extrinsic factors such as grain boundaries, impurities and twinning.

An individual grain boundary is a place where the ordinary lattice structure is interrupted. In the simplest case, there is only a misalignment of atoms. In dirtier grain boundaries, there may be impurities present; there may be changes in the charge density, or deficiencies of oxygen, etc. All large single crystals of cuprates have grain boundaries. In order to reduce the effect of extrinsic factors in measurements, it is better to use small-size single crystals. Thin films can be used too, but almost all high-T_c superconducting thin films are mosaics, not single thin crystals. Nevertheless, the highest critical current J_c in cuprates is obtained in thin films. The effect of impurities on superconducting properties of cuprates will be discussed separately (see Section 3.17).

In YBCO, the slight difference in length between the a and b lattice constants breaks the symmetry of the crystal lattice. During crystal growth, it is energetically quite easy for propagation to switch from a to b, or vice versa. The switching is called *twinning*, and results in a minor irregularity in an otherwise uniform single crystal. Twinning occurs accidentally as a single crystal grows, creating twin boundaries. In practice, it is possible to de-twin single crystals; however, the procedure is very delicate.

Finally, one has take into account that high-T_c superconductors change their properties when exposed to moisture or moist air. As a consequence, the surface properties of cuprates are not necessarily representative of the interior material. For example, YBCO is extremely sensitive to moisture, while, the bismuth compounds are less affected.

2. Doping and charge distribution

The simplest copper oxide perovskites are insulators. To become superconducting they have to be doped by charge carriers. It is generally agreed that the effect of doping has the most profound influence on superconducting properties of cuprates. Basically, there are two ways to increase the number of charge carriers in cuprates chemically: (i) to substitute metallic atoms in charge reservoirs by higher-valence atoms, and/or (ii) to change the number of oxygen atoms. Doping increases the number of electrons or holes at the Fermi level. The concentration of charge carriers in cuprate is low, $\sim 5 \times 10^{21}$, in comparison with that of conventional superconductors, $\sim 5 \times 10^{22}$–10^{23}. Due to the large coherence length in conventional superconductors, only a 10^{-4} part (= $\Delta(0)/E_F$) of the electrons located near the Fermi surface participate in pairing. In cuprates, $\sim$10% of all conduction electrons (holes) form the Cooper pairs.

2.1 Charge doping and T_c

In conventional superconductors, the critical temperature rises monotonically with the rise of charge-carrier concentration, $T_c(p) \propto p$. In cuprates, the $T_c(p)$ dependence is nonmonotonic, as shown in Figs 3.1 and 3.11a. In most hole-doped cuprates (but not in all), the $T_c(p)$ dependence has the bell-like shape and can be approximated [14] by the empirical expression

$$T_c(p) \simeq T_{c,max}[1 - 82.6(p - 0.16)^2], \tag{3.1}$$

where $T_{c,max}$ is the maximum critical temperature for a given compound. Superconductivity occurs within the limits $0.05 \leq p \leq 0.27$ which vary slightly in different cuprates. Different doping regions of the superconducting phase are mainly known as the *underdoped*, *optimally doped* and *overdoped* regions, as shown in Fig. 3.13. The insulating phase at $p < 0.05$ is usually called the *undoped* region. These designations are used in the remainder of the book. Above 0.27, cuprates become metallic.

Theoretically, an oxygen atom acquires two electrons from another atom. In practice, the situation is slightly more complicated. If the variation of hole concentration is achieved by changing the oxygen content, then the number of doped holes created by an added oxygen atom in different doping regions is

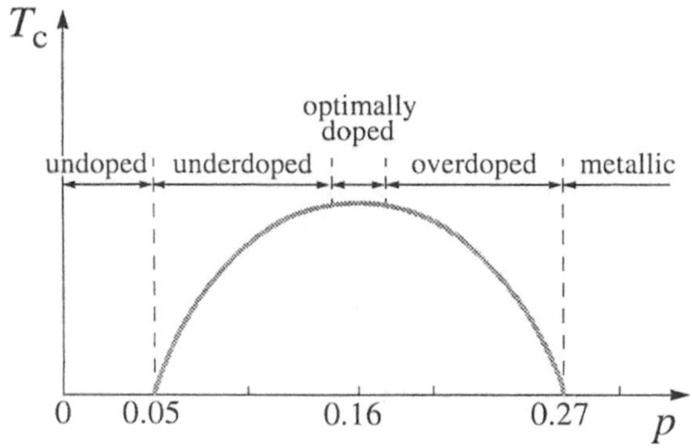

Figure 3.13. Critical temperature as a function of doping. In most cuprates, the superconducting phase appears when the doping level is between $0.05 \leq p \leq 0.27$. The phase diagram is schematically divided into five regions: undoped ($p < 0.05$), underdoped ($0.05 \leq p \leq 0.14$), optimally doped ($0.14 < p < 0.18$), overdoped ($0.18 \leq p \leq 0.27$), and metallic ($0.27 < p$).

not exactly two [15]. In the underdoped region, each extra oxygen atom contributes slightly more than two holes to the overall charge-carrier density. In the optimally doped region, the number of holes created by an added oxygen atom is precisely two. In the overdoped region, each extra oxygen atom contributes only one hole to the overall charge-carrier density. Thus, the $p(n_O)$ dependence, where n_O is the number of extra oxygen atoms, is nonlinear.

In three- and four-layer cuprates, there are two types of CuO_2 planes: outer and inner. In each three(four)-layer block, the outer CuO_2 planes are connected directly to charge reservoirs. At the same time, the inner CuO_2 planes in each three(four)-layer block are isolated from the charge reservoirs. As a consequence, the outer and inner CuO_2 planes have nonequivalent hole doping [16]. In the overdoped region, the carriers are mostly doped in the outer planes and the inner planes remain at the optimum doping level. This is probably the reason why the critical temperature of four-layer compounds is lower than that of the three-layer cuprates of the same family: in four-layer cuprates, the optimal doping is never achieved in all four planes simultaneously.

Most of superconducting cuprates are hole-doped. There are only a few electron-doped compounds which exhibit superconductivity. The ratio between the maximum critical temperatures in hole-doped and electron-doped cuprates is 135 K/24 K = 5.6. The same ratio obtained in hole-doped and electron-doped C_{60} (in a field-effect transistor configuration) is very similar and equals 4.7 [17]. The $T_c(p)$ dependences for electron- and hole-doped C_{60} have bell-like shapes as do those in cuprates. It seems that the electron-hole asymmetry is fundamental: superconducting hole-doped compounds will always have the critical temperature a few times higher than T_c of their electron-doped partners. This issue will be discussed again in Chapters 8 and 10.

In fact, it is possible to dope electrons or holes into initially underdoped cuprates by using a field-effect transistor configuration [18]. The more complex thallium compounds can also be made into electron carriers. A variant upon 1212 configuration is $TlSr_2Ca_{1-x}A_xCu_2O_{7+\delta}$, where A denotes a rare earth element. This material is chemically tunable from hole-doped to electron-doped by varying x. In a field-effect transistor configuration, the infinite-layer compound $CaCuO_2$ which is the nominally insulating material can be easily doped either by electrons or holes. The maximum T_c values of 89 K and 34 K are observed respectively for hole- and electron-type doping of around 0.15 charge carriers per CuO_2.

2.2 Charge inhomogeneities

In the simple model of a metal, conduction electrons form a "sea" of weakly interacting, negatively charged particles which are distributed homogeneously inside the metal. So, the electrons are free to travel in all three directions. In cuprates, CuO_2 planes are two-dimensional, so the analogue of the electron-sea model for cuprates would be a model in which charge carriers are homogeneously distributed into two-dimensional copper oxide planes. This, however, is not the case. In cuprates and in many other compounds with low dimensionality, the distribution of charge carriers is inhomogeneous. Moreover, in cuprates, the charge-carrier distribution is inhomogeneous on a *microscopic* scale and a *macroscopic* scale, as schematically shown in Fig. 3.14.

In the undoped region ($p < 0.05$) of hole-doped cuprates, doped holes prefer not to be distributed homogeneously into CuO_2 planes but to form dynamical one-dimensional "objects" called *charge stripes*. In the undoped region, charge stripes are diagonal, i.e. they run not along –O–Cu–O–Cu– bonds (see Fig. 3.2b) but along the diagonal –Cu–Cu–Cu– direction, as shown in Fig. 3.14. Since, in undoped cuprates, the hole concentration is low, the distance between charge stripes separated by two-dimensional insulating antiferromagnetic domains is large. The charge-stripe phase is not distributed homogeneously: there are two types of small islands containing either the intact antiferromagnetic phase or the diagonal charge-stripe phase.

In the underdoped region ($0.05 < p < 0.13$), charge stripes are vertical (or horizontal) and packed closer, as shown in Fig. 3.14. In this doping region, the average distance between charge stripes d_s is approximately proportional to $1/p$, and saturates near $p = \frac{1}{8}$, as shown in Fig. 3.15. Above $p = \frac{1}{8}$, the distance between stripes is practically unchanged. The dynamical charge stripes are separated by two-dimensional magnetic stripes which can be considered as the local memory effect of the antiferromagnetic insulating phase. As p increases, the fraction of intact antiferromagnetic domains decreases, as shown in Fig. 3.16a, but the two types of islands containing the intact antiferromagnetic phase and the vertical charge-stripe phase still coexist.

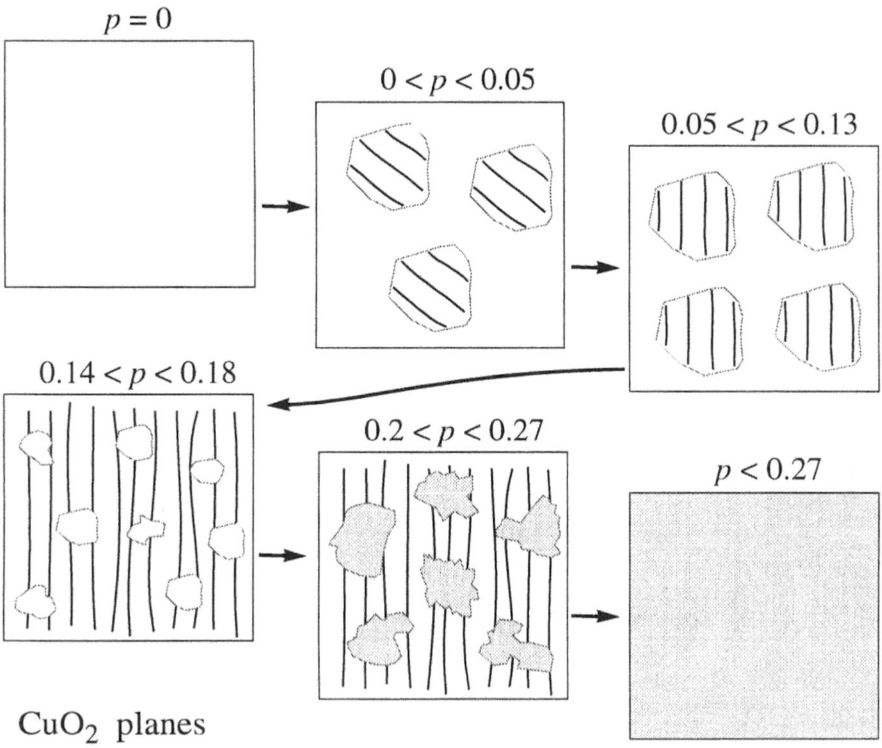

Figure 3.14. Sketch of charge distribution in CuO_2 planes as a function of doping. The antiferromagnetic and metallic phases are shown in white and grey, respectively. The lines show charge stripes (for more details, see text).

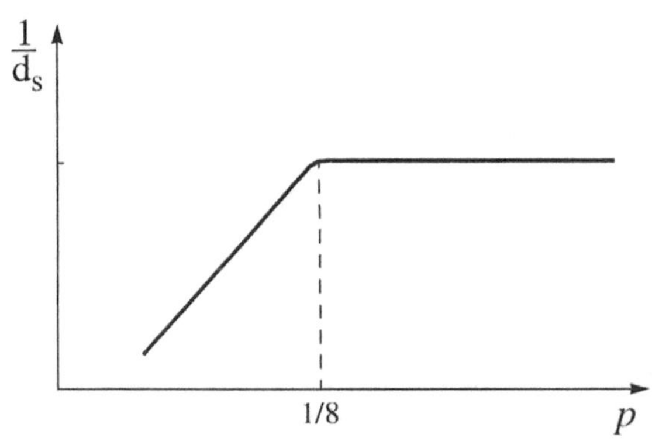

Figure 3.15. Doping dependence of incommensurability δ ($\propto 1/d_s$) of spin fluctuations, where d_s is the distance between charge stripes [19].

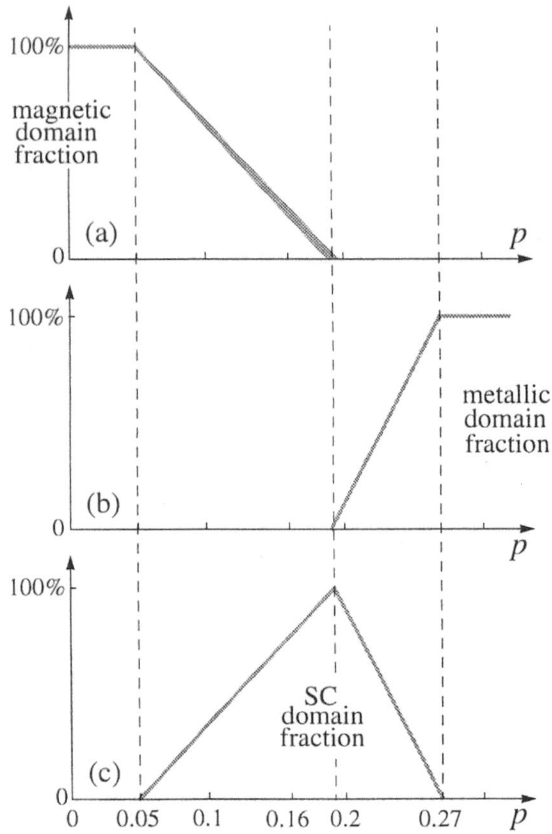

Figure 3.16 (a) Magnetic domain fraction in LSCO and YBCO [20, 21]. (b) Metallic domain fraction derived from plot (a) and Fig. 3.15. (c) Superconducting (SC) domain fraction at $T < T_c$, which can be associated with the vertical charge-stripe phase.

The length of a separate charge stripe is about 100 Å. The charge stripes are dynamical: they meander and can move in the transverse direction. Consequently, they are not strictly one-dimensional but *quasi*-one-dimensional. The stripes are half-filled: that is one positive electron charge per two copper sites along the stripes (in nickelates, the charge density along the stripes is 1 hole per Ni site). The physics of the stripe phase, including the magnetic order between the charge stripes, will be considered in detail separately (see Section 3.19).

In the near optimally doped region ($p \sim 0.16$) and in the overdoped region ($0.2 < p < 0.27$), the average distance between charge stripes remains almost unchanged, as shown in Figs 3.14 and 3.15. Therefore, as the doping level increases, new doped holes take over the virgin antiferromagnetic islands which completely vanish at $p = 0.19$, as shown in Fig. 3.16a. Above $p = 0.19$, small metallic islands start appearing, as schematically depicted in Fig. 3.16b.

Above $p = 0.27$, the charge-carrier distribution is homogeneous in two-dimensional CuO_2 planes, and cuprates become non-superconducting metals.

The data shown in Fig. 3.16a do not imply that magnetic correlations disappear above $p \sim 0.19$. In fact, magnetic relaxation is still dominant in the highly overdoped region [22]. The superconducting phase fraction as a function of doping is schematically shown in Fig. 3.16c. The superconducting phase can be associated with the vertical charge-stripe phase. The fraction of the superconducting phase has a maximum in the slightly overdoped region at $p \sim 0.19$, and not at $p_{opt} = 0.16$.

3. Superconducting properties

In spite of an unconventional type of superconductivity in cuprates, the latter display a number of properties that are in many ways similar to those of conventional BCS superconductors. First, superconductivity in cuprates occur due to electron pairing: earlier studies, performed in 1987, showed that the charge of charge carriers in LSCO and YBCO equals $2e$, where e is the electron charge. So, the main cause of superconductivity in conventional and high-T_c superconductors is the same—the electron pairing. Second, there is an energy gap in the electronic excitation spectrum below the critical temperature. Thus, in high-T_c superconductors, the electron pairing also leads to the appearance of energy gap. Hence, the basic principles underlying the phenomenon of superconductivity in different materials are the same.

Just as in conventional superconductors, there is a jump in the heat capacity at T_c, which indicates that the origins of the phase transitions are similar. Doped cuprates do have the Fermi level, so they are not insulators. The nonmonotonic dependence of $T_c(p)$ (see Fig. 3.13) is analogous to the nonmonotonic behavior of $T_c(p)$ in superconducting semiconductors. The isotope effect also exist in cuprates; however, it is strongly dependent on hole concentration.

Thus, in spite of an unconventional type of superconductivity in cuprates and in some other materials, the basic principles behind our understanding the phenomenon of superconductivity, described in the BCS theory, can be applied to any superconducting system (not to the Bose-Einstein condensate).

Here we consider some characteristics above and below the transition temperature. First, we focus on the superconducting state, and then we discuss normal-state properties of cuprates, which are definitely more unconventional than those below T_c. Tunneling measurements in cuprates are presented separately in Chapter 12.

3.1 The isotope effect

The isotope effect is one of the main indicators of the BCS mechanism of superconductivity. Experimentally, the isotope effect is found to be extremely small in cuprates at the optimal doping level $p_{opt} = 0.16$. This fact was initially taken as evidence against the BCS mechanism of high-T_c superconductivity

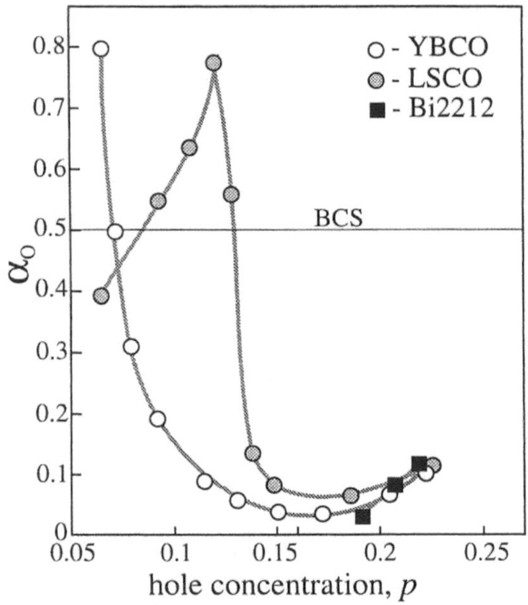

Figure 3.17. Oxygen isotope effect in LSCO, YBCO and Bi2212 [23, 24].

and, mistakenly, against the phonon-pairing mechanism. If the pairing mechanism is different from the BCS mechanism, *this does not mean that phonons are irrelevant.*

In fact, there is a huge isotope effect in cuprates. Figure 3.17 shows the oxygen (^{16}O vs ^{18}O) isotope-effect coefficient $\alpha_O \equiv d\ln(T_c)/d\ln(M)$, where M is the isotopic mass, as a function of doping level in LSCO, YBCO and Bi2212. Fist, one can see in Fig. 3.17 that the oxygen-isotope effect in cuprates is not universal: it is system- and doping-dependent. Second, in the underdoped region, α_O can be much larger than the BCS value of 0.5. According to the BCS theory, the isotope effect cannot be larger than 0.5. Third, at optimal dopings and in the slightly overdoped region, the oxygen-isotope effect is small, confirming the earlier results. In YBCO and probably in Bi2212, the $\alpha_O(p)$ dependence has an inverted bell-like shape. At the same time, in LSCO, the maximum isotope effect is located near $p = \frac{1}{8}$.

The copper (^{63}Cu vs ^{65}Cu) isotope effect has also been studied in LSCO and YBCO [23, 25]. In LSCO, the copper-isotope effect is similar to the oxygen-isotope effect, shown in Fig. 3.17. The copper-isotope effect in YBCO is small (even at low dopings), and can be even negative.

All the data of oxygen and copper isotope effects in the cuprates clearly demonstrate that the mechanism of high-T_c superconductivity is *different* from the BCS type.

It is worth mentioning that in LSCO there is a huge oxygen-isotope effect on lattice fluctuations above T_c [26], A large oxygen-isotope effect on the in-plane penetration depth is observed in LSCO [27] and Bi2223 [28]. In HoBCO, there is a large oxygen-isotope effect on the onset pseudogap temperature T^* [29].

Thus, it is obvious that in cuprates the underlying lattice strongly affects the electronic properties below and above T_c, although the effect is different from the predictions of the BCS theory.

3.2 Absence of the correlation between $\Delta(0)$ and T_c

According to the BCS theory, the energy gap at $T = 0$ and the critical temperature relate to each other as $2\Delta(0) = 3.52k_BT_c$. Thus, $T_c = \frac{\Delta(0)}{1.76} \simeq \frac{\Delta}{2}$ ($k_B = 1$).

In high-T_c superconductors, the energy gap obtained in tunneling and angle-resolved photoemission (ARPES) measurements is always larger than $2T_c$, particularly in the underdoped region. For example, in the optimally doped region of Bi2212, thus near $T_{c,max} = 95$ K, the magnitude of the energy gap obtained in tunneling and ARPES measurements is about 35–40 meV (= 400–460 K) which gives the ratio $\frac{\Delta}{T_c} = 4.2$–4.8. In other cuprates (LSCO, NCCO, Bi2201 etc.), the ratio $\frac{\Delta}{T_c}$ is also larger than 2, reaching as large as 30 in the underdoped region. Hence, in cuprates, the energy gap and the critical temperature do not correlate with each other. This means that in cuprates *the electron pairing and the onset of the long-range phase coherence have different mechanisms.* It is worth while to recall that, even in conventional superconductors, the pairing and the phase coherence are two different phenomena.

3.3 Effective mass anisotropy

Because of the layered structure of high-T_c superconductors, electrons move easily into CuO_2 planes, and with difficulty between the planes. Thus, anisotropy of the crystal structure of cuprates affects transport properties. To account for the anisotropy, it is conventionally agreed that the effective mass changes with crystal direction. Instead of being a single-valued scalar m, the effective electron mass becomes a tensor. In cuprates, to a good approximation the effective electron mass is a diagonal tensor, and the in-plane effective masses have similar values, $m_a \approx m_b$. The value of the in-plane effective mass in cuprates is larger than the electron mass m_e by a factor of between four and five: $m_{ab} \simeq (4\text{–}5)m_e$.

Anisotropy is defined by the ratio of the effective mass of the electron in the various directions, $\gamma^2 = \frac{m_c}{m_a}$. In YBCO, the effective mass ratio is about $\gamma^2 \approx 30$. LSCO exhibits somewhat higher anisotropy, $\gamma^2 \approx 200$, while Bi- and Tl-based cuprates are much more anisotropic than YBCO: in Bi- and Tl-compounds, the ratio is about 50 000. Such a large anisotropy, which is totally foreign to conventional superconductors, means that electrons can barely move in the c-axis direction, and the cuprates are effectively two-dimensional.

3.4 Resistivity and the effect of magnetic field

Consider changes of the temperature dependences of in-plane and c-axis resistivities obtained in hole-doped cuprates by increasing doping level. As shown schematically in Fig. 3.18, in the undoped region, the in-plane ρ_{ab} and out-of-plane ρ_c resistivities both are semiconducting; that is, the resistivities first fall with decreasing temperature, attaining their minimum values, and then sharply increase at low temperatures. The steep rise of the resistivities is due to charge ordering along charge stripes [30]. The difference between the absolute values of ρ_{ab} and ρ_c is a few orders of magnitude. For example, in YBCO, $\rho_c/\rho_{ab} \sim 10^3$. It is important noting that ρ_{ab} and ρ_c attain their minimum values at different temperatures.

In the underdoped region, the in-plane and out-of-plane resistivities passing through the minimum values both attain a maximum and then fall, vanishing below T_c, as shown in Fig. 3.18. The absolute values of the resistivities decrease in comparison with those in the undoped region, and the ratio ρ_c/ρ_{ab} decreases as well. The latter means that, as the doping level increases, two-dimensional cuprates become quasi-two-dimensional. For example, in LSCO (x = 0.06), $\rho_c/\rho_{ab} = 4\times10^3$ and decreases to $\rho_c/\rho_{ab} \sim 10^2$ at x = 0.28. The sharp fall in ρ_c and ρ_{ab} due to the transition to the superconducting state literally interrupts the rise corresponding to the insulating charge-ordering state.

In Fig. 3.18, near the optimally doped region, the in-plane resistivity above the critical temperature is now almost linear. However, the out-of-plane resistivity remains similar to that in the underdoped region, shifting to high temperatures and to low absolute values. Thus, in the optimally doped region, the in-plane resistivity is almost metallic while the out-of-plane resistivity still exhibits the semiconducting behavior. In high-quality single crystals, the extension of a $\rho_{ab}(T)$ dependence passes through zero as in the graph in Fig. 3.19a.

In the overdoped region, both ρ_{ab} and ρ_c become metallic, as shown in Fig. 3.18, probably, with the only exception from the general tendency being ρ_c in Bi2212. Figure 3.19 shows the in-plane and out-of-plane resistivities in an overdoped Bi2212 single crystal. As one can see in Fig. 3.19b, above $T_c \simeq 80$ K the out-of-plane resistivity remains semiconducting, even, in the overdoped region.

In cuprates there is also a weak in-plane anisotropy: ρ_a and ρ_b are not exactly the same. If, in YBCO, the in-plane anisotropy ρ_a/ρ_b at 300 K varying from 1.23 in the underdoped region to 2.5 in the optimally doped region is mainly, but not only, due to the CuO chains, in other cuprates, the weak in-plane anisotropy is due to self-organized charge stripes. For example, in

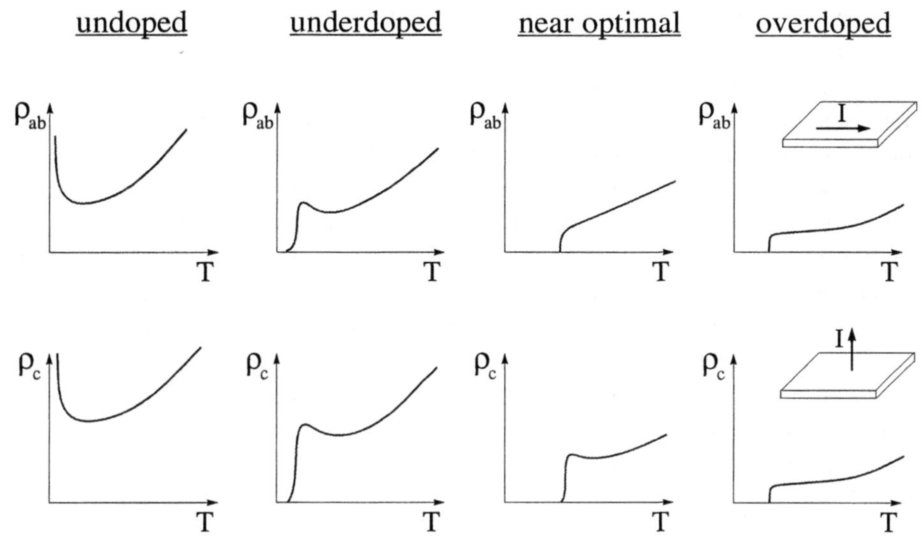

Figure 3.18. Schematic overview of transport properties of cuprates at different dopings. The in-plane resistivity is shown at the top, and the out-of-plane resistivity at the bottom. The insets depict the direction of the current in each case.

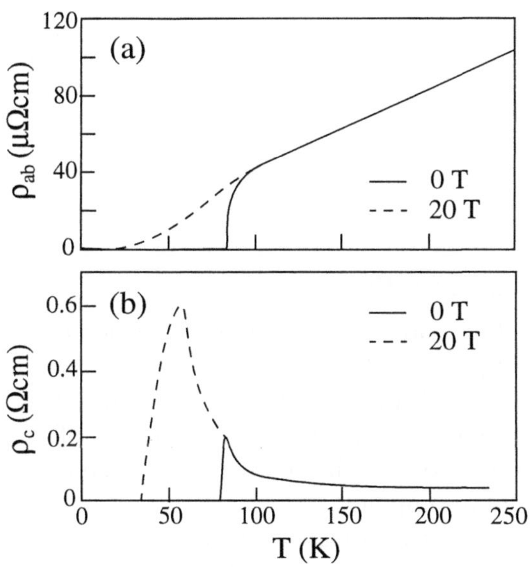

Figure 3.19. Temperature dependences of (a) in-plane and (b) out-of-plane resistivities in slightly overdoped Bi2212 with $T_c \simeq 80$ K. The solid lines show the resistivities in zero magnetic filed, and the dashed lines in a *dc* magnetic field of 20 Tesla [31].

undoped LSCO, the ρ_b/ρ_a ratio increases with lowering temperature attaining 1.4 at 4.2 K [32].

The effect of an applied magnetic field on in-plane and out-of-plane resistivities is shown in Fig. 3.19. An applied magnetic field having the *high* magnitude reveals the parts of ρ_{ab} and ρ_c at low temperatures, camouflaged by the onset of the superconducting phase in zero magnetic field. This is similar to the behavior of ρ_c in overdoped Bi2212, as shown in Fig. 3.19b. An applied magnetic field having the *low* magnitude smears the transition to the superconducting state, as shown in Fig. 3.19a. By contrast, in conventional superconductors, by applying magnetic field, the superconducting transition remains sharp, simply shifting to low temperatures. On the other hand, such behavior of the resistivity in conventional superconductors can be easily understood by taking into account the fact that, in metallic superconductors, there is no metal-insulator transition as there is in cuprates.

Table 3.2. Characteristics of optimally doped cuprates: T_c, the coherence length ξ_i, the penetration depth λ_i and the upper critical magnetic fields B^i_{c2} (i = ab or c).

Compound	T_c (K)	ξ_{ab} (Å)	ξ_c (Å)	λ_{ab} (Å)	λ_c (Å)	B^{ab}_{c2} (T)	B^c_{c2} (T)
NCCO	24	70–80	~15	1200	260 000	7	-
LSCO	38	33	2.5	2000	20 000	80	15
YBCO	93	13	2	1450	6000	150	40
Bi2212	95	15	1	1800	7000	120	30
Bi2223	110	13	1	2000	10 000	250	30
Tl1224	128	14	1	1500	–	160	-
Hg1223	135	13	2	1770	30 000	190	-

3.5 Coherence length

In superconducting cuprates, the average distance between two electrons (holes) in a Cooper pair, i.e. the coherence length, is very short in comparison with the coherence length in conventional superconductors, 400–10^4 Å. The short coherence length in cuprates is a consequence of the large energy gap and the small Fermi velocity. The shortness of the coherence length is a very important feature of the cuprates.

Moreover, due to the two-dimensional structure of cuprates, the coherence length depends on the crystal direction: the coherence length along the c axis, ξ_c, is much smaller than the in-plane coherence length ξ_{ab}. In different hole-doped cuprates, the in-plane coherence length varies between 10 and 35 Å, while the out-of-plane coherence length is only 1–5 Å. In general, the in-plane coherence length in cuprates with low T_c is larger than that of cuprates with high T_c: the data for selected cuprates are listed in Table 3.2. In electron-doped NCCO, the in-plane coherence length is a few times larger than that of in hole-doped cuprates, $\xi_{ab} \approx 80$ Å.

The small values of c-axis coherence length means that the transport along the c axis is incoherent, even in the superconducting state. For example in Bi2212, $\xi_c \sim 1$ Å which is a few times smaller than the interlayer distance.

3.6 Penetration depth and superfluid density

The penetration depth λ is one of the most important parameters of the superconducting state because λ directly relates to the superfluid density as $n_s \propto 1/\lambda^2$. Thus, penetration-depth measurements are a probe of *coherence* properties of the condensate. The penetration depth can be obtained by different techniques such as microwave, infrared, muon-Spin-Relaxation (μSR), ac-susceptibility, inductance measurements etc. Table 3.2 lists the penetration-depth data for some cuprates. In Table 3.2, one can see that in cuprates λ is very large, particularly in the c-axis direction, meaning that n_s is very low.

Uemura and co-workers showed that, in underdoped cuprates, T_c depends linearly on the superfluid density and this dependence is universal for all superconducting cuprates, as shown in Fig. 3.20. Moreover, this remarkable de-

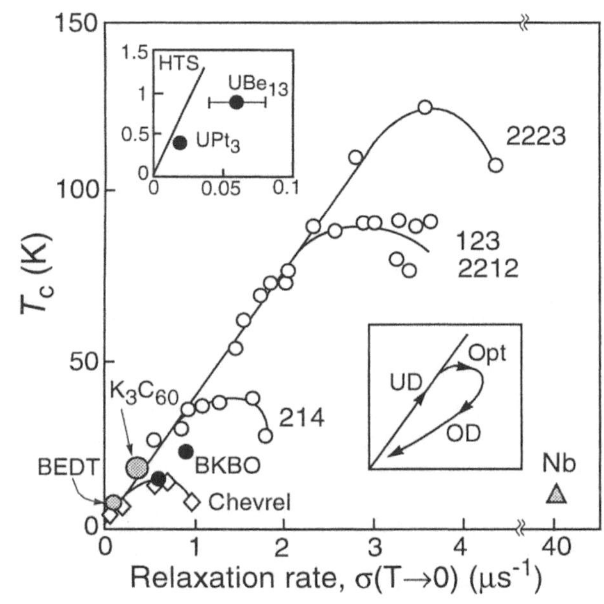

Figure 3.20. Critical temperature versus muon spin relaxation rate $\sigma(T \rightarrow 0)$ for various superconductors ($\sigma \propto 1/\lambda^2 \propto n_s/m^*$). Heavy fermions are shown in the upper inset. BEDT is a quasi-one-dimensional organic superconductor. The lower inset shows the "boomerang" path with increasing doping: the underdoped (UD), optimally doped (Opt) and overdoped (OD) regions [33].

pendence is manifest not only in cuprates but also in quasi-one dimensional organic superconductors, C_{60}, Chevrel phases, heavy fermions and $Ba_{1-x}K_xBiO_3$ (BKBO). Consequently, "all of these systems belong to a special group of superconductors with some fundamental features in common and, possibly, share a common condensation mechanism" [33]. At optimal doping, the critical temperature starts to saturate and, in the overdoped region, decreases, making the "boomerang path", as shown in the lower inset of Fig. 3.20. Thus, in the overdoped region, the superfluid density falls with increase of hole concentration: "the suppression of T_c in this region is therefore *not* due to the disappearance of superconducting carriers but rather should be attributed to other causes" [33]. Indeed, in the overdoped region, additional charge carriers gather into metallic islands, as shown in Fig. 3.16b.

The Uemura plot depicted in Fig. 3.20 also shows that in cuprates, the concentration of charge carriers is more than one order of magnitude lower than that in metallic Nb superconductor. Thus, this "encourages development of a theory of superconductivity in low carrier densities and strong pairing interactions which interpolates to the BCS theory and the Bose-Einstein condensation" [33]. The plot also indicates that heavy fermions have relatively high T_c as scaled with their low superfluid density, which is a consequence of their heavy effective mass m^*. For cuprates, the Uemura plot suggests that the optimum doping is different in different cuprates, confirming that in each compound the maximum T_c is determined by a set of parameters, principally by the doping level and structural parameters.

Figure 3.21a shows the "inverted" Uemura plot—the doping dependence of penetration depth in LSCO. From Figs 3.20 and 3.21a, the penetration depth as a function of doping attains a minimum not in the optimally doped region but in the slightly overdoped region. In Bi2212, at different doping levels, the minimum of in-plane penetration depth is also located in the slightly overdoped region [36]. In Fig. 3.21a, one can see that, in LSCO, by varying p, $\lambda_{ab}(p)$ and $\lambda_c(p)$ are approximately proportional to each other.

In Fig. 3.21a, the dashed line schematically shows the behavior of in-plane and out-of-plane penetration depths in the heavily overdoped region, as *anticipated* from the Uemura plot in Fig. 3.20.

Finally, it is worth mentioning that, in YBCO, due to the presence of chains, the penetration depth is smaller along the chains than the penetration depth along the a axis: in YBCO with T_c = 93 K, λ_a = 1550–1600 Å and λ_b = 800–1000 Å. The value λ_{ab} = 1450 Å, listed in Table 3.2 for YBCO, presents the value $\sqrt{\lambda_a \lambda_b}$.

3.7 Electronic specific heat and the condensation energy

The normal–superconducting transition is a second-order phase transition accompanied by a jump in the heat capacity. In conventional superconductors,

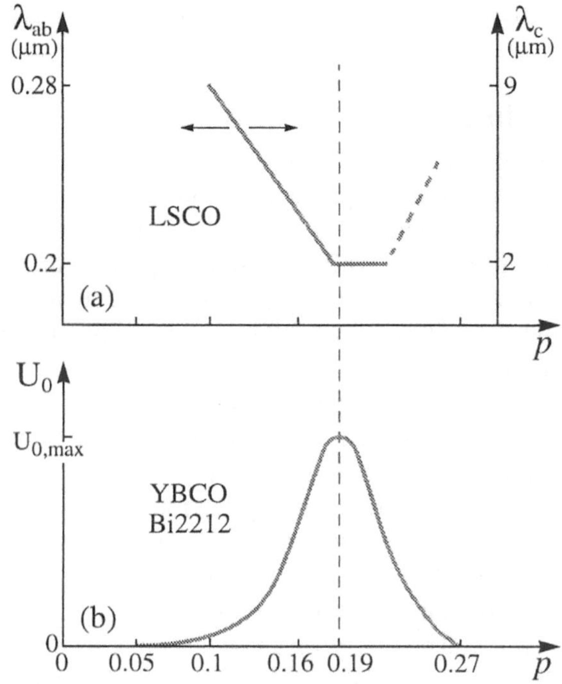

Figure 3.21. (a) Absolute values of the in-plane λ_{ab} and out-of-plane λ_c magnetic penetration depth as a function of doping for LSCO [34]. The dashed line shows an expected behavior of λ_i ($i = ab, c$) at high doping, as inferred from Fig. 3.20. (b) Superconducting condensation energy U_0 as a function of doping for Bi2212 and YBCO [35].

on cooling the heat capacity of a superconductor has a discontinuous jump at T_c. The jump is a measure of the pair density, but is not, in general, a $T = 0$ property. As noted in Chapter 2, the heat capacity of cuprates also has a jump at T_c.

By studying the thermodynamic properties of cuprates, it is possible to obtain the condensation energy, U_0, defined by

$$U_0 = \int_0^{T_c} (S_{NS} - S_{SC})dT, \tag{3.2}$$

where S_{NS} is the extrapolated normal-state electronic entropy and S_{SC} is the superconducting-state electronic entropy. Figure 3.21b shows U_0 as a function of doping, obtained in YBCO and Bi2212. In Fig. 3.21b, one can see that superconductivity in the cuprates is most robust in the slightly overdoped region, at $p \simeq 0.19$, and not at optimal doping ($p \simeq 0.16$). This is in good agreement with the penetration-depth data depicted in Fig. 3.21a. From Fig. 3.21, superconductivity in the underdoped region is very weak. The maximum values of

U_0 in YBCO and Bi2212 are $U_{0,max}$ = 2.6 J/g atoms and $U_{0,max}$ = 2 J/g atoms, respectively.

3.8 Effect of impurities

One of the crucial tests for the superconducting state, which arises due to a specific mechanism, is how magnetic and non-magnetic impurities affect it. In conventional superconductors, non-magnetic impurities have a weak effect on superconducting properties, while magnetic impurities drastically affect the superconducting state. It is already an experimental fact that magnetic and non-magnetic impurities have the opposite effect on superconductivity mediated by magnetic fluctuations [37, 38].

The coherence length of conventional superconductors is very large; therefore the effect of an impurity on superconductivity on a microscopic scale compared to that on a macroscopic scale is practically the same. This, however, is not the case for superconducting cuprates, the coherence length of which is very short. Consequently, it is necessary to consider separately how magnetic and non-magnetic impurities affect high-T_c superconductivity on a macroscopic scale and on a microscopic scale.

Macroscopic scale. Surprisingly, on a macroscopic scale, magnetic and non-magnetic impurities have a *similar* effect on superconductivity in cuprates. The partial substitution of Fe, Ni and Zn for Cu affects T_c similarly with $dT_c/dx \approx$ -4–5 K/at.%, independently of the substitutional element [39]. An exception to this rule is Zn-doped YBCO, where Zn suppresses T_c three times faster (-12 K/at.%) than, for example, Ni does. Experimentally, Zn atoms occupy not only Cu sites in CuO_2 planes, but also Cu-chain sites. The effect of Zn-on-chain location is that Zn interrupts the coherence between nearest CuO_2 planes.

The superfluid density is suppressed similarly by a magnetic and a non-magnetic impurity. However, the superfluid density is suppressed much faster than T_c: a 10% reduction in T_c results in a 27% reduction in n_s.

Microscopic scale. Despite their similar effect on T_c, magnetic and non-magnetic impurities cause very different effects on their local environment. Tunneling measurements performed above Zn and Ni impurities situated in CuO_2 planes show that Zn creates voids around it, locally suppressing the superconductivity [38]. The voids in swiss cheese are reminiscent of the local Zn effect on high-T_c superconductivity. In contrast, a magnetic Ni atom has surprisingly little effect on its local environment in CuO_2 planes: superconductivity is not interrupted at a Ni site [37]. The main conclusion from these remarkable results is that, in CuO_2 planes, there is magnetically-mediated superconductivity.

Another surprising effect produced locally by a nonmagnetic Zn atom is that Zn induces a local magnetic moment of $0.8\mu_B$ either in hole-doped cuprates or electron-doped NCCO, where μ_B is the Bohr magneton. Magnetic moments on all Cu sites around a Zn atom have a staggered order; thus a nonmagnetic Zn atom does not destroy local antiferromagnetic correlations, but enhances them. Logically, magnetically mediated superconductivity should be enhanced around Zn atoms too; however, this is not the case. Magnetic Fe and Ni atoms locally induce an effective magnetic moment of $4.9\mu_B$ and $0.6\mu_B$ in hole-doped cuprates, and $2.2\mu_B$ and $2\mu_B$ in electron doped NCCO, respectively. Thus, in hole-doped cuprates, Ni located in CuO_2 planes reduces the effective magnetic moments on neighboring Cu spins.

3.9 Critical magnetic fields and critical current J_c

All superconducting cuprates are type-II superconductors. So, they have two critical magnetic fields: $B_{c1}(0)$ and $B_{c2}(0)$. Moreover, due to the highly anisotropic structure of cuprates, there is a huge anisotropy of the critical magnetic fields applied parallel and perpendicular to the CuO_2 planes. Thus, in cuprates, there are four different critical magnetic fields: $B_{c1}^{ab}(0)$, $B_{c1}^{c}(0)$, $B_{c2}^{ab}(0)$ and $B_{c2}^{c}(0)$. The upper symbols ab and c denote the critical value of B applied parallel to the ab planes and parallel to the c axis, respectively. B_{c2} is much larger when the field is applied parallel to the CuO_2 planes than when it is applied perpendicular to those planes. This is because most of the conduction is in the planes: a magnetic field applied parallel to the planes is not very effective in destroying superconductivity within the planes. The values of B_{c2}^{ab} and B_{c2}^{c} for some cuprates, extrapolated from the resistivity data, are listed in Table 3.2. These values are very approximate since the resistivity measurements can only be done near T_c, where the magnetic fields are accessible in the laboratory. The extrapolation to zero temperature, made on the basis of near-T_c data, is inaccurate because the $B_{c2}(T)$ dependence is not linear. Magnetization measurements give substantially higher values than those listed in Table 3.2. As an example, magnetization B_{c2} data for YBCO are $B_{c2}^{ab} \approx 670$ T and $B_{c2}^{c} \approx 125$ T.

In cuprates, the lower critical fields $B_{c1}^{ab}(0)$ and $B_{c1}^{c}(0)$ are very small: in YBCO, $\sim 2 \times 10^{-2}$ T and $\sim 5 \times 10^{-2}$ T, respectively. In Hg1223, $B_{c1}^{ab}(0) \approx 3 \times 10^{-2}$ T. It is interesting that the anisotropy in B_{c1} has the sign opposite to that in B_{c2}: $B_{c2}^{c} < B_{c2}^{ab}$ but $B_{c1}^{c} > B_{c1}^{ab}$.

In metallic superconductors described by the BCS theory, $B_{c2} \propto T_c^2$. In cuprates the relation is different: $B_{c2} \propto T_c^{\sqrt{2}}$, found in cuprates with low T_c.

As for their transport properties—the critical current in layered cuprates is also very anisotropic. The highest values of the critical current J_c are obtained in epitaxial thin films of YBCO. At liquid helium temperature, the critical current in the ab plane is almost 10^8 A/cm^2 and, along the c axis of order 10^5 A/cm^2.

3.10 Phase stiffness

In quantum mechanics, whenever two electrons interact, they do so by having their wave functions overlap. Denoting the two wave functions by Ψ_1 and Ψ_2 and the interaction potential by H, the interaction depends upon the overlap integral

$$\int_{-\infty}^{+\infty} \Psi_1 H \Psi_2 dV, \tag{3.3}$$

where dV is an element of volume, and the integration is carried out over all space.

The superconducting state requires the electron pairing *and* the onset of long-range phase coherence. As discussed in Chapter 2, in conventional superconductors described by the BCS theory, two electrons create a Cooper pair at T_c due to electron-phonon interactions. The phase coherence among the Cooper pairs is also established at T_c due to the overlap of Copper-pair wave functions. The overlap of the wave functions occurs immediately at T_c, because the superfluid density is relatively high and, as a consequence, the average distance between the Cooper pairs is much smaller than the coherence length (the size of a Cooper pair). In other words, the phase stiffness in metallic superconductors is very high. In cuprates, depending on the doping level, the size of a Cooper pair is smaller than or similar to the distance between the Cooper pairs. So, at such conditions, the Cooper pairs can exist well above T_c, particularly, in the underdoped region.

As defined in Chapter 2, the ratio T_{PO}/T_c characterizes the phase stiffness, where T_{PO} is the temperature at which the phase order would disappear if the disordering effects of all degrees of freedom were ignored. In conventional superconductors, the ratio T_{PO}/T_c is very high, and lies between 2×10^2 and 2×10^5. At the same time, in cuprates this ratio is very small, 0.7–16, indicating that the pairing may occur well above T_c, and with fluctuating phase coherence. Indeed, a few experiments show phase-coherence fluctuations just above T_c [40]. The presence of the Cooper pairs above T_c is discussed below (see Section 3.13).

It is important to keep in mind that, in conventional (as well as in unconventional) superconductors, the pairing mechanism and the mechanism of the establishment of long-range phase coherence are different.

3.11 Phase coherence along the *c* axis

Superconductivity in cuprates is two-dimensional. In cuprates (and other layered superconductors), T_c is the temperature at which superconducting CuO_2 planes establish the phase coherence along the c axis.

It is mistakenly believed that the phase coherence along the c axis is established due to the Josephson coupling between the planes. Measurements of the in-plane (ρ_{ab}) and out-of-plane (ρ_c) resistivities in Tl2212, as a function of applied pressure, show that $\rho_c(T)$ shifts smoothly down with increase of pressure; however T_c first increases and then *decreases* [41]. This result can not be explained by the interlayer Josephson-coupling mechanism. Thus, "any model that associates high-T_c with the interplane Josephson coupling should therefore be revisited" [41].

Chapter 9 is devoted to the mechanism of phase coherence in cuprates: the mechanism of interplane hopping is magnetic, due to spin fluctuations.

3.12 Two energy scales: pairing and phase-coherence

The energy gap is one of the most important characteristics of the superconducting state. Since the discovery of high-T_c superconductivity in cuprates they have been extensively studied by different techniques. Surprisingly, in cuprates different techniques measure different values of the energy gap. The discrepancy remained a mystery until it was realized that different experimental methods probe two different energy gaps. Moreover, the magnitudes of the two gaps strongly depend on the doping level.

Figure 3.22 shows the two energy scales in cuprates as a function of hole concentration [42, 43]. Tunneling and angle-resolved photoemission (ARPES) spectroscopies probe the pairing energy gap Δ_p, since these techniques are

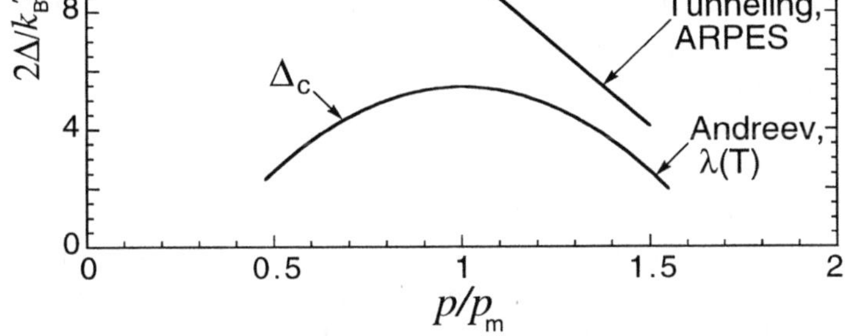

Figure 3.22. Low-temperature phase diagram of superconducting cuprates: the pairing energy scale Δ_p is obtained in tunneling and photoemission (ARPES) measurements, while the phase-coherence energy scale Δ_c in Andreev-reflection and penetration-depth measurements (p_m = 0.16).

sensitive to the single-quasiparticle excitation energy. Penetration-depth and Andreev-reflection measurements, which are exclusively sensitive to the coherence properties of the condensate, probe the coherence energy range Δ_c. In Fig. 3.22, Δ_p increases linearly as p decreases, whereas Δ_c has a parabolic dependence on p since Δ_c is proportional to T_c as $2\Delta_c/k_BT_c \simeq 5.4$. Electron-doped cuprates also have two energy scales [44].

In a BCS condensate, Δ_p and Δ_c are identical due to high phase stiffness. In cuprates, two electron-like or hole-like quasiparticles couple with each other above T_c, but the *steady* phase coherence among the pairs is established at T_c. Just above T_c, there are fluctuations of phase coherence.

It is important realizing that the pairing Δ_p scale is the in-plane energy scale Δ_p^{ab}, while the coherence Δ_c scale is mainly the c-axis energy scale Δ_c^c.

In Fig. 3.22, $\Delta_p(p)$ and $\Delta_c(p)$ have different dependences on p and do not correlate with each other. This fact reveals that the pairing and the phase coherence are most likely governed by two different mechanisms.

3.13 Cooper pairs above T_c

The presence of pairing and phase-coherence energy scales shown in Fig. 3.22 automatically implies the presence of Cooper pairs above T_c. Indeed, a few experiments revealed traces of the Cooper pairs above T_c. For example, in slightly overdoped Bi2212, the sign of the pairs has been seen in tunneling measurements 20 K above T_c (see Chapter 12). In underdoped Bi2212, fluctuations of phase coherence have been observed 20 K above T_c [40]. In fact, in highly underdoped and undoped regions, incoherent Cooper pairs exist in cuprates at room temperature (see Fig. 3.34).

3.14 Symmetry of the order parameter: s-wave vs d-wave

In conventional superconductors, each electron of a Cooper pair has opposite momentum and spin compared to the other. The order parameter Ψ corresponding to such a superconducting ground state has an s-wave symmetry (an angular momentum $\ell = 0$), as shown in Fig. 3.23. In an s-wave condensate, the energy gap has no nodes. Theoretically, two electrons can also be bound in a singlet state with an angular momentum $\ell = 2$ (d-wave). Figure 3.23 also shows a d-wave ($d_{x^2-y^2}$) symmetry of the order parameter. A key feature distinguishing the $d_{x^2-y^2}$ symmetry is that it has two positive and two negative lobes. The d-wave symmetry of the order parameter was first attributed to the superconducting ground state of heavy fermions just before the discovery of high-T_c superconductors. In the case of an anisotropic d-wave state, the energy gap has four nodes, as shown in Fig. 3.23.

The situation in cuprates is *very* peculiar. All phase-sensitive techniques show that hole-doped and electron-doped cuprates have the $d_{x^2-y^2}$ supercon-

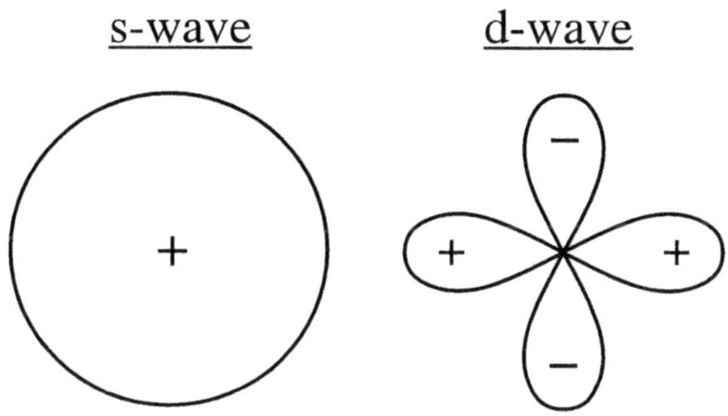

Figure 3.23. Two possible symmetries of the superconducting order parameter, shown in real space: isotropic s-wave and $d_{x^2-y^2}$ wave (d-wave).

ducting ground state. ARPES measurements also observe four-fold symmetry of energy gap. At the same time, all tunneling measurements performed in hole-doped cuprates along the c axis show that the superconducting condensate has an s-wave symmetry [45]. Theoretically, tunneling between a s-wave superconductor and a d-wave superconductor in the direction perpendicular to the plane where the d-wave lobes are located is not expected because the coupling of the positive lobes should cancel that of the negative lobes. In [45], tunneling along the c axis was observed between two parts of a Bi2212 single crystal, turned relatively to each other by 45°. In the latter case, if the symmetry of the superconducting ground state of Bi2212 is a d-wave, the tunneling current has to be zero. This, however, is not the case. Tunneling measurements in electron-doped cuprates also show the presence of an s-wave condensate [44].

Knowledge of the symmetry of the superconducting ground state of cuprates places constraints on the theory of high-T_c superconductivity. For example, pair fields with d-wave symmetry are not favored by phonons. In contrast, spin fluctuations exclusively mediate superconductivity with a $d_{x^2-y^2}$ ground state (or g-wave).

We return to discuss this issue in Chapter 10.

3.15 Phonons in cuprates

In optimally doped cuprates, the isotope effect is very weak (see Fig. 3.17). This fact was initially taken as evidence against the BCS mechanism of high-T_c superconductivity and, mistakenly, against the phonon-pairing mechanism. If the pairing mechanism is different from the BCS mechanism, this does not mean that phonons are irrelevant. In fact, phonons are an essential part of the mechanism of high-T_c superconductivity.

First, the lattice and charge dynamics are inexorably mixed in superconducting cuprates [26–30, 46–49]. Second, phonons are responsible for the formation of charge stripes in cuprates [50–52]. Third, as shown above, the structural transitions precede the pseudogap formation and the transition into the superconducting state. Finally, the lattice and spin degrees of freedom are strongly coupled below T_c (see Section 3.13).

Thus, the role of phonons in superconducting cuprates is difficult to underestimate; however, it is different from that in conventional superconductors. In BCS-type superconductors, T_c increases with lattice softening. In high-T_c superconductors, T_c increases with lattice stiffening, as shown in Fig. 3.24. A logical assumption is that phonons, as a component of the superconducting mechanism, act differently in cuprates compared to metallic superconductors.

Inelastic neutron measurements (INS) show that two phonon branches seem to be important in cuprates: 55 meV and 70 meV [47, 49]. Both phonon branches are associated with oxygen vibrations: the 55 meV phonons are attributed either to in-plane or out-of-plane Cu–O bond-bending vibrations. The 70 meV branch is associated with half-breathing-like oxygen phonon modes that propagate in the CuO_2 plane. Independently of the origin of these phonon modes, their energies are unusually high, indicating that the electron-phonon coupling is very strong. The energy range of these phonon modes corresponds to an optical branch of lattice vibrations. Kresin and Wolf [54] proposed that in two-dimensional cuprates, the optical branch has a shape similar to the acoustic branch, and is elevated to about 15 meV at $k = 0$. It is called a *soft* optical mode.

The phonon dispersion curves of stoichiometric La_2NiO_4 (LNO) are found to closely resemble those of stoichiometric LSCO ($x = 0$) [55]. Doping by extra oxygen or cation deficiency leads to phonon anomalies at high energies similar to those observed in doped LSCO. Thus, the lattice dynamics for the corresponding doped and pure LNO and LSCO shows a surprising similarity: not only are the phonon dispersion curves of the two undoped materials very similar, but also the same phonon anomalies develop upon doping. It is worth noting that, in nickelates, doped charge carriers form one-dimensional stripes similar to those of cuprates.

It is interesting that in new MgB_2 superconductor, high-energy phonons peaking at 36 meV (acoustic branch) and at 54, 78, 89 and 97 meV (optical branch) seem to be important for superconductivity, as revealed by inelastic neutron scattering measurements [56]. Phonons are responsible for electron pairing in MgB_2 as well as in cuprates (see Chapters 8, 10 and 11).

3.16 Magnetic properties

Superconductivity and magnetism were earlier regarded as mutually exclusive phenomena. However, after the discovery of superconductivity in heavy

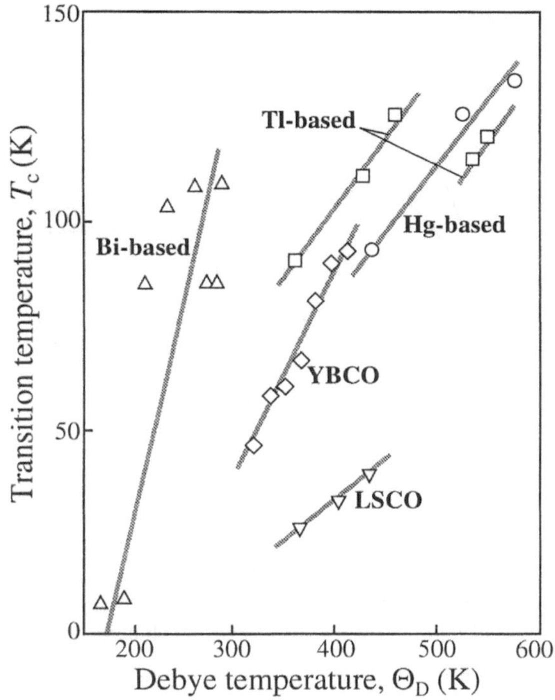

Figure 3.24. Critical temperature as a function of Debye temperature for different cuprates [53].

fermions, antiferromagnetic fluctuations were thought to be responsible for mediating the electron pairing in these materials. Indeed, recent research showed that spin fluctuations participate in mediating superconductivity in heavy fer-mions. This intimate relationship between superconductivity and magnetism also appear to be crucial in cuprates which inherited their magnetic properties from the parent compounds—antiferromagnetic Mott insulators.

Inelastic neutron scattering (INS) measurements show that cuprates display a wide variety of magnetic properties. Because INS measurements require large-size homogeneous single crystals, INS studies have been performed only in a few cuprates: YBCO, LSCO, Bi2212 and Tl2201. Generally, all the cuprates exhibit common features of magnetic interactions. At the same time, there are some particularities of magnetic correlations in each cuprate. We consider first magnetic properties of YBCO since it exhibits both commensurate and incommensurate peaks.

YBCO is a double-layer cuprate and, in first approximation, it can be modeled as a set of weakly coupled CuO_2 bilayers. By neglecting the local anisotropy and other smaller interaction terms, the high-frequency spin dynamics can

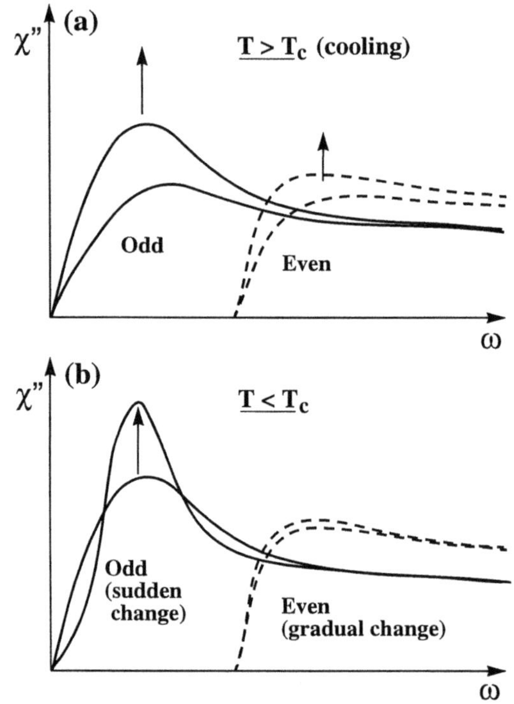

Figure 3.25. Schematic diagram summarizing the temperature evolution of the magnetic excitation spectra of the underdoped YBCO compound (a) in the normal state, and (b) across the superconducting transition. The excitations in the even channel evolve smoothly with temperature. The excitations in the odd channel undergo an abrupt sharpening at the resonance energy across T_c.

be described using the Heisenberg Hamiltonian for a single bilayer

$$H = \sum_{ij} J_{\|} S_i \cdot S_j + \sum_{ij'} J_{\perp} S_i \cdot S_{j'}, \tag{3.4}$$

where $J_{\|}$ and $J_{\perp}$ are the intralayer and interlayer superexchange constants, respectively. The first term in the expression represents the nearest-neighbor coupling between Cu spins S_i in the same plane and the second the nearest-neighbor coupling between Cu spins in different planes. In YBCO, $J_{\|} \simeq 120$ meV and $J_{\perp} \simeq 12$ meV. INS studies reveal that, in YBCO and other cuprates, spin-wave excitations present up to $2J_{\|}$.

By analogy with phonon-excitation spectra (see Fig. 2.4), spin-wave excitations in *bilayer* YBCO are split into two channels: acoustic and optical. In the acoustic (odd) channel, pairs of neighboring spins in adjacent planes rotate in the same sense about their average direction. In the optical (even) channel, the

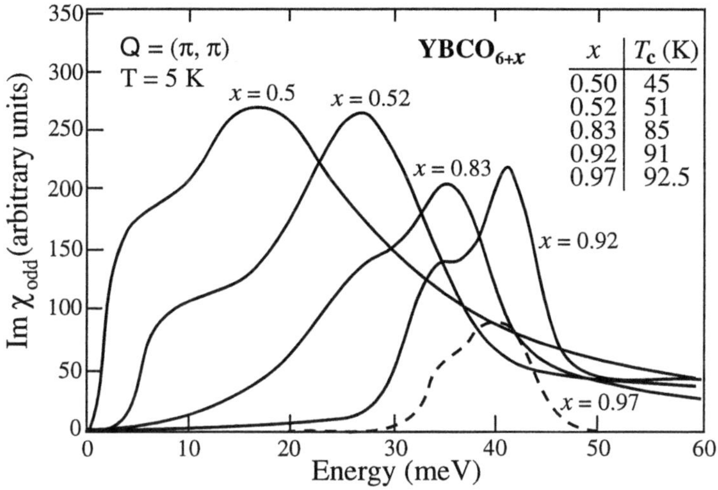

Figure 3.26. Magnetic resonance peak in YBCO versus doping level (the spin susceptibility in the odd channel at 5 K).

spins in adjacent planes rotate in opposite directions, thus sensing the restoring force from the interplane coupling $J_\perp$.

Figure 3.25 schematically summarizes the temperature dependence of the magnetic excitation spectra of underdoped YBCO. At any doping level of YBCO, in the even channel, there is a gap in the spin-wave excitation spectrum. The even (optical) excitations are almost unchanged across the superconducting transition, as shown in Fig. 3.25. The odd (acoustic) excitations undergo an abrupt sharpening on cooling through T_c. This sharp mode in the odd channel is called the *magnetic resonance peak*, which appears at the antiferromagnetic wave vector $Q = (\pi, \pi)$ exclusively below T_c. The resonance peak corresponds to a collective spin excitation. The energy at which the resonance peak appears, E_r, scales with doping level. Figure 3.26 shows the resonance peak as function of doping: E_r scales linearly with T_c. The antiferromagnetic correlations weaken in the overdoped region; however, the magnetic relaxation is still dominant in the highly overdoped region [22].

Apart from the resonance peak which is commensurate, INS measurements have found four incommensurate peaks at some energy transfers, which appear in YBCO below T_c. Each spin scattering peak occurs at an incommensurate wave vector $Q \pm \delta$, as schematically shown in Fig. 3.27. In Fig. 3.27a, upon doping, commensurate antiferromagnetic Bragg peaks (grey stars) caused by antiferromagnetic ordering in ideal Mott insulator disappear and are replaced by four broadened incommensurate dynamic peaks (black circles and squares). This indicates that spin fluctuations are displaced from the commensurate peak by a small amount δ, related in the underdoped region to the doping p by

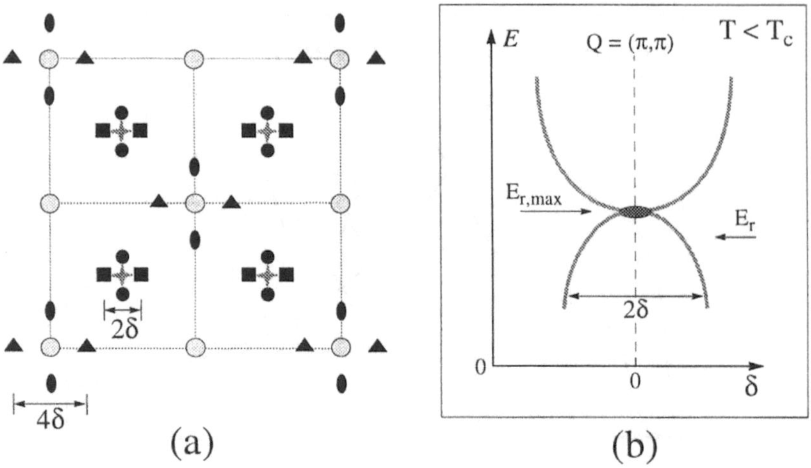

Figure 3.27. (a) Neuron scattering data in the reciprocal (momentum) space, obtained in LSCO. The commensurate antiferromagnetic Bragg peaks, obtained in undoped LSCO at wave vector $Q = (\pi, \pi)$, are shown by grey stars. Upon doping, the commensurate peaks disappear and are replaced by four broadened incommensurate dynamic peaks (black squares and circles) with incommensurability δ. The fundamental lattice Bragg peaks are shown by large grey dots. Upon doping, new peaks (black triangles and ovals) are observed, which are displaced by 2δ from the fundamental lattice peaks. (b) Energy dependence of the peak position of the incommensurate peaks versus incommensurability δ in YBCO. There are two legs below the maximum resonance energy $E_{r,max}$ = 41 meV. Above the resonance, the signal is also split. In the underdoped region, the resonance peak appears at $E_r < E_{r,max}$.

$\delta = p$. The incommensurate peaks can be caused by a sinusoidal spin density wave (SDW) slowly fluctuating in space and time. At the same time, scattering at $Q \pm \delta$ can also arise from a spin wave that is locally commensurate, but whose phase jumps by π at a periodic array of domain walls, termed antiphase boundaries, or charge stripes. The charge stripes will be discussed below (see the following Section).

At all dopings, the incommensurability and the commensurate resonance appear to be inseparable parts of the general features of the spin dynamics in YBCO, as shown in Fig. 3.27b. In optimally doped YBCO, the resonance peak appears at the maximum resonance energy $E_{r,max}$ = 41 meV and, below 41 meV, there are well-separated incommensurate peaks. At low energies (< 30 meV), the intensity of the incommensurate peaks is strongly reduced. This can result from the opening of a spin gap below T_c. In Fig. 3.27b, above $E_{r,max}$, the peaks are separated again, and the separation gradually increases with increasing energy up to $2J_{\parallel}$, although the peaks have a broader width and a much weaker intensity than those below $E_{r,max}$. In the underdoped region, the commensurate resonance peak is shifted to a lower energy $E_r < E_{r,max}$, scaling with T_c, and the incommensurate peaks appear below E_r. On heating through T_c, the incommensurate peaks are strongly renormalized upon approaching T_c and vanish in the normal state.

In LSCO, the low energy magnetic excitations have been extensively studied, and the observed spin fluctuations are only characterized by incommensurate peaks. The magnetic resonance peak has never been observed in LSCO. Incommensurability in LSCO is consistent with that in YBCO with the same hole doping, but, in LSCO it persists in the normal state. In LSCO, the peak position is unchanged across the transition temperature. Only around room temperature does the incommensurate structure begin to disappear. In LSCO, δ is energy-independent but depends strongly on the doping level. In single-layer LSCO, spin excitations do not split into the acoustic and optical channels as do those in double-layer YBCO.

The similarity of spin dynamics in two different cuprates, LSCO and YBCO, demonstrate that the spin dynamics do not depend on the details of the Fermi surface, but have an analogous form to that for the stripe phase. This will be further considered below (see the next Section).

The resonance mode observed in YBCO has also been found in double-layer Bi2212 and in single-layer Tl2201 [57]. Incommensurability has not yet been seen in Bi2212 and Tl2201, but it is expected to be observed.

In the studied double-layer cuprates with different dopings, the position of the magnetic resonance peak scales with T_c as $E_r/k_BT_c \simeq 5.4$. In single-layer Tl2201, this ratio is slightly larger $E_r/k_BT_c \simeq 5.9$ [57]. In YBCO, with different dopings, the gap in the spin-wave excitation spectrum, observed in the even channel and shown in Fig. 3.25, surprisingly coincides with the pairing energy scale Δ_p shown in Fig. 3.22 [58]. What is even more surprising is that, in the *undoped* region where the resonance mode in the odd channel is obviously absent, the gap in the spin-wave excitation spectrum in the even channel remains and continues to scale with Δ_p, increasing as p decreases. This means that the existence of the pairing Δ_p scale is a necessary but *not* the sufficient condition for appearance of superconductivity in cuprates. Second, the magnetic resonance peak *directly* relates to the establishment of the phase coherence. The latter means that the resonance mode is either a consequence or the mediator of the phase coherence.

Finally, a question which is often overlooked is whether the incommensurate response is a part of a continuous dispersion below the resonance peak. Or, is the resonance peak a consequence of the incommensurability? In fact, they are inseparable parts of the same phenomenon. The case of LSCO, however, shows that the incommensurability can exist without the presence of the commensurate peak, but not vice versa.

3.17 Stripe phase

As discussed above, charge carriers doped into insulating CuO_2 planes are distributed inhomogeneously either on a macroscopic scale or on a microscopic scale, as shown in Fig. 3.14. Here we consider the charge distribution and the magnetic order in CuO_2 planes on a microscopic scale (the nanoscale)—the stripe phase.

In Fig. 3.27a, we discussed only the disappearance of the commensurate antiferromagnetic Bragg peaks (grey stars) upon doping, which are replaced by four broadened incommensurate dynamic peaks (black circles and squares). The gray circles in Fig. 3.27a are the fundamental lattice Bragg peaks in ideal square CuO_2. In Nd-doped LSCO, upon doping, the long-range order occurs, which is manifest by the appearance of four new peaks (black ovals and triangles) displaced by 2δ from the fundamental lattice peaks (grey circles), as shown in Fig. 3.27a. The magnitudes of the displacement of the charge and spin peaks indicate that the charge modulation is twice larger than the spin modulation. The picture which is consistent with such a charge/spin ordering is shown in Fig. 3.28a. In this structure, the vacant sites introduced by doping reside at antiphase boundaries. The spin direction in antiferromagnetic domains rotates by 180° on crossing a domain wall.

The charge stripes in cuprates are half-filled, that is, one positive electron charge per two copper sites along the stripes (in nickelates, the charge density along the stripes is 1 hole per Ni site). The stripe phase shown in Fig. 3.28a, which is nearly one-dimensional, would not cause the appearance of four incommensurate antiferromagnetic peaks (black circles and squares in Fig. 3.27a), but only two of them (either black circles or squares). It is assumed that the appearance of the four incommensurate peaks is the consequence of stripe orientation in adjacent CuO_2 layers: the stripe orientation is alternately rotated by 90° layer by layer, as schematically shown in Fig. 3.28b.

At dopings $p < 0.13$, the average distance between charge stripes decreases and saturates near $p = \frac{1}{8}$, as shown in Fig. 3.15. Above $p = \frac{1}{8}$, the distance between stripes is practically unchanged. The length of a separate charge stripe is about 100 Å. The charge stripes are dynamical: they meander and can move in the transverse direction. In YBCO, the fluctuating charge stripes exist over the whole doping range, and become visible below a temperature somewhat higher than the pseudogap temperature [59], which is discussed below. The fluctuating charge stripes are insulating on a microscopic scale [60, 19], that is, there is a charge order on the stripes which is similar to, but **not** the conventional charge-density-wave (CDW) order. The latter fact means that there is a charge gap on stripes. In addition to the charge gap on stripes, there is a spin gap in magnetic domains which are located between charge stripes, as shown in Fig. 3.28a.

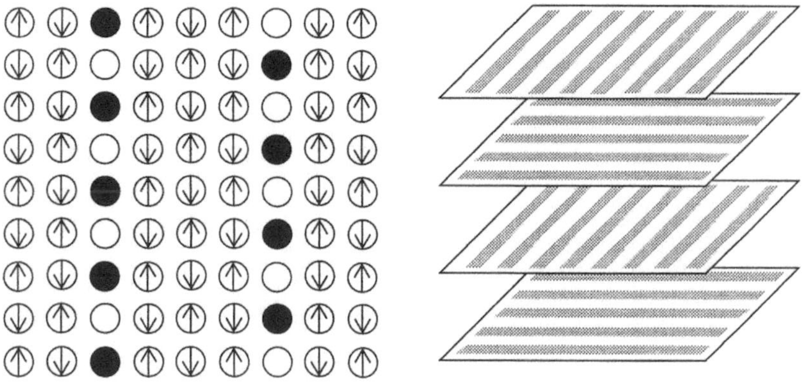

(a) (b)

Figure 3.28. (a) Idealized diagram of the spin and charge stripe pattern within a CuO_2 plane with hole density of 1/8. Only the Cu atoms are shown; the oxygen atoms, which are located between the copper atoms, have been omitted. Arrows indicate the spin orientations. Holes (filled circles) are situated at the anti-phase domain boundaries. (b) The stripe orientation in adjacent CuO_2 layers is assumed being alternately rotated by 90°.

A similar stripe order is observed in nickelates. However, the charge orientation in nickelates is diagonal, as it is in undoped cuprates. In addition, holes in the doped nickelates are less destructive to the antiferromagnetic background than they are in the isostructural cuprates. The phase diagram of $La_{2-x}Sr_xNiO_4$ (LSNO) is very similar to the phase diagram of the isostructural LSCO, having a series of phases which are closely related to the HTT, LTO and LTT phases of LSCO. In nickelates, the charge stripes are also insulating, and the charge gap on the stripes is much easier to observe because the charge stripes in nickelates are *less* dynamic compared to those in cuprates. Thus, there are many similarities between the stripe phases of cuprates and nickelates. The stripe order is also observed in manganites.

Charge stripe ordering in cuprates, nickelates and manganites is evidence for *extremely large electron-lattice coupling*. Such inhomogeneous electronic states are strongly connected to underlying lattice instabilities.

As an example, Figure 3.29 shows the neutron-diffraction data obtained in Nd-doped LSCO [2]. In Fig. 3.29, one can see that, on cooling, the charge-stripe order appears somewhat below the structural transformation, $T_{CO} < T_d$, where T_{CO} (T_d) is the onset temperature of the charge order (the structural transition). Spins in hole-pure regions become antiferromagnetically aligned at much lower temperature T_{MO}. In nickelates and manganites, the magnetic order always appears after charge ordering. For example, in LSNO having x = 0.29, 0.33 and 0.39, $T_{MO} \simeq$ 115, 180 and 150 K and $T_{CO} \simeq$ 135, 230 and 210 K, respectively. In LSNO, a clear charge gap is formed below T_{CO}. In

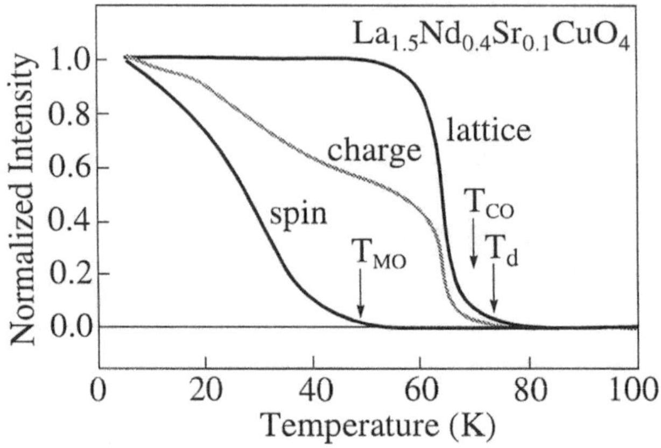

Figure 3.29. Temperature evolutions of lattice, charge and spin superlattice peaks in neutron diffraction measurements in Nd-doped LSCO. The magnetic, charge and lattice orderings are marked by T_{MO}, T_{CO} and T_d, respectively. The lattice transformation corresponds to the LTO→LTT (or LTLO) transition. The backgrounds are subtracted [2].

$La_{0.35}Ca_{0.65}MnO_3$, $T_{MO} \simeq 140$ K and $T_{CO} \simeq 260$ K. In the latter case, the magnitude of the charge gap $2\Delta(0)/k_B T_{CO} \sim 13$ is too large for a conventional CDW order on charge stripes. In nickelates and manganates, the structural transitions which precede charge ordering have been observed in acoustic measurements.

Figure 3.30 summarizes the discussion in the previous paragraph, and schematically shows the sequence of onset temperatures of charge/spin order and structural/superconducting transition in cuprates and nickelates. In Fig. 3.30, one can see that the magnetic order helps to stabilize the charge order.

In LSCO ($x = 0.15$), charge stripes are associated with the LTT phase, and magnetic domains with the LTO phase [61]. Thus, in LSCO, the LTT and LTO phases alternate, having width of about 8 and 16 Å, respectively. These results suggest that, in LSCO, charge, spin and lattice degrees of freedom at low temperature are coupled; and that the underlying lattice is responsible for charge/spin-stripe effects. The LTT phase can be considered as moving domain walls into the LTO phase.

3.18 Chains in YBCO

On the nanoscale, chains in YBCO are insulating at low temperatures, having a well-defined $2k_F$-modulated charge-density wave (CDW) order, where k_F is the momentum at the Fermi surface. The CDW is clearly observed in tunneling and nuclear quadrupole resonance (NQR) measurements. On a macroscopic scale, the chains conduct electrical current by solitons which are

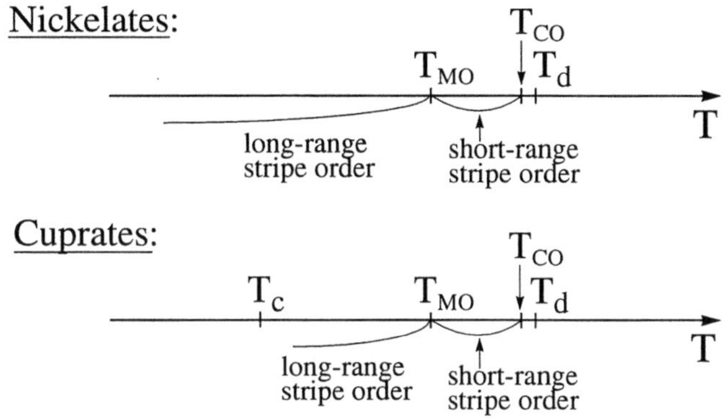

Figure 3.30. Characteristic temperatures in nickelates and cuprates: a structural transition occurs at T_d; a charge ordering at T_{CO}, and a spin ordering at T_{MO}. The short-range stripe order appears below T_{CO} and develops into the long-range stripe order at T_{MO}. The difference between nickelates and cuprates is that cuprates become superconducting at T_c.

directly observed on chains in tunneling measurements. Solitons are discussed in detail in Chapters 5 and 6.

3.19 Acoustic measurements in cuprates

Acoustic or ultrasound measurements provide direct information not only about lattice properties but also about the electron-phonon interaction and electronic properties. Although the technique is able to supply very useful information, acoustic measurements are surprisingly under-utilized in cuprates. This is the reason why acoustic measurements are discussed separately. The ultrasound technique is described in Chapter 2.

Acoustic-resonance spectroscopy measurements performed in cuprates show that superconductivity in cuprates is not of the BCS type. In conventional superconductors, when the sample is cooled to the superconducting state, the elastic coefficients always *decrease* (softening). In high-T_c superconductors, the elastic coefficient always *increase* (stiffening), as shown in Fig. 3.31. As an example, Figure 3.31 shows the elastic transverse coefficient $(C_{11} - C_{12})/2$ obtained in LSCO (x = 0.14) as a function of temperature [62]. Other elastic coefficients for various modes in LSCO reveal not only anisotropic lattice properties in the normal state, but also anisotropic coupling between superconductivity and lattice deformation [8, 63]. In LSCO and other cuprates, the elastic coefficients C_{66} and C_{11} display a large jump at the structural transition temperature $T_d \simeq 200$ K (see Fig. 3.9), while the coefficients $(C_{11} - C_{12})/2$, C_{44} and C_{33} show only a small kink at T_d. The drop in the in-plane shear elastic coefficient C_{66}, upon cooling through T_d more than by half, is attributed to the formation of domain walls [64].

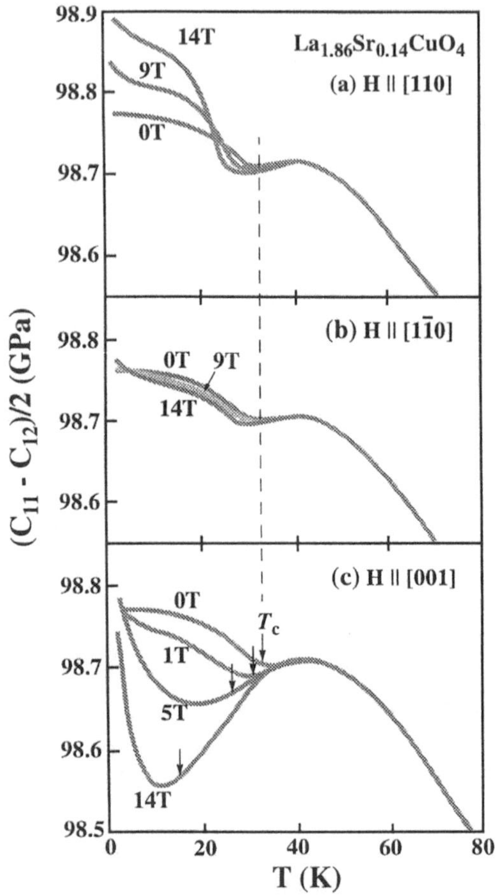

Figure 3.31 Temperature dependence of the transverse elastic coefficient $(C_{11} - C_{12})/2$ in LSCO (x = 0.14) in applied magnetic fields. The field direction is (a) $H \parallel [110] \perp u$, (b) $H \parallel [1\bar{1}0] \parallel u$, and (c) $H \parallel [001] \perp u$, where u is the displacement. Arrows in (c) indicate T_c suppressed by the applied magnetic field.

At T_c, the coefficient C_{33} reflecting the properties along the c axis has a jump which is very similar to that observed in conventional superconductors at T_c. The latter means that, in cuprates, the phase coherence is established at T_c along the c axis. The C_{11} coefficient also exhibits a small jump at T_c.

Acoustic measurements performed in undoped and underdoped LSCO, YBCO, NCCO, Bi-and Tl-based compounds show that the elastic coefficients display some kind of structural transition at maximum $T_{c,max}$ for each compound, although some of them either are not superconducting or have low T_c. This means that these structural transitions, which occur at $T_{c,max}$, do not require the presence of superconductivity. Consequently, the maximum critical temperature, $T_{c,max}$, for each compound is determined by the lattice! Again, the underlying (unstable) lattice is the driving force for high-T_c superconductivity in cuprates.

In Fig. 3.31, the transverse elastic coefficient $(C_{11} - C_{12})/2$ starts to soften at a temperature around 50 K, which is substantially higher than T_c = 36 K;

but begins to harden just below T_c. This is different from the behavior of other elastic coefficients, which do not show any corresponding anomaly just above T_c. The softening can be the result of the LTO→LTT structural transition. In this case, according to Gor'kov, the softening as a function of temperature has to be logarithmic [65]. The softening in the $(C_{11} - C_{12})/2$ coefficient has also been observed in YBCO and Bi2212.

What is the most striking in Fig. 3.31 is the effect of the magnetic field applied along various axes on $(C_{11} - C_{12})/2$ and on some other elastic coefficients. They exhibit a strong and anisotropic dependence on the direction and magnitude of the applied magnetic field for temperatures below T_c, as shown in Fig. 3.31. Above T_c, the applied magnetic field has no effect on the elastic coefficients, independently of the magnitude and the orientation of the field. This clearly shows that, below T_c, *the spin and lattice degrees of freedom are anisotropically and strongly coupled.*

All these unusual elastic properties of cuprates can naturally be understood in the framework of a scenario for high-T_c superconductivity described in Chapter 10: an interpretation of these acoustic measurements is presented in Chapter 8.

3.20 Effect of pressure

The pressure coefficient of T_c shows a nearly universal correlation in cuprates, when plotted as a function of T_c and normalized to its largest value. The coefficient is zero in optimally doped cuprates and becomes negative in overdoped materials. Among studied cuprates, LSCO is an exception to this universal scaling behavior.

The highest measured T_c = 164 K is observed in Hg1223 under a pressure of about 30 GPa.

4. Normal-state properties

There is an interesting contrast between the development of the physics of cuprates and that of the physics of conventional superconductors. Just before the creation of the BCS theory, the normal-state properties of conventional metals were very well understood; however superconductivity was not. The situation with the cuprates was just the opposite: at the time high-T_c superconductivity was discovered, there already existed a good understanding of the phenomenon of superconductivity, but the normal-state properties of the cuprates were practically unknown. The understanding of the normal-state properties of cuprates is crucial for the understanding of the mechanism of high-T_c superconductivity.

The normal-state properties of cuprates are abnormal in comparison with those of conventional superconductors. In a sense, it is ironic to use the word "normal" in the description of physical properties of cuprates above T_c. The odd behavior of the cuprates above the critical temperature is caused by the presence of a partial energy gap called the pseudogap.

4.1 Pseudogap

The pseudogap is a depletion of the density of states above the critical temperature. Cuprates do have a connected Fermi surface that appears to be consistent with conventional band theory. At all dopings, the pseudogap is pinned to the Fermi level, and therefore dominates the normal-state low-energy excitations. The pseudogap was observed for the first time in nuclear magnetic resonance (NMR) measurements, and therefore mistakenly, interpreted as a spin gap. Later, photoemission, tunneling, Raman, specific-heat and infrared measurements also provided evidence for a gap-like structure in electronic excitation spectra. Thus, it became clear that the pseudogap is not a spin gap but a gap to both spin and charge excitations; or alternatively, there are two spatially separated pseudogaps: one is a spin gap and the second is a charge gap. The magnitude of pseudogap(s) depends on doping level, p: it is large in the underdoped region and decreases as p increases.

The pseudogap is a normal-state gap; however phase-coherence fluctuations above T_c also contribute to the pseudogap below T_{pair}, the temperature at which the Cooper pairs start to form. The temperature at which the pseudogap vanishes is usually labeled by T^*. Figure 3.32 schematically shows how T_c, T_{pair} and T^* relate to each other in the underdoped region. The *magnitude* of the normal-state gap is not affected much by cooling through T_c; however, there is a re-organization of excitations inside the pseudogap (in charge and spin sectors). Below the critical temperature, the superconducting gap is predominant.

As an example, Figure 3.33 depicts the pseudogap in tunneling and resistivity measurements. In tunneling measurements above the critical temperature, conductances $dI(V)/dV$ directly relate to the density of states. In Fig. 3.33a, tunneling measurements performed by the break-junction technique in

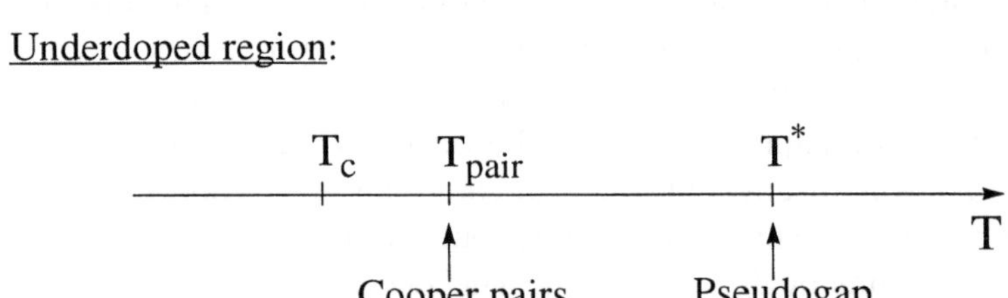

Figure 3.32. Characteristic temperatures in the underdoped region: a pseudogap appears at T^*; the pairing occurs at T_{pair}, and the sample becomes superconducting at T_c.

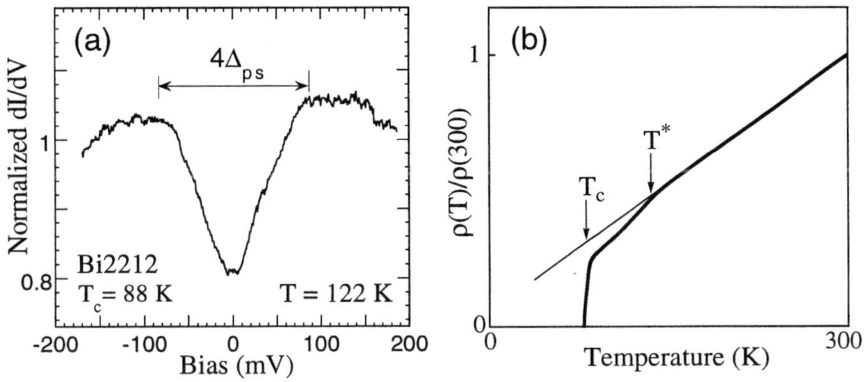

Figure 3.33. (a) Pseudogap in slightly overdoped Bi2212, obtained in tunneling measurements. (b) Pseudogap in resistivity measurements: T^* is the temperature at which the resistivity deviates on cooling from a linear dependence.

overdoped Bi2212 clearly demonstrate the presence of a slightly asymmetrical pseudogap in the density of states. In the measurements, the temperature 122 K is chosen because, in slightly overdoped Bi2212, $T_{pair} \simeq 116$ K (see Chapter 12). Thus, the temperature 122 K is above T_{pair}, consequently the spectrum shown in Fig. 3.33a represents a pure normal-state gap which disappears upon increasing the temperature. The normal-state pseudogap observed in tunneling measurements corresponds to a charge gap on charge stripes, because in order to pull out an electron (hole) from a charge stripe where the carriers reside, it is necessary to overcome the charge gap.

Figure 3.33b shows how the pseudogap temperature T^* is determined in resistivity measurements: T^* is the temperature at which resistivity deviates from a linear temperature dependence. The pseudogap determined in transport measurements directly relates to a spin gap in insulating antiferromagnetic domains (see Chapter 10).

4.2 Pseudogap temperature T^*

All measurements in cuprates show that the magnitude of pseudogaps decreases with increasing doping level. However, there is a clear discrepancy between the phase diagrams inferred from transport measurements, on the one hand, and from tunneling measurements, on the other. Transport and NMR measurements show that, in the overdoped region, the pseudogap is absent above T_c, thus $T^*_{transp} < T_c$. At the same time, in tunneling measurements, the pseudogap is observed well above T_c, indicating that $T_c < T^*_{tunnel}$.

Figure 3.34 shows the doping dependence of T_c, T_{pair} and two T^*, one of which is obtained in tunneling measurements in Bi2212, T^*_{tunnel}, while the

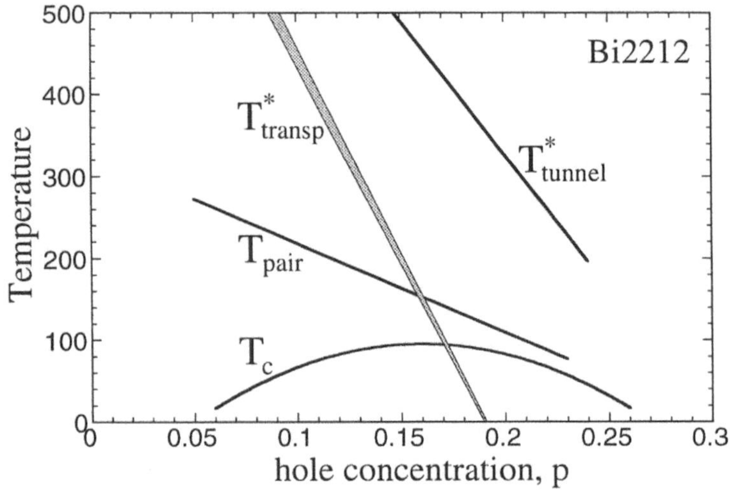

Figure 3.34. Phase diagram of Bi2212: T_c is the transition temperature; T_{pair} is the pairing temperature (see Fig. 3.32), and T^*_{tunnel} and T^*_{transp} are the pseudogap temperatures obtained in tunneling (see Fig. 3.33a) and transport (see Fig. 3.33b) measurements, respectively.

second emerges in transport and NMR measurements, T^*_{transp}. In Fig. 3.34, the temperatures T_c and T_{pair} directly relate to the energy scales Δ_c and Δ_p, respectively, which are shown in Fig. 3.22. The critical temperature T_c as a function of doping has no correlations with the other three temperatures shown in Fig. 3.34. T_{pair} is proportional to T^*_{tunnel} as $3T_{pair} \approx T^*_{tunnel}$. It is not surprising that T^*_{transp} and T^*_{tunnel}, shown in Fig. 3.34, don't coincide because they correspond to two different pseudogaps: a spin gap in insulating magnetic domains and a charge gap on charge stripes, respectively. In Fig. 3.34, T^*_{transp} has a similar doping dependence as that of the fraction of antiferromagnetic domains shown in Fig. 3.16a. In cuprates, the point $p = 0.19$ is called *the critical quantum point.* In general, in a quantum critical point, the magnetic order is about forming/disappearing. We shall discuss the quantum critical point $p = 0.19$ as well as the origins of the two pseudogaps in Chapters 9 and 10.

4.3 Structural transitions above T_c

We already considered structural transitions, most of which occur in the normal state. In LSCO, there are at least two structural transitions before an LSCO sample becomes superconducting, as shown in Fig. 3.9. These transitions were observed in X-ray, neutron-scattering and acoustic measurements.

In YBCO, there are at least three structural transitions which occur at T_c, 140–150 K and 220–250 K [9]. The transition at 220–250 K is close to that shown in Fig. 3.5; therefore it can be associated with the HTT→LTO transition. Unfortunately, the other two structural transitions in YBCO are unknown: the measurements performed in YBCO with different dopings ($0.55 \leq x \leq 1$) by ion channeling spectrometry cannot identify the type of these structural transitions. Heat-capacity measurements in YBCO (x = 0.85–0.95) show the presence of three anomalies in the temperature dependence of heat capacity, occurring in each of the regular intervals 100–200 K, 205–230 K and 260–290 K [66].

In Bi2212 (T_c = 84.5 K), three structural phase transitions are observed in acoustic measurements: at 95 K, 150 K and 250 K [10]. Even tunneling measurements in slightly overdoped Bi2212 ($p \simeq 0.19$), besides identifying $T_{pair} \approx$ 110-120 K, show that there are notable changes at 205–215 K (see Chapter 12).

The CuO-chain oxygen ions in $YBCO_8$ (124) undergo a correlated displacement in the a direction (perpendicular to the chains) of about 0.1 Å, with the onset of correlations occurring near 150 K. A similar effect is observed in YBCO.

It is worth recalling that observations made on the basis of transport measurements suggest that the LTO→LTT transition induces intra-gap electronic states in the middle of the pseudogap [6]. Indeed, the LTT structure seems to attract charge carriers, forming charge stripes [61].

As emphasized above, in the underdoped and optimally doped regions, some of these transitions are purely structure dependent, and almost independent of hole doping. In contrast, the critical temperature $T_c \propto p^2$. Acoustic measurements performed in undoped and underdoped LSCO, YBCO, NCCO, Bi-and Tl-based compounds show that the elastic coefficients display some kind of structural transition at maximum $T_{c,max}$ for each compound, although some of these either are not superconducting or have low T_c. This suggests that these structural transitions do not require the presence of superconductivity. Consequently, in each compound, $T_{c,max}$ is determined by the underlying (unstable) lattice which is the driving force for high-T_c superconductivity in cuprates.

4.4 Magnetic ordering in the undoped region

The analysis of a great deal of data shows that, in *all* layered non-superconducting compounds, including undoped cuprates, the long-range antiferromagnetic (ferromagnetic) order develops at T_N (T_C) along the c axis, where T_N (T_C) is the Néel (Curie) temperature [67]. At the same time, in-plane magnetic correlations exist above T_N (T_C).

This issue will be discussed in more detail in Chapter 9.

5. Theory

There is no theory of high-T_c superconductivity. "The problem is not that there is no theory of high-temperature superconductivity, there are too damn many of them" [68].

This book provides a basic description of the mechanism of high-T_c superconductivity, but doesn't provide a complete theory of the phenomenon. The combination of two existing theories—namely, the bisoliton theory and the theory based on spin fluctuations, can describe in first approximation the phenomenon of superconductivity in cuprates. However, there are a few details which cannot be accounted for by these two theories. For example, neither of the theories consider the presence of charge stripes. Thus, more is needed on the theoretical front.

As mentioned in Introduction, there are more than 100 different theoretical models of high-T_c superconductivity. It is senseless to attempt to consider all the models: we only need two of them. The bisoliton theory will be discussed in detail in Chapter 7. Here we concentrate on the theory of superconductivity based on spin-fluctuation mechanism. The reader who is interested in various theories of high-T_c superconductivity is referred to the theoretical reviews [69–72].

The magnetic-fluctuation mechanism of superconductivity was first proposed as an explanation of superconductivity in heavy fermions, which were discovered in 1979 [73]. This model is based on a short-range Coulomb interaction leading to an exchange coupling $J \times \vec{S}_i \vec{S}_j$ between near-neighbor copper spins, $\vec{S}_i$ and $\vec{S}_j$, and strong magnetic spin fluctuations. The superexchange constant is denoted by J. In cuprates, it has an extremely high magnitude, $J \sim$ 120–130 meV.

The underlying microscopic physics can be described by the $t - J$ model defined by the Hamiltonian

$$H = H_t + H_J = -t \sum_{\langle nm \rangle \sigma} (d^\dagger_{n\sigma} d_{m\sigma} + H.c.) + J \sum_{\langle nm \rangle} S_n S_m, \tag{3.5}$$

where $d^\dagger_{n\sigma}$ is the creation operator of a hole with spin σ ($\sigma = \uparrow, \downarrow$) at site n on a two-dimensional square lattice. The $d^\dagger_{n\sigma}$ operators act in the Hilbert space with no double electron occupancy. The spin operator is $S_n = \frac{1}{2} d^\dagger_{n\sigma} \sigma_{\alpha\beta} d_{n\beta}$, and $\langle nm \rangle$ are the nearest-neighbor sites on the lattice. At half-filling (one hole per site) the $t - J$ model is equivalent to the Heisenberg antiferromagnetic model, which has long-range Néel order in the ground state. Under doping, the long-range antiferromagnetic order is destroyed; however, local antiferromagnetic order is preserved.

The model gives the critical temperature in reasonable agreement with experimental data for high-T_c superconductors. The spin-fluctuation mechanism of superconductivity results in the $d_{x^2-y^2}$ symmetry of superconducting order parameter.

6. Applications

Soon after the discovery of high-T_c superconductors, it was realized that, at liquid nitrogen temperatures, enormous savings are possible. For example, if cryogenic liquids are used for cooling, a litre of liquid helium costs approximately \$25, as opposed to \$0.60 for a litre of liquid nitrogen. The difference in the cost of electricity, if electrical cryocoolers are used at 4.2 K and 77 K, is similar.

The main problem in using cuprates for applications is that, like all ceramics, high-T_c superconductors are very brittle and very difficult to shape and handle, while long, flexible, superconducting wires are necessary for many large-scale applications. Large supercurrents can only flow along CuO_2 planes, and only a small fraction of the material in a completed device is likely to be correctly oriented. The grain boundaries attract impurities, leading to weak links, which reduce the inter-grain current density and provide an easy path for flux vortices to enter the material. Flux creep or vortex penetration into high-T_c superconductors is unusually rapid. The coherence length or diameter of a vortex core tends to be very small. This is a problem because pinning is most effective if the defect or impurity is of the same size as the coherence length.

6.1 Small-scale applications

The small-size devices based on thin films of cuprates are now commercially available. Progress is now largely limited by refrigeration packages, not by materials, films or junctions. Thin films of YBCO are widely used for small-size high-T_c superconducting devices because YBCO has a high critical temperature and can accommodate a high current density. The most success has been achieved with SQUIDs (Superconducting QUantum Interference Devices), which are based on the Josephson effect. The SQUID sensitivity is so high that it can detect magnetic fields 100 billion times smaller than the Earth's magnetic field. Other devices that have reached commercial availability are high-T_c superconducting passive RF and microwave filters for wide-band communications and radars. These are based on conventional microstrip and cavity designs. They have the advantages of very low noise and much higher selectivity and efficiency than conventional filters. They are now used at mobile-phone base stations. Finally, high-T_c superconducting thin films are also widely used for bolometric detection of radiation.

6.2 Large-scale applications

Large-scale applications for high-T_c superconductors present a major challenge to the materials scientists. Compared with the small-scale applications, a large-scale application generally requires much larger currents and lengths of superconductor in a working environment where the magnetic field may be several Teslas. The most important applications under consideration are in magnets, power transmission cables, current leads, fault current limiters, transformers, generators, motors, and energy storage. Applications related to magnet technology are probably among the most significant that are under development at the present time. These include magnetic energy storage, Maglev trains (relaying on repulsion between magnets mounted on the train and the guideway) and magnets for MRI (Magnetic Resonance Imaging) and other medical applications. In all these cases the superconductor must not only carry a large current with zero resistance under a high magnetic field, but it also must be possible to fabricate it in long lengths with high flexibility and a high packing density. Research on large-scale applications of high-T_c superconductors has focused on the Bi-based family because it is difficult to grow YBCO in bulk. Bi2212/Bi2223 powder is packed into a silver tube, which is drawn fine and goes through a sintering, rolling and annealing process. The major remaining barrier to wider use is cost. However, in some cases the extra cost is justified: high-T_c superconducting underground power transmission cables, which can carry 3 to 5 times the current of a copper cable of the same diameter, are already coming into commercial use in cities such as Detroit. The use of high-T_c superconductors is now a multi-billion-dollar growing business: more than 50 companies around the world have set out to commercialize high-T_c superconductors over the past 10 years.

7. A final remark

After reading this Chapter, the keywords which have to be associated with cuprates and high-T_c superconductivity should be the following expressions:

- an unstable lattice and structural phase transitions;
- a strong electron-phonon coupling;
- a metal-insulator transition;
- domain walls (charge stripes);
- spin fluctuations;
- a charge gap;
- a spin gap

- phase-coherence fluctuations;
- a pairing gap;
- a phase-coherence gap, and
- strongly correlated materials.

CUPRATES: ANOMALY IN TUNNELING SPECTRA

If you do not expect the unexpected, you will not find it.
—Heraclitus [of Ephesus] (ca 550–475 BC)

This Chapter is the shortest and, at the same time, the most important in the book. The pairing mechanism of high-T_c superconductivity is based on an anomaly found in the tunneling spectra of Bi2212 [1, 2], which is described here. This Chapter is crucial for understanding the mechanism of high-T_c superconductivity, and written for a reader not familiar with tunneling measurements. The main purpose of this Chapter is to present "pure" experimental data. They will be discussed later in the framework of a theoretical model in Chapter 6.

Experimental details of the growth and characterization of Bi2212 single crystals used here can be found in Appendix A of Chapter 12. Tunneling measurements were performed by break junctions and in-plane point contacts. The description of break-junction and point-contact setups can be found in Appendix B of Chapter 12.

1. Tunneling measurements in Bi2212

Tunneling current-voltage $I(V)$ characteristics obtained in superconductor-insulator-normal (SIN) metal junctions of conventional superconductors are well described by the so-called BTK (Blonder-Tinkham-Klapwijk) theory [3].

In Chapter 2, Figure 2.13 shows theoretical $I(V)$ characteristics for an SIN junction at different temperatures. In the *tunneling* regime, these theoretical $I(V)$ curves at high positive bias, depending on the normal resistance of the junction, always lie somewhat below the normal-state curve (the straight line in Fig. 2.13), while at low negative bias above the line. In conventional superconductors, the BTK predictions are successfully verified by tunneling experiments. However, in cuprates, the BTK theory is violated.

Figure 4.1 shows two $I(V)$ curves obtained at low temperatures in underdoped (Fig. 4.1a) and overdoped (Fig. 4.1b) Bi2212 single crystals in superconductor-insulator-superconductor (SIS) junctions. In Fig. 4.1, one can see that the $I(V)$ curves at high positive (low negative) bias do not pass below (above) the dashed lines which are parallel to the $I(V)$ curves at high bias but far above (below) them. This result cannot be explained by the d-wave symmetry of the order parameter (in conventional superconductors, the order parameter has an s-wave symmetry). Upon a visual inspection of Figs. 2.13 and 4.1, it is obvious that something is "wrong" with the tunneling $I(V)$ characteristics of Bi2212. The only difference between the two $I(V)$ characteristics in the two plots of Figs. 4.1 is the bias scale. This is because, in the underdoped region (Fig. 4.1a), the gap magnitude Δ is larger than that in the overdoped region (Fig. 4.1b).

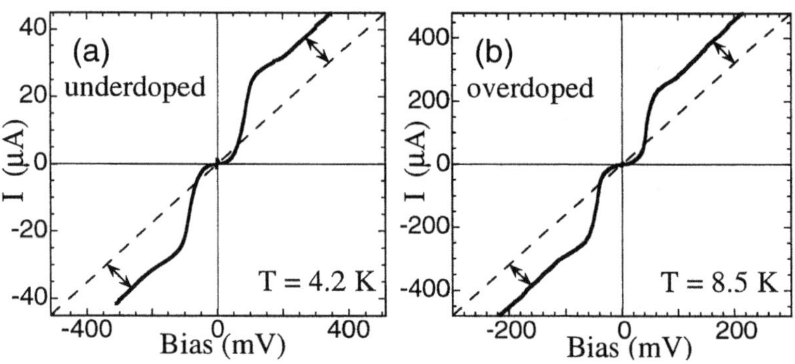

Figure 4.1. Tunneling current-voltage $I(V)$ characteristics measured in SIS junctions of (a) an underdoped Bi2212 single crystal with the critical temperature T_c = 83 K [4], and (b) an overdoped Bi2212 single crystal with T_c = 88 K. The data are obtained at temperatures of 4.2 K and 8.5 K, respectively. In both plots, the dashed lines are parallel to the $I(V)$ curves at high bias, and the arrows indicate the offset from the lines.

As discussed in Chapter 3, the pseudogap observed in tunneling measurements in high-T_c superconductors is a charge gap (on charge stripes). Intrinsic c-axis tunneling data obtained in Bi2212 micron-size mesas reveal that, below the critical temperature, tunneling spectra consist of a contribution from the pseudogap and a contribution from the superconducting condensate. The contribution from the superconducting condensate literally appears above the contribution from the pseudogap [5]. So, in order to understand the origin of the anomaly in the tunneling $I(V)$ characteristics of Bi2212, it is necessary to separate the two contributions. The anomaly originates either from the superconducting condensate or from the pseudogap.

The contribution from the pseudogap can be directly obtained from tunneling measurements carried out in the normal state of Bi2212. From the tunneling characteristics obtained below the critical temperature and those obtained in the normal state, by taking the difference between the spectra, it is then possible to estimate the contribution resulting from the superconducting condensate. This is equivalent to the procedure usually used in neutron scattering measurements. In our case, however, this procedure only leads to an *estimation* of the contribution because by subtracting the spectra we assume that the pseudogap crosses T_c without modification. This is not necessary the case. In fact, on cooling through T_c, as we shall see in Chapter 12, the magnitude of the pseudogap remains almost unchanged; however, the quasiparticle excitations inside the pseudogap, thus at low bias, are strongly regrouped. We will first discuss tunneling data obtained in the superconducting state.

1.1 Measurements below T_c

Figure 4.2a shows tunneling spectra measured by a break junction (thus, in an SIS junction) at low temperature $T = 8.5$ K in an overdoped Bi2212 single crystal with a critical temperature $T_c = 88$ K. The conductance $dI(V)/dV$ (grey curve in Fig. 4.2a) exhibits typical features of SIS-junction conductances obtained in Bi2212: well-defined quasiparticle peaks, a zero-bias peak due to the Josephson current, dips and humps outside the gap structure. The $I(V)$ characteristic in Fig. 4.2a (black curve) was already discussed—it is shown in Fig. 4.1b. Figure 4.2b presents a set of tunneling $I(V)$ characteristics obtained in different overdoped Bi2212 single crystals with $T_c = 87$–89 K. Considering the $I(V)$ curves shown in Fig. 4.2b is evident that the anomaly is a robust feature of tunneling $I(V)$ characteristics obtained in SIS junctions of Bi2212.

Figure 4.2c depicts tunneling spectra which are similar to those in Fig. 4.2a, but obtained in an SIN junction of Bi2212. Because of the SIN junction, quasiparticle peaks in the conductance shown in Fig. 4.2c appear at a bias two times smaller than the peak bias in Fig. 4.2a. The $I(V)$ curve in Fig. 4.2c clearly indicates that the observed anomaly is not an SIS-junction effect but an intrinsic feature of tunneling $I(V)$ characteristics obtained in Bi2212.

1.2 Measurements above T_c

In order to estimate the contribution from the superconducting condensate in the tunneling $I(V)$ characteristics, we need to know the contribution from the pseudogap, which can be directly obtained in tunneling measurements in the normal state of Bi2212. The question is how to obtain tunneling $I(V)$ characteristics in the normal state?

There are, at least, three different solutions. The first two consist in applying a magnetic field below T_c, while the third solution is to measure $I(V)$

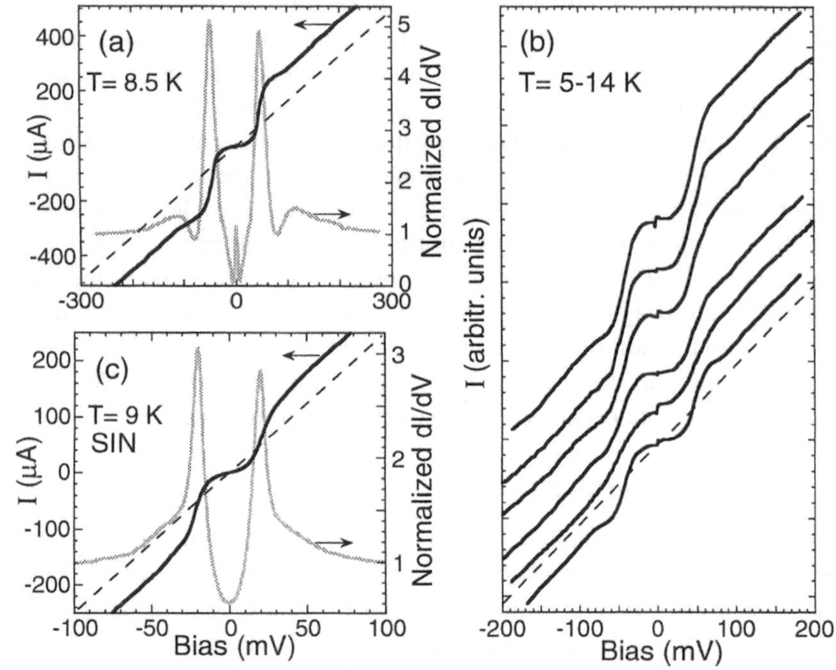

Figure 4.2. (a) Tunneling $I(V)$ (black curve) and $dI(V)/dV$ (grey curve) characteristics obtained at T = 8.5 K in an SIS junction of an overdoped Bi2212 single crystal with T_c = 88 K (The $I(V)$ curve is also shown in Fig. 4.1b). (b) Set of $I(V)$ curves obtained at T = 5–14 K in different overdoped Bi2212 single crystals with T_c = 87–89 K. The curves are offset for clarity. (c) $I(V)$ (black curve) and $dI(V)/dV$ (grey curve) characteristics obtained at T = 9 K in an SIN junction of an overdoped Bi2212 single crystal with T_c = 87.5 K. In all plots, the dashed lines are parallel to the $I(V)$ curves at high bias. In plot (a), the label of the horizontal axis is the same as in the other plots. In plot (b), the vertical scale is similar to that in plot (a).

somewhat above T_c. In the first case, a magnetic field with a magnitude larger than the upper critical magnetic field B_{c2} in Bi2212 renders the whole sample normal. In the second case, by applying a magnetic field with a magnitude larger than the lower critical field B_{c1}, vortices will enter the sample. In the latter case, the normal-state characteristics can be obtained inside vortex cores. Because B_{c2} in Bi2212 is very large, the first solution cannot be realized in laboratory conditions. The second solution seems to be suitable, however, tunneling spectra obtained inside vortex cores have subgap structures which are usually interpreted as a manifestation of bound states (see Chapter 12). Moreover, technically, it is not easy to realize such measurements. So, we are left with a straightforward solution: to measure $I(V)$ characteristics somewhat above T_c. The main disadvantage of measurements performed above T_c is the presence of substantial thermal fluctuations.

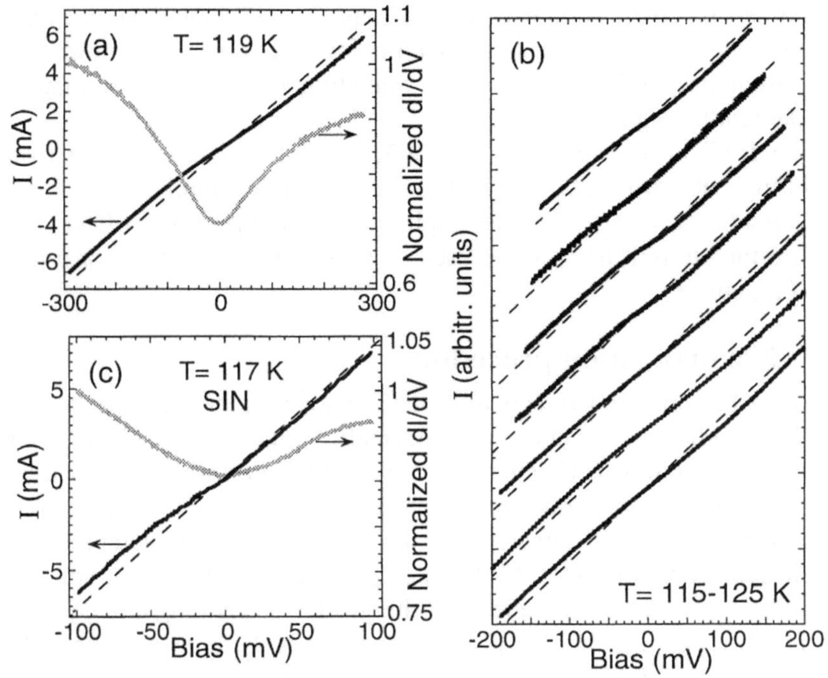

Figure 4.3. (a) $I(V)$ (black curve) and $dI(V)/dV$ (grey curve) characteristics obtained at T = 119 K in the same SIS junction as those in Fig 4.2a. (b) Set of $I(V)$ curves obtained at T = 115–125 K in different overdoped Bi2212 single crystals with T_c = 87–89 K. The curves are offset for clarity. (c) $I(V)$ (black curve) and $dI(V)/dV$ (grey curve) characteristics obtained at T = 117 K in the same SIN junction as those in Fig. 4.2c. In all plots, the dashed lines are parallel to the $I(V)$ curves at high bias. In plot (a), the label of the horizontal axis is the same as in the other plots. In plot (b), the vertical scale is similar to that in plot (a).

Figure 4.3a depicts tunneling spectra obtained at T = 119 K in the same Bi2212 single crystal as those in Fig. 4.2a. Figure 4.3b presents a set of $I(V)$ curves measured at T = 115–125 K in different overdoped Bi2212 single crystals. The data in Figs. 4.3a and 4.3b are obtained in SIS junctions. Figure 4.3c shows tunneling spectra obtained in an SIN junction at T = 117 K in the same Bi2212 single crystal as those in Fig. 4.2c. The temperatures between 115 and 125 K presented in Fig. 4.3 were not chosen by accident: we shall see in Chapter 12 that, in slightly overdoped Bi2212, the onset of superconductivity occurs at 110–120 K. So, the temperatures 115–125 K are above the onset of superconductivity in these Bi2212 samples and, as a consequence, the data shown in Fig. 4.3 provide exclusively the contribution from the normal-state pseudogap.

Considering the $I(V)$ characteristics shown in Fig. 4.3, one can see that, at high positive (low negative) bias, they pass somewhat below (above) the dashed lines which are parallel to the $I(V)$ curves at high bias. So, the anomaly in tunneling $I(V)$ characteristics, found below T_c, vanishes somewhat above T_c. Consequently, this anomaly in the tunneling $I(V)$ curves originates from the superconducting condensate.

In the $I(V)$ curves shown in Fig. 4.3, a small "negative" offset from the straight lines above T_c is caused by the pseudogap. We shall discuss this "negative" offset in Chapter 12. The amplitude of the "negative" offset is small, therefore the straight line can be used as a first approximation for a normal-state $I(V)$ curve.

1.3 Normalization procedure

In order to compare two sets of tunneling spectra, they have to be normalized. The conductances can be easily normalized at high bias. How to normalize the corresponding $I(V)$ curves is not a trivial question. The conductance curves at high bias, thus, far away from the gap structure, are almost constant. Consequently, for an SIS junction, by normalizing two conductance curves at high bias, the equation $(dI(V)/dV)_{1,norm} \simeq (dI(V)/dV)_{2,norm}$ holds at bias $|V| \gg 2\Delta/e$, where e is the electron charge. By integrating this equation we have $I(V)_{1,norm} \simeq I(V)_{2,norm} + C$, where C is the constant, meaning that the corresponding $I(V)_{i,norm}$ curves are parallel to each other at high bias.

The two conductances shown in Figs. 4.2a and 4.3a, obtained in the same Bi2212 single crystal. Normalized by their values at the lowest negative bias, these two conductances are presented in Fig. 4.4a. In Fig. 4.4a, the conductance measured in the normal state of Bi2212 is shown in grey, while the black curve represents the conductance obtained in the superconducting state. Figure

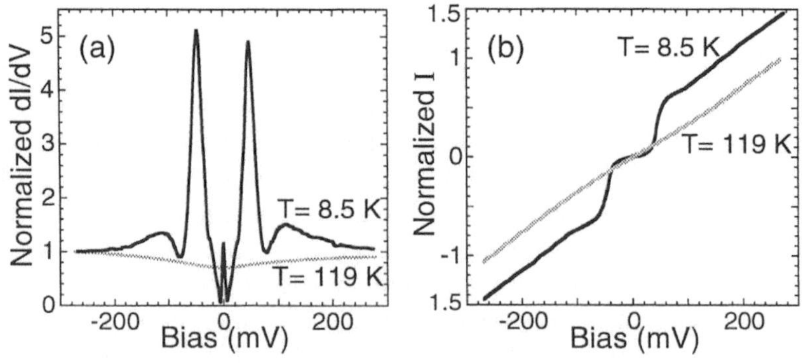

Figure 4.4. Normalized tunneling characteristics from Figs. 4.2a and 4.3a, which are obtained in the same Bi2212 single crystal (T_c = 88 K) at different temperatures, T= 8.5 K (black curves) and T = 119 K (grey curves): (a) the conductances, and (b) the $I(V)$ curves. In plot (b), the grey curve is normalized by its value at minimum negative bias, and the black curve is adjusted to be parallel at high bias to the grey curve (see text for more details).

4.4b depicts the two $I(V)$ characteristics normalized as described in the previous paragraph: the normal-state $I(V)$ curve from Fig. 4.3a (the grey curve in Fig. 4.4b) is normalized by its value at minimum negative bias, and the $I(V)$ curve from Fig. 4.2a (the black curve in Fig. 4.4b) is adjusted to be parallel at high bias to the normalized normal-state curve. We are in a position to estimate the contribution from the superconducting condensate.

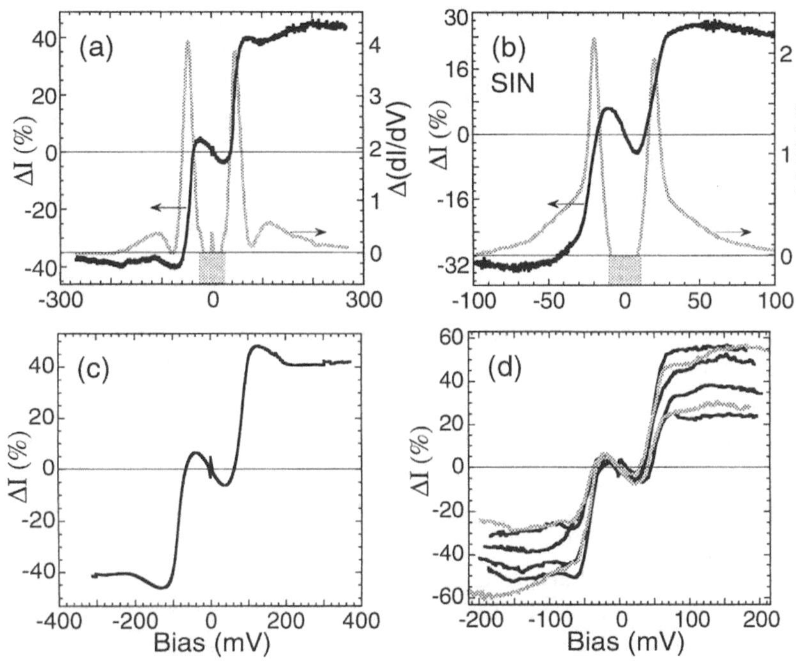

Figure 4.5. (a) Difference between the two normalized $I(V)$ curves presented in Fig. 4.4b (black curve), and the difference between the normalized conductances shown in Fig. 4.4a (grey curve). (b) Difference between the two $I(V)$ curves from Figs. 4.2c and 4.3c, normalized before subtraction as those in Fig 4.4b (black curve), and the difference between their normalized conductances (grey curve). The data are obtained in the same SIN junction. (c) Difference between the $I(V)$ curve and the dashed line, shown in Fig. 4.1a, normalized before subtraction as those in Fig. 4.4b. The dashed line is used as an estimate of the normal-state curve (see text for more details). (d) Differences between $I(V)$ characteristics shown in Fig. 4.2b and their corresponding normal-state curves, normalized before subtraction as those in Fig. 4.4b. Two curves are shown in grey for clarity. In all plots, the current differences are presented in %. In plots (a) and (b), the labels of the horizontal axes are the same as those in plots (c) and (d). In plots (a) and (b), the grey boxes cover the parts of the conductances, which are below zero and have no physical meaning.

1.4 Contribution from the superconducting condensate

Figure 4.5a depicts two differences: one is the difference between the conductances shown in Fig. 4.4a (grey curve), and the second is the difference between the $I(V)$ characteristics presented in Fig. 4.4.b (black curve). In first approximation, the two curves in Fig. 4.5a correspond to the "pure" contribution from the superconducting condensate in the tunneling spectra.

The same procedure of normalization has been done for the tunneling spectra shown in Figs. 4.2c and 4.3c, which are obtained in the same SIN junction. The differences are shown in Fig. 4.5b. Upon a visual inspection of Figs. 4.5a and 4.5b, one can see that the data in both plots are very similar. The grey boxes in Figs 4.5a and 4.5b cover the parts of the conductances with negative values, which have no physical meaning. At this stage, the conductance data are not very important.

The $I(V)$ characteristic shown in Fig. 4.1a, which is obtained in an *underdoped* Bi2212 single crystal, went through the same normalization procedure, and the result is presented in Fig. 4.5c. Since the data in Fig. 4.1a were obtained elsewhere [4], a normal-state $I(V)$ characteristic was not measured. So, in this particular case, we use the straight line (the dashed line in Fig. 4.1a) as the normal-state curve. The $I(V)$ characteristics shown in Fig. 4.2b and their corresponding normal-state curves were normalized in the same manner, and the obtained differences are presented in Fig. 4.5d. Considering all the $I(V)$ curves in Fig. 4.5, one can see that the contribution from the superconducting condensate in *these* $I(V)$ characteristics at high bias is 25–55 % above the contribution from the pseudogap.

Analyzing the $I(V)$ characteristics shown in Fig. 4.5, which, in first approximation, represent the contribution from the superconducting condensate, it is easy to observe the general trends of the curves. First, at high bias, the curves reach a plateau value. Second, at the gap bias, the curves rise/fall sharply. Last, at low bias, the curves go to zero. In Fig. 4.5, the negative slope of the curves at low bias represents the negative differential resistance and is an artifact. This artifact is simply a consequence of the rough estimation used here. Figure 4.6a depicts an idealized $I(V)$ characteristic summarizing the observed tendencies. The first derivative of the $I(V)$ characteristic is shown in Fig. 4.6b.

Comparing Figs. 2.13 and 4.6a, it is evident that the physics of high-T_c superconductors is *different* from the physics of conventional superconductors. The abnormality found in the tunneling $I(V)$ characteristics of Bi2212 will be discussed later in Chapter 6.

2. Tunneling measurements in YBCO

It is important to show that the anomaly found in Bi2212 is inherent to other cuprates too.

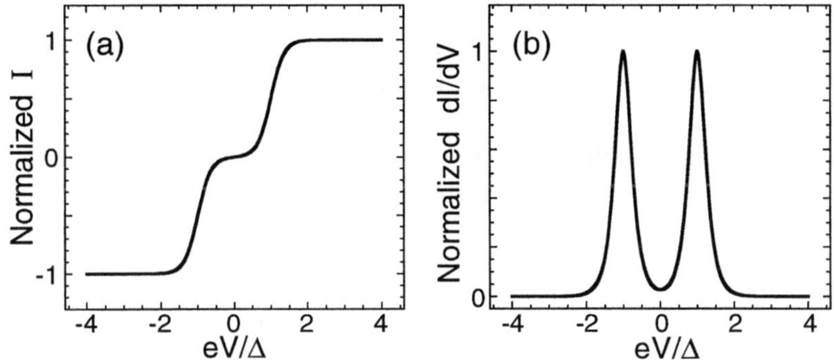

Figure 4.6. (a) Idealized contribution from the superconducting condensate to tunneling $I(V)$ characteristics obtained in SIN junctions of Bi2212. (b) The first derivative of the idealized $I(V)$ characteristic. In both plots, the curves are normalized by their maximum values.

Figure 4.7 shows the $I(V)$ characteristic obtained in an in-plane SIS tunneling junction formed between Pb and near optimally doped YBCO. In Fig. 4.7, one can see that the $I(V)$ characteristic at positive (negative) bias lies somewhat above (below) the straight (dashed) line. If we subtract this linear contribution from the $I(V)$ curve, it is evident that the result of the subtraction will look similar to that in Fig. 4.5c. Since the anomaly is not inherent to conventional superconductors, the superconducting condensate of Pb cannot be responsible for this offset. Therefore, the anomaly is caused by the supercon-

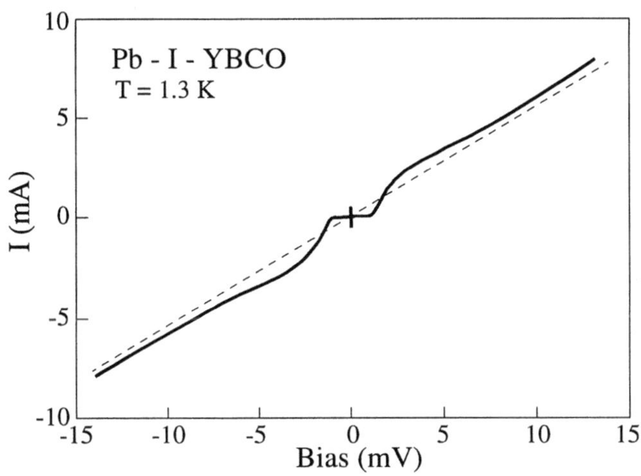

Figure 4.7. Tunneling current-voltage $I(V)$ characteristic obtained in an in-plane Pb-I-YBCO junction at T = 1.3 K, where I stands for insulator. The near optimally doped YBCO has a critical temperature of $T_c \simeq 92$ K [6]. The dashed line is parallel to the $I(V)$ curve at high bias.

ducting condensate of YBCO. Since the $I(V)$ characteristic in Fig. 4.7 is not obtained in an YBCO-insulator-YBCO junction but in a Pb-insulator-YBCO junction, the amplitude of the offset from the straight line is smaller than those in Fig. 4.1.

Finally, one can conclude that the anomaly found in tunneling $I(V)$ characteristics is inherent, at least, to two members of high-T_c superconductors—Bi2212 and YBCO.

Chapter 5

NONLINEAR EXCITATIONS: SOLITONS

True laws of Nature cannot be linear.

—Albert Einstein

One may wonder what this Chapter is doing in a book describing the phenomenon of superconductivity. The main purpose of this Chapter is to present a brief introduction to nonlinear excitations, and to underline the soliton concept. Acquaintance with nonlinear excitations is necessary since the Cooper pairs in high-T_c superconductors are pairs of soliton-like excitations. I am sure that, in 15 years or so, the presence of this Chapter in a book describing the phenomenon of superconductivity will be considered absolutely natural. There are excellent books devoted to the description of solitons and related phenomena; the reader who is interested knowing more on the soliton issues is referred to a few books (see Appendix).

1. Introduction

For a long time linear equations have been used for describing different phenomena. For example, Newton's, Maxwell's and Schrödinger's equations are linear, and they take into account only a linear response of a system to an external disturbance. However, the majority of real systems are *nonlinear*. Most of the theoretical models are still relying on a *linear* description, corrected as much as possible for nonlinearities which are treated as small perturbations. It is well known that such an approach can be absolutely wrong. The linear approach can sometimes miss completely some essential behaviors of the system.

Nonlinearity has to do with *thresholds*, with multi-stability, with hysteresis, with phenomena which are changed *qualitatively* as the excitations are changed.

In a linear system, the ultimate effect of the combined action of two different causes is merely the superposition of the effects of each cause taken individually. But in a nonlinear system adding two elementary actions to one another can induce dramatic new effects reflecting the onset of cooperativity between the constituent elements.

To understand *nonlinearity*, one should first understand *linearity*. Consider linear waves. In general, a *wave* may be defined as a progression through matter of a state of motion. Characteristic properties of any linear wave are: (i) the shape and velocity of a linear wave are independent of its amplitude; (ii) the sum of two linear waves is also a linear wave; and (iii) small amplitude waves are linear. Figure 5.1a shows an example of a periodic linear wave. Large amplitude waves may become nonlinear.

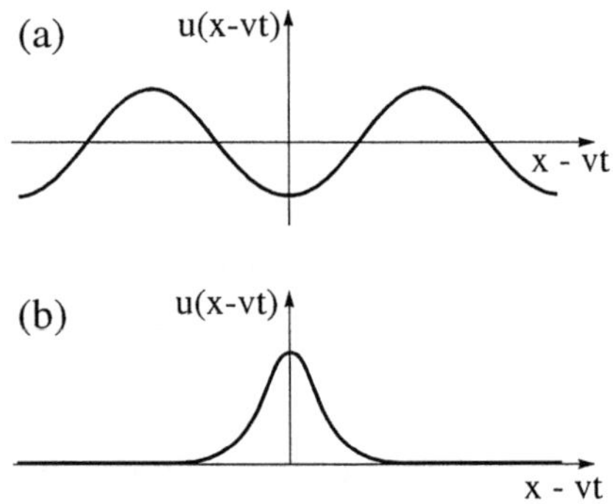

Figure 5.1. Sketch of (a) a periodic linear wave, and (b) a solitary wave.

The fate of a wave travelling in a medium is determined by properties of the medium. *Nonlinearity* results in the distortion of the shape of large amplitude waves, for example, in turbulence. However, there is another source of distortion—the *dispersion* of a wave. More than 100 years ago the mathematical equations describing *solitary waves* were solved, at which point it was recognized that the solitary wave, shown in Fig. 5.1b, may exist due to a precise balance between the effects of nonlinearity and dispersion. Nonlinearity tends to make the hill steeper (see Fig. 5.1b), while dispersion flattens it. The solitary wave lives "between" these two dangerous, destructive "forces." Thus, *the balance between nonlinearity and dispersion is responsible for the existence of the solitary waves.* As a consequence, the solitary waves are extremely robust.

Solitary waves or *solitons* cannot be described by using linear equations. Unlike ordinary waves which represent a spatial periodical repetition of elevations and hollows on a water surface, or condensations and rarefactions of a density, or deviations from a mean value of various physical quantities, solitons are single elevations, such as thickenings etc., which propagate as a unique entity with a given velocity. The transformation and motion of solitons are described by nonlinear equations of mathematical physics.

The history of solitary waves or solitons is unique. The first *scientific* observation of the solitary wave was made by Russell in 1834 on the water surface. One of the first mathematical equations describing solitary waves was formulated in 1895. And only in 1965 were solitary waves fully understood! Moreover, many phenomena which were well known before 1965 turned out to be solitons! Only after 1965 was it realized that solitary waves on the water surface, nerve pulse, vortices, tornados and many others belong to the same category: they are all solitons! That is not all, the most striking property of solitons is that they behave like particles!

Mathematically, there is a difference between "solitons" and "solitary waves." *Solitons* are localized solutions of integrable equations, while *solitary waves* are localized solutions of non-integrable equations. Another characteristic feature of *solitons* is that they are solitary waves that are not deformed after collision with other solitons. Thus the variety of *solitary waves* is much wider than the variety of the "true" solitons. Some solitary waves, for example, vortices and tornados are hard to consider as *waves*. For this reason, they are sometimes called *soliton-like excitations*. To avoid this bulky expression we shall often use the term *soliton* in all cases. This is not dangerous when we are talking about general properties of soliton-like excitations.

2. Russell's discovery

The first observation of the solitary wave or soliton was made by John Scott Russell near Edinburgh (Scotland) in August 1834. He was observing a boat moving on a shallow channel and noticed that, when the boat suddenly stopped, the wave that it was pushing at its prow "rolled forward with great velocity, assuming the form of a large solitary elevation, a rounded, smooth and well defined heap of water which continued its course along the channel apparently without change of form or diminution of speed" [1]. He followed the wave along the channel for more than a mile. The shape of the solitary wave observed by Russell is similar to that in Fig. 5.1b.

Russell published the first report of this event in 1838. He called this solitary wave the *Wave of Translation*. A more detailed account of it and of successive experiments was published in his *Report on Waves* in 1844 [1].

Between 1834 and 1844, Russell performed numerous experiments in natural environments—on canals, rivers and lakes—as well as in his "laboratory," which was a specially designed small reservoir in his garden. In these studies, he found the following main properties of the solitary wave:

- An isolated solitary wave has a constant velocity and does not change its shape.
- The dependence of the velocity v on the canal depth h and the height of the wave, u, is given by the relation

$$v = \sqrt{g(h+u)},$$

where g is the gravity acceleration. This formula is valid for $u < h$.

- A high enough solitary wave decays into two or more smaller solitary waves. The new "born" solitary waves have different heights and, as a consequence, their velocities are different.
- There exist only solitary elevations (humps); solitary cavities (depressions) are never met.

In his report, Russell made a remark that "the great primary waves of translations cross each other without change of any kind in the same manner as the small oscillations produced on the surface of a pool by a falling stone". Researchers were very much puzzled by this striking phenomenon which was understood only 130 years later.

Russell was not only an exceptionally observant scientist and excellent experimenter but a first class theorist. However, the attitude to solitons of some most prominent theoretical experts at that time was quite different, and his colleagues neither saw its significance nor shared his enthusiasm. Moreover, there were even a few papers, including one written by Stokes, which concluded that the solitary wave cannot exist even in liquids with vanishing viscosity.

Unfortunately, such a situation happens often in science. As noted in Chapter 1, Abrikosov could not publish his famous paper during 5 years. The results described in the paper seemed to other physicists very strange. Even after 1957, when it was published, the results were accepted only after experimental proof of several predicted effects. Everyone who is really active in science experienced in his (her) lifetime the same attitude from other scientists. "I know how helpless an individual is against the spirit of his time," – Ludwig Boltzmann.

So, this was really a bad time for the solitary wave. Between 1844, when Russell's report was published, and 1965, fewer than two dozen papers relating to the solitary wave were published.

3. Korteweg–de Vries equation

Fortunately, Russell lived long enough to see the solitary wave "acquited" (he died in 1882). Boussinesq published a paper in 1871, in which he showed that Russell's solitary wave may exist and approximately calculated its shape and velocity. The final verdict to the existence of the solitary wave was "announced" by Korteweg and de Vries in 1895. They reexamined all previous considerations and introduced a new equation for describing solitary waves that we now call the *Korteweg–de Vries* (KdV) equation [2].

Let $u(x,t)$ denote the height of the wave above the free surface at equilibrium, where t is the time and x is the coordinate along propagation of the solitary wave, the KdV equation is

$$\frac{\partial u}{\partial t} + \frac{3}{2h} u \frac{\partial u}{\partial x} + \frac{h^2}{6} \frac{\partial^3 u}{\partial x^3} = 0, \tag{5.1}$$

where h is the depth of the water in the canal. This equation is written in a frame moving at the speed of long-wave linear disturbances of the surface. The exact solution of the KdV equation describing the soliton is

$$u(x,t) = u_0 \times sech^2 \left(\frac{x - vt}{\ell} \right), \tag{5.2}$$

where u_0 is the initial height of the soliton; $v = \sqrt{gh}(1 + \frac{u_0}{2h})$, and 2ℓ is the "width" of the soliton and defined by the equation

$$S \equiv \frac{3}{4} \frac{u_0 \ell^2}{h^3} = 1. \tag{5.3}$$

Equation (3) is the mathematical condition expressing the balance between the dispersion and nonlinearity effects in the soliton. Although the parameter S was known for decades, its role in soliton theory became clear fairly recently. If S is much larger than 1, nonlinearity effects will prevail (if the hump is smooth enough). They will strongly deform the hump and it will eventually break into several pieces that will probably give rise to several solitons. If $S < 1$, dispersion prevails, and the hump will gradually diffuse. For $S \approx 1$ the hump has a soliton-like shape and, if its velocity is close to the soliton velocity, it will slightly deform and gradually become the real soliton, described by the KdV equation.

The KdV equation played a crucial role in our times in resurrecting the soliton. For pure mathematicians, the history of solitons begins with the KdV equation. However, mathematicians did not fully realize the importance of the shallow water equation and neglected the KdV equation until the physicists returned it to life after 70 years of oblivion. In fact, Korteweg and de Vries themselves had no idea of the brilliant future awaiting their equation.

4. Numerical simulations

With the appearance of computers, it became possible to simulate the behavior of nonlinear systems. Here we shall consider two first computer "experiments": the first was performed by Fermi, Pasta and Ulam in 1955 [3], and the second ten years later—in 1965 by Zabusky and Kruskal [4]. Historically, both simulations were very important for understanding the behavior of nonlinear systems described by the KdV equation.

In 1955, Fermi, Pasta and Ulam (FPU) decided to investigate the behavior of a one-dimensional chain of 64 particles of mass m, bound by massless springs. They accounted for nonlinear forces by assuming that stretching the spring by $\Delta\ell$ generates the force $k\ell + \alpha(\Delta\ell)^2$. The nonlinear correction to Hooke's law, $\alpha(\Delta\ell)^2$, was assumed to be small as compared to the linear force, $k\Delta\ell$. Thus, the computer had to solve the following equations

$$m\ddot{u}_n = k(u_{n+1} - u_n) - k(u_n - u_{n-1}), \tag{5.4}$$

with additional nonlinear terms in the right-hand side:

$$\alpha[(u_{n+1} - u_n)^2 - (u_n - u_{n-1})^2]. \tag{5.5}$$

In the harmonic case ($\alpha = 0$), the energy stored initially in a given mode stays in that mode and the system does not approach thermal equilibrium. For $\alpha \neq 0$, Fermi, Pasta and Ulam started their simulation by exciting the lowest mode ($n = 1$), i.e. by choosing a non-localized initial condition having the shape of the plane wave with wavelength equal to the size of the system. At the beginning of the simulation, they observed that the energy slowly goes into other modes, so the modes 2, 3, 4, ... became gradually excited. But, to their surprise, after about 157 periods of the fundamental mode, almost all the energy was back in the lowest mode! After this recurrence time, the initial state was almost restored. This remarkable result, known as the FPU paradox, shows that introducing nonlinearity in a system does not guarantee an equipartition of energy. They also showed that nonlinear excitations can emerge not only from localized initial conditions but also from non-localized initial conditions or from thermal excitations.

When Kruskal and Zabusky heard of the results from the authors, they immediately started similar computer experiments: they considered movements of continuous nonlinear strings. After many attempts, they came to a striking conclusion: for small amplitudes, vibrations of the continuous FPU string are best described by the KdV equation!

The KdV equation describes a variety of nonlinear waves, and is suitable for small amplitude waves in materials with weak dispersion.

Zabusky and Kruskal solved the FPU paradox in 1965 because they plotted the displacements in *real space* instead of looking at the Fourier modes. Figure 5.2 shows the result of their simulations. They considered the evolution of a simple harmonic wave $u(0, x) = \cos(\pi x)$ shown by the grey curve in Fig. 5.2. At some moment T_B (= $1/\pi$), a characteristic step is formed (see the dashed curve), which later at time $3.6T_B$ gives rise to a sequence of solitons, presented by the solid curve. In Fig. 5.2, the solitons are enumerated in descending order, the first having the largest amplitude. The first soliton is also the fastest one, and it is running down other solitons and eventually colliding with them. Shortly, after each collision, the solitons reappear virtually unaffected in size and shape. Thus, Zabusky and Kruskal proved that the KdV solitons are not changed in collisions, like rigid bodies. This property of solitary waves was mentioned in Russell's report. Zabusky and Kruskal also found that, as two solitons pass through each other, they accelerate. As a consequence, their trajectories deviate from straight lines, meaning that the solitons have particle-like properties. We shall discuss this remarkable feature in the next Section.

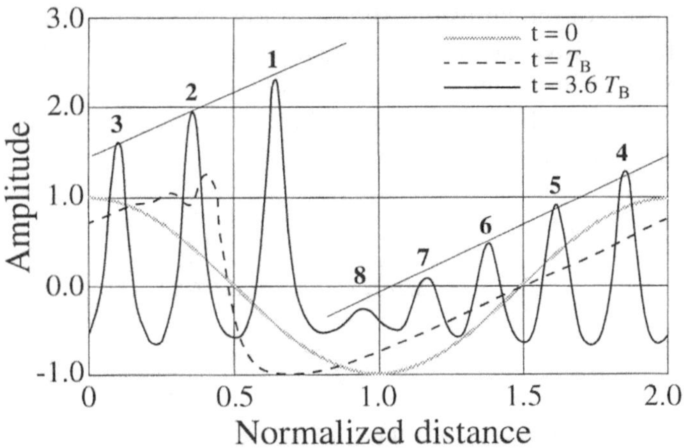

Figure 5.2. The temporal development of the wave form $u(x)$ [4].

For the results of the simulation, shown in Fig. 5.2, the recurrence time T_R was found to be $T_R = 30.4T_B$. The FPU recurrence is simply a manifestation of the stability of the solitons. In fact, Zabusky and Kruskal studied waves in a plasma, which satisfy the KdV equation, and they coined the term *soliton*, 131 years after of its discovery.

5. Particle-like properties

Intuitively, it is natural to consider solitary waves are *waves*. However, what Zabusky and Kruskal discovered in their computer simulation is that solitons behave like particles. Historically, neither Russell nor other scientists who studied the solitary wave more than a century after him noticed its striking resemblance to a particle. As mentioned above, Russell was aware of the particle-like property of two colliding solitary waves—after collision both solitary waves preserve their shapes and velocities. However, Russell didn't notice that, when the higher wave overtake the lower one, it appears that the first simply goes through the second and moves on ahead. Russell obviously thought that this was true. In reality this is not the case. If Russell could have had a video camera, he would have seen that the process of the collision is somewhat different.

When the larger solitary wave touches the smaller solitary wave, which move slower, the larger one slows down and diminishes while the smaller one accelerates and increases. When the smaller wave becomes as large as the original one, the waves detach and the ex-smaller one, which now is higher and faster, goes forward while the ex-bigger one is now moving behind at lower speed. Figure 5.3 schematically shows a collision of two solitary waves. In Fig. 5.3, two solitary waves are depicted at different periods of time. The observable result of the collision, shown in Fig. 5.3, is that the larger wave appears to have **shifted** ahead of the position that it would have occupied in uniform motion (without the collision). While the smaller one appears correspondingly to have **shifted** backwards. Thus, the waves do not penetrate each other, but rather collide like tennis balls.

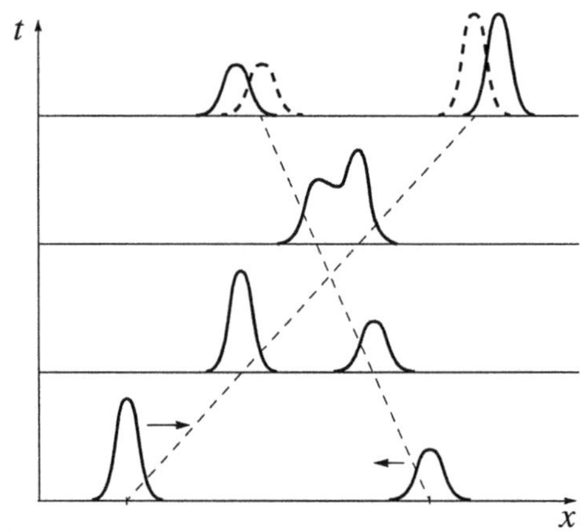

Figure 5.3. A schematic representation of a collision between two solitary waves.

In order to understand this analogy, let us consider a collision of two tennis balls uniformly moving along the axis x. Suppose that the balls are identical, and they do not rotate. Let the velocities of the balls be v_1 and v_2. In the *center of mass system* of coordinates, which moves with the velocity $v = (v_1 - v_2)/2$, the balls have velocities v and -v. Figure 5.4 shows two balls at different periods of time. Assume that the balls touch each other at time t_1 and detach at time t_2. Between t_1 and t_2, the balls suffer first squeezing and then expanding. What is interesting that, at the end of the collision, thus at t_2, the centers of the balls are slightly ahead (behind) of the positions O_1' and O_2' which correspond to the balls positions if there were no collision. This sort of shift occurs if the

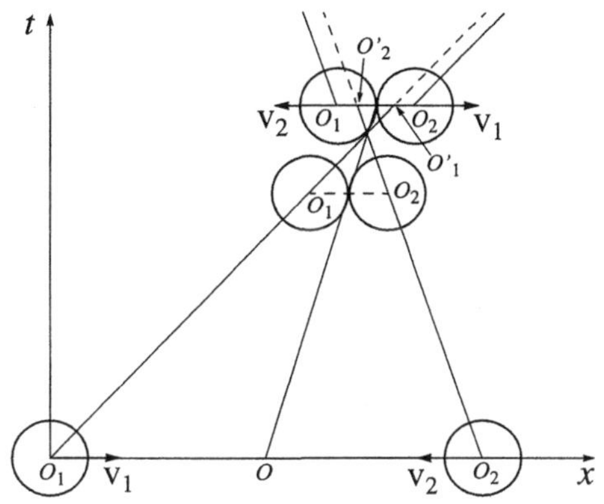

Figure 5.4. A schematic representation of a collision between two tennis balls [5].

time of the collision $(t_2 - t_1)$ is smaller than the "characteristic time" $2R/v$, where R is the radius of the balls.

This experiment can be performed at home, if there is a fast-speed video camera. The question is why had no one noticed this striking particle-like property of the solitary wave? If Russell could not see this subtle effect because he didn't have proper equipment, it is more difficult to understand why experimenters using modern cinema equipment failed to observe the shift in 1952. The only reasonable explanation of this blindness of scientists is that everybody, including Russell, perceived the solitary wave as a *wave*. Even it was clear that this wave is very unusual, nobody could imagine regarding it as a particle. As soon as Zabusky and Kruskal found in their computer simulations that the solitary wave has much in common with particles, they cut off the word "wave" and gave the new name *soliton*, by analogy with electron, proton and other elementary particles.

6. Frenkel-Kontorova solitons

To finish historical introduction to the soliton, let us consider one more event which is important in the history of the soliton.

All the examples of solitons, considered in the previous sections, are described by the KdV equation, and all these solitons belong to the same class of solitons: they are *nontopological*. The nontopological nature of these solitons can be easily understood because the system returns to its initial state after the passage of the wave. However, there are other types of solitons. The second group of solitons are so-called *topological* solitons, meaning that, after the passage of a topological soliton, the system is in a state which is different from its initial state. The topological stability can be explained by the analogy of the impossibility of untying a knot on an infinite rope without cutting it.

Frenkel and Kontorova theoretically predicted in 1939 topological solitons [6]. In fact, they found a special sort of a defect, called a *dislocation*, which exist in the crystalline structure of solids. The dislocations are not immobile—they can move inside the crystal.

Frenkel and Kontorova studied the simple possible model of a crystal which is shown in Fig. 5.5. In this one-dimensional model, atoms (black balls in Fig. 5.5) are distributed in a periodic sequence of hills and hollows which represent the substrate periodic potential. The balls rest at the bottom of each hollow because of the gravitational force. In Fig. 5.5, the springs which connect the balls represent the forces acting between atoms. It is evident that, in this model, a situation when one of the hollows is empty while all the balls are resting at the bottoms is impossible because of the springs. If one of the balls leaves a hollow (for example, n-th ball), the neighboring balls (n-1 and n+1) will be forced to follow. This creates an excitation which propagates further. The length of the excitation (or dislocation) is much larger than the interatomic

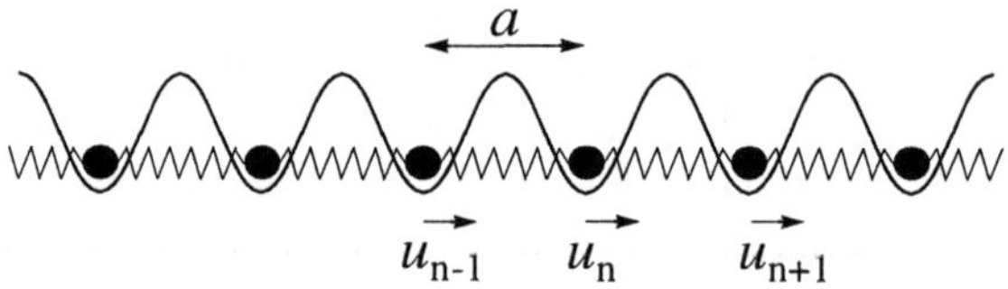

Figure 5.5. The Frenkel-Kontorova model of one-dimensional atomic chain: a chain of atoms with linear coupling (springs) interacting with a periodic nonlinear substrate potential.

distance, a. The long dislocation is mobile because small shifts of each atom do not require a noticeable energy supply. So, the dislocations in an ideal crystal freely move without changing its shape. However, if the crystal is imperfect, the dislocations will be attracted or repulsed by the defects. It is not difficult to understand that two dislocations repel each other, while dislocations are attracted by antidislocations.

The evolution of the dislocations and other topological solitons is described by the so-called *sine-Gordon* equation which we shall discuss in the next Section. The dislocation has the shape of a kink, shown in Fig. 5.6a. It has the *tanh*-like shape. The amplitude of the kink-soliton is independent of its velocity. As a consequence, topological solitons can be entirely static. The energy density of the kink-soliton is shown in Fig. 5.6b. In Fig. 5.6b, one can see that most of the energy of the kink, having the $sech^2$ shape, is localized in its core (in the middle of the kink). Consequently, *the soliton is a localized packet of energy*. The energy of the soliton falls exponentially away from its center.

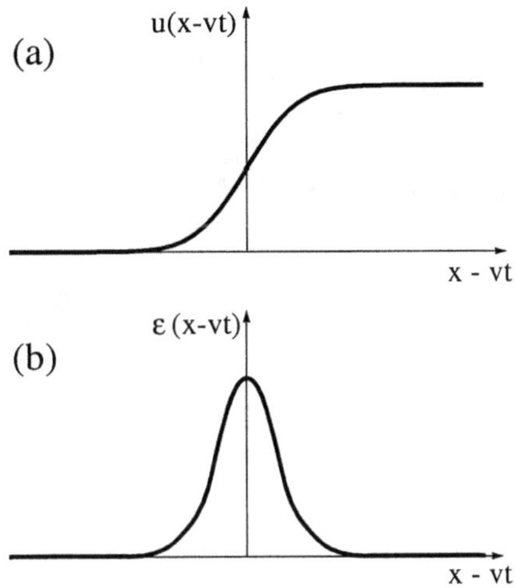

Figure 5.6. Schematic plots of (a) a kink-soliton solution, and (b) the energy density of the kink-soliton.

7. Topological solitons in a chain of pendulums

In order to understand better the nature of topological solitons, let us consider the propagation of a soliton in a chain of pendulums coupled by torsional springs. This mechanical transmission line is, probably, the simplest and one of the most efficient system for observing topological solitons and for studying their remarkable properties.

Figure 5.7 shows 21 pendulums, each pendulum being elastically connected to its neighbors by springs. If the first pendulum is displaced by a small angle θ, this disturbance propagates as a small amplitude linear wave from one pendulum to the next through the torsional coupling. As it moves along the chain, the small amplitude localized perturbation spreads over a larger and larger domain due to dispersive effects. The soliton is much more spectacular to observe. It is generated by moving the first pendulum by a full turn. This 2π rotation propagates along the whole pendulum chain and, even, reflects at a free end to come back unchanged.

The pendulum chain involves only simple mechanics and it is easy to write its equations of motion. Its energy is the sum of the rotational kinetic energy, the elastic energy of the torsional string connecting two pendulums, and the gravitational potential energy. Denoting by θ_n the angle of deviation of pendulum n from its vertical equilibrium position, by a the distance between pendulums along the axis, by m its mass and L the distance between the rotation axis and its center of mass, the expression of the energy is then

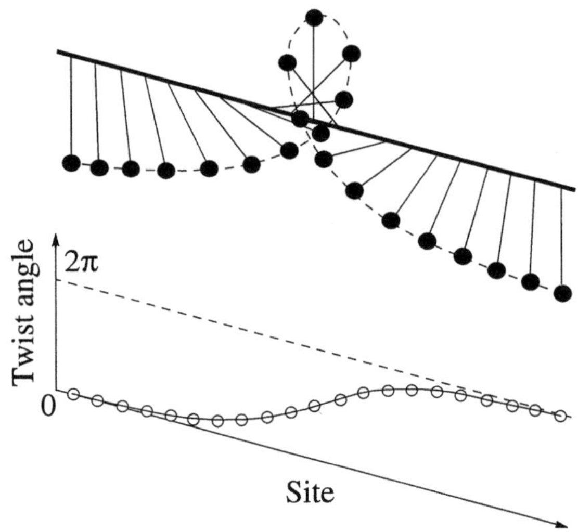

Figure 5.7. A soliton in a chain of pendulums coupled by a torsional spring. The soliton is a 2π rotation. The lower figure shows the rotation angle of the pendulums as a function of their position along the chain.

$$H = \sum_n \frac{1}{2} I \left(\frac{d\theta_n}{dt} \right)^2 + \frac{1}{2} \beta (\theta_n - \theta_{n-1})^2 + mgL(1 - \cos\theta_n), \tag{5.6}$$

where I is the momentum of inertia of a pendulum around the axis, and β is the torsional coupling constant of the springs. The equations of motion which can be deduced from the hamiltonian are

$$\frac{d^2\theta_n}{dt^2} - \frac{c_0^2}{a^2}(\theta_{n+1} + \theta_{n-1} - 2\theta_n) + \omega_0^2 \sin\theta_n = 0, \tag{5.7}$$

where $\omega_0^2 = mgL/I$ and $c_0^2 = \beta a^2/I$.

This system of nonlinear differential equations, often called the *discrete sine-Gordon* equation, cannot be solved analytically. The common attitude in solving nonlinear equations is to *linearize* the equations. In our case, it can be done by replacing $\sin\theta_n$ by its small amplitude expression $\sin\theta_n \approx \theta_n$. Then the system of equations is easy to solve but *essential physics has been lost.* The linearized equations have no localized solutions and they have no chances to describe the soliton even approximately because $\theta_n = 2\pi$ is not a small angle. Thus, *by linearizing a set of nonlinear equations to get an approximate solution is not always a good idea.*

There is however another possibility to solve approximately this set of equations, while preserving their full nonlinearity, if the coupling between the pendulums is strong enough, i.e. $\beta \gg mgL$ (or $c_0^2/a^2 \gg \omega_0^2$). In this case, adjacent pendulums have similar motions and the discrete set of variables $\theta_n(t)$ can be replaced by a single function of two variables, $\theta(x,t)$ such that $\theta_n(t) = \theta(x = n, t)$. A Tylor expansion of $\theta(n+1,t)$ and $\theta(n-1,t)$ around $\theta(n,t)$ turns the discrete sine-Gordon equation into the partial differential equation

$$\frac{\partial^2\theta(x,t)}{\partial t^2} - c_0^2\frac{\partial^2\theta(x,t)}{\partial x^2} + \omega_0^2 \sin\theta = 0. \tag{5.8}$$

The equation is called the *sine-Gordon* equation, and it has been extensively studied in soliton theory because it has exceptional mathematical properties. In particular, it has a soliton solution

$$u(x-vt) = 4\arctan\left[\exp\left(\pm\frac{\omega_0}{c_0}\frac{x-vt}{\sqrt{1-v^2/c_0^2}}\right)\right], \tag{5.9}$$

which is plotted in Figs. 5.7 and 5.8. The ($\pm$) signs correspond to localized soliton solutions which travel with the opposite screw senses: they are respectively called a *kink* soliton and an *antikink* soliton. They are shown in Fig. 5.8: the pendulums rotate from 0 to 2π for the kink and from 0 to -2π for the antikink.

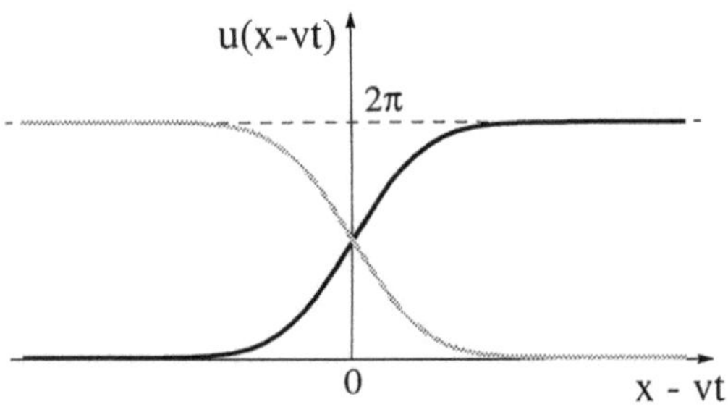

Figure 5.8. A schematic plot of a kink-soliton and antikink-soliton solutions for the chain of pendulums shown in Fig. 5.7.

As the KdV equation, the sine-Gordon equation also contains dispersion and nonlinearity. However, in the sine-Gordon equation, both dispersion and nonlinearity appear in the same term $\omega_0^2 \sin\theta$, where ω_0 is a characteristic frequency of a system.

The topological nature of solitons in the chain of pendulums can be easily demonstrated by plotting the gravity potential acting on the pendulums versus θ and the position x of the pendulum. As one can see in Fig. 5.9, each kink joints two successive equilibrium states (potential minima). Thus, the kink-soliton can be considered as a *domain wall* between two degenerate energy

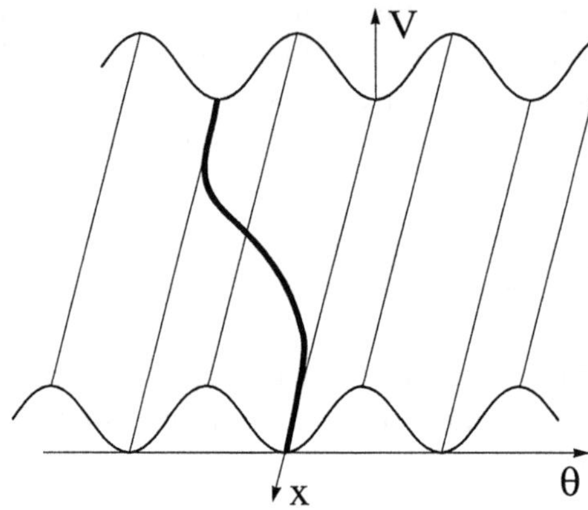

Figure 5.9. Sketch of the sinusoidal potential of the pendulum chain. The solid curve shows the trajectory of the kink-soliton which can be considered as a domain wall between two degenerate energy minima.

minima. The topological soliton is an excitation which interpolates between these minima.

In the expression of the soliton solution, one can see that a topological soliton cannot travel faster than c_0 which represents the velocity of linear waves (in solids, the longitudinal sound velocity). As the soliton velocity v approaches c_0, the soliton remains constant but its width gets narrower owing to the Lorentz contraction of its profile, given by $d\sqrt{1 - v^2/c_0^2}$, where $d = c_0/\omega_0$ is a *discreteness* parameter. Thus, the topological solitons behave like relativistic particles.

From the condition $c_0^2/a^2 \gg \omega_0^2$ (see above), the discreteness parameter $d = c_0/\omega_0$ is much larger than the distance between pendulums, $d \gg a$. If $d \simeq a$, the angle of rotation varies abruptly from one pendulum to the next, and the continuum (long-wavelength) approximation cannot be used.

Comparing topological and nontopological solitons, it is worth remarking that the amplitude of the kink is independent of its velocity and, when the velocity $v = 0$, the soliton solution reduces to

$$u(x) = 4\arctan[\exp(\pm x/d)]. \tag{5.10}$$

Thus, by contrast to nontopological solitons, *the kink soliton may be entirely static*, losing its wave character. The second feature of topological solitons (kinks) is that they have antisolitons (antikinks) which are analogous to antiparticles. In contrast, for nontopological solitons, there are no antisolitons.

By using such a simple mechanical device shown in Fig. 5.7, it is easy to check experimentally the exceptional properties of the topological solitons. Launching a soliton and keeping on agitating the first pendulum, it is possible to test the ability of the soliton to propagate over a sea of linear waves. The particle-like properties appear clearly if one static soliton is created in the middle of the chain and then a second one is sent. The collision looks to the observer exactly similar to a shock between elastic marbles. Thus, the pendulum chain provides an experimental device which convincingly demonstrate the unique properties of the soliton.

In addition to the kink and antikink solutions, the sine-Gordon equation also has solutions in the form of oscillating pulses called a *breather* or *bion* (meaning a "living particle"). Breathers can be considered as a soliton-antisoliton bound state. An example of a breather is shown in Fig. 5.10. As other topological solitons, breathers can move with a constant velocity or be entirely static. Theoretically, breathers interact with other breathers and other solitons in the same manner as all topological solitons do. However, in reality, breather-type solitons can be easily destroyed by almost any type of perturbation.

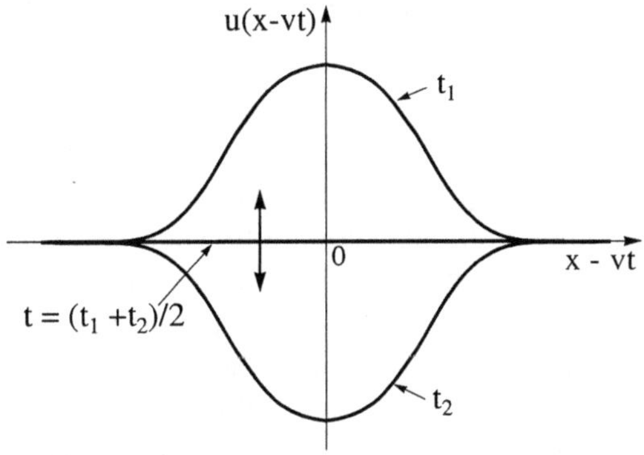

Figure 5.10. Sketch of a sine-Gordon breather or a bion at different times t.

Stationary breathers are pulsating objects. If, in the chain of pendulums shown in Fig. 5.7, one launches a kink and antikink (with the opposite screw sense) with sufficiently low velocity, one can observe a bounded pair that is a breather solution. Nevertheless, owing to dissipation effects present on the real line, only a few breathing oscillations can be observed. The oscillations then decrease with time and energy is radiated onto the line.

8. Different categories of solitons

There are a few ways to classify solitons. For example, as discussed above, there are *topological* and *nontopological* solitons. Independently of the topological nature of solitons, all solitons can be divided into two groups by taking into account their profiles: *permanent* and *time-dependent*. For example, kink solitons have a permanent profile (in ideal systems), while all breathers have an internal dynamics, even, if they are static. So, their shape oscillates in time.

The third way to classify the solitons is in accordance with nonlinear equations which describe their evolution. Here we discuss common properties of solitons on the basis of the third classification, i.e. in accordance with nonlinear equations which describe the soliton solutions.

Up to now we have considered two nonlinear equations which are used to describe soliton solutions: the KdV equation and the sine-Gordon equation. There is the third equation which exhibits true solitons—it is called the *nonlinear Schrödinger* (NLS) equation. We now summarize soliton properties on the basis of these three equations, namely, the Korteweg-de Vries equation:

$$u_t = 6uu_x - u_{xxx}; \tag{5.11}$$

the sine-Gordon equation:

$$u_{tt} = u_{xx} - \sin u, \tag{5.12}$$

and the nonlinear Schrödinger equation:

$$iu_t = -u_{xx} \pm |u|^2 u, \tag{5.13}$$

where u_z means $\frac{\partial u}{\partial z}$. For simplicity, the equations are written for the dimensionless function u depending on the dimensionless time and space variables.

There are *many* other nonlinear equations (i.e. the Boussinesq equation) which can be used for evaluating solitary waves, however, these three equations are particularly important for physical applications. They exhibit the most famous solitons: the KdV (pulse) solitons, the sine-Gordon (topological) solitons and the *envelope* (or NLS) solitons. All the solitons are one-dimensional (or quasi-one-dimensional). Figure 5.11 schematically shows these three types of solitons. Let us summarize common features and individual differences of the three most important solitons.

8.1 The KdV solitons

The exact solution of the KdV equation is given by Eq. (5.2). The basic properties of the KdV soliton, shown in Fig. 5.11a, can be summarized as follows [7]:

i. Its amplitude increases with its velocity (and vice versa). Thus, they cannot exist at rest.

ii. Its width is inversely proportional to the square root of its velocity.

iii. It is a unidirectional wave pulse, i.e. its velocity cannot be negative for solutions of the KdV equation.

iv. The sign of the soliton solution depends on the sign of the nonlinear coefficient in the KdV equation.

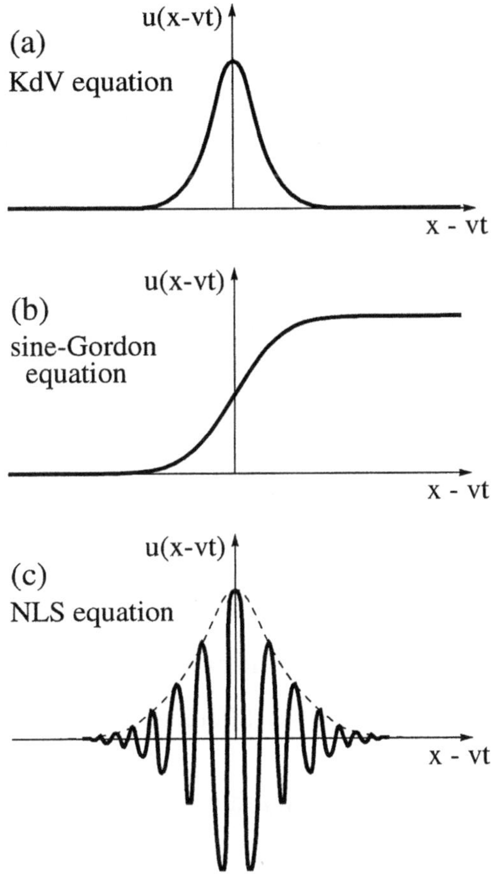

Figure 5.11. Schematic plots of the soliton solutions of: (a) the Korteweg–de Vries equation; (b) the sine-Gordon equation, and (c) the nonlinear Schrödinger equation.

The conservation laws for the KdV solitons represent the conservation of mass

$$M = \frac{1}{2}\int u dx; \tag{5.14}$$

momentum

$$P = -\frac{1}{2}\int u^2 dx; \tag{5.15}$$

energy

$$E = \frac{1}{2}\int (2u^2 + u_x^2) dx, \tag{5.16}$$

and center of mass

$$M x_s(t) = \frac{1}{6}\int x u dx = tP + const. \tag{5.17}$$

The KdV solitons are nontopological, and they exist in physical systems with weakly nonlinear and with weakly dispersive waves. When a wave impulse breaks up into several KdV solitons, they all move in the same direction (see, for example, Fig. 5.2). The collision of two KdV solitons is schematically shown in Fig. 5.3. Under certain conditions, the KdV solitons may be regarded as particles, obeying the standard laws of Newton's mechanics. In the presence of dissipative effects (friction), the KdV solitons gradually decelerate and become smaller and longer, thus, they are "mortal."

8.2 The topological solitons

The exact solution of the sine-Gordon equation is given by Eq. (5.9). The basic properties of a topological (kink) soliton shown in Fig. 5.11b can be summarized as follows [7]:

i. Its amplitude is independent of its velocity—it is constant and remains the same for zero velocity, thus the kink may be static.

ii. Its width gets narrower as its velocity increases, owing to Lorentz contraction.

iii. It has the properties of a relativistic particle.

iv. The topological kink which has a different screw sense is called an *antikink*.

For the chain of pendulums shown in Fig. 5.7, the energy of the topological (kink) soliton, E_K, is determined by

$$E_K = \frac{m_0 c_0^2}{\sqrt{1 - \frac{v^2}{c_0^2}}}, \tag{5.18}$$

where c_0 is the velocity of *linear* waves, and the soliton mass m_0 is given by

$$m_0 = 8\frac{I}{a}\frac{\omega_0}{c_0} = 8\frac{I}{a}\frac{1}{d}. \tag{5.19}$$

One can also introduce the relativistic momentum

$$p_K = \frac{m_0 v}{\sqrt{1 - \frac{v^2}{c_0^2}}}. \tag{5.20}$$

At rest ($v = 0$) one has $p = 0$, and the static soliton energy is

$$E_{0,K} = m_0 c_0^2. \tag{5.21}$$

Topological solitons are extremely stable. Under the influence of friction, these solitons only slow down and eventually stop and, at rest, they can live "eternally." In an infinite system, the topological soliton can only be destroyed by moving a semi-infinite segment of the system above a potential maximum. This would require an infinite energy. However, the topological soliton can be annihilated in a collision between a soliton and an anti-soliton. In an integrable system having exact soliton solutions, solitons and anti-solitons simply pass through each other with a phase shift, as all solitons do, but in a real system like the pendulum chain which has some dissipation of energy, the soliton-antisoliton equation may destroy the nonlinear excitations. Figure 5.12 schematically shows a collision of a kink and an antikink in an integrable system which has soliton solutions. In integrable systems, the soliton-breather and breather-breather collisions are similar to the kink-antikink collision shown in Fig. 5.12.

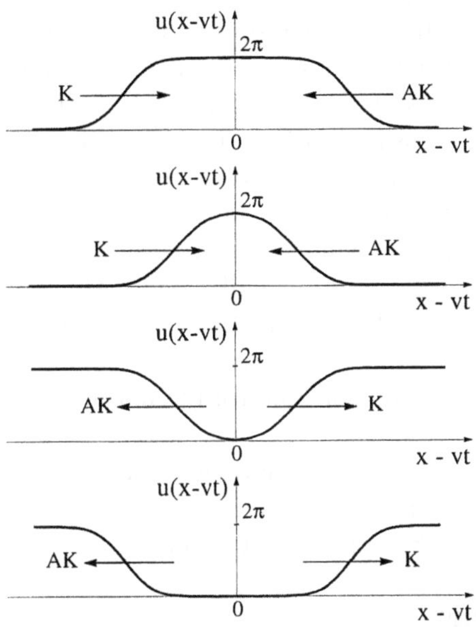

Figure 5.12 Sketch of a collision between a kink (K) and an antikink (AK). The phase shift after the collision is not indicated.

The sine-Gordon equation has almost become ubiquitous in the theory of condensed matter, since it is the simplest nonlinear wave equation in a periodic medium.

8.3 The envelope solitons

The NLS equation is called the *nonlinear* Schrödinger equation because it is formally similar to the Schrödinger equation of quantum mechanics

$$\left[i\hbar\frac{\partial}{\partial t} + \frac{\hbar^2}{2m}\frac{\partial^2}{\partial x^2} - U\right]\Psi(x,t) = 0, \tag{5.22}$$

where U is the potential, and $\Psi(x,t)$ is the wave function.

The NLS equation describes self-focusing phenomena in nonlinear optics, one-dimensional self-modulation of monochromatic waves, in nonlinear plasma etc. In the NLS equation, the potential U is replaced by $|u|^2$ which brings into the system self-interaction. The second term of the NLS equation is responsible for the dispersion, and the third one for the nonlinearity. A solution of the NLS equation is schematically shown in Fig. 5.11c. The shape of the enveloping curve (the dashed line in Fig. 5.11c) is given by

$$u(x,t) = u_0 \times sech((x - vt)/\ell), \tag{5.23}$$

where 2ℓ determines the width of the soliton. Its amplitude u_0 depends on ℓ, but the velocity v is *independent* of the amplitude, distinct from the KdV soliton. The shapes of the envelope and KdV solitons are also different: the KdV soliton has a $sech^2$ shape. Thus, the envelope soliton has a slightly wider shape. However, other properties of the envelope solitons are similar to the KdV solitons, thus, they are "mortal" and can be regarded as particles. The interaction between two envelope solitons is similar to the interactions between two KdV solitons (or two topological solitons), as shown in Fig. 5.3.

In the envelope soliton, the stable groups have normally from 14 to 20 humps under the envelope, the central one being the highest one. The groups with more humps are unstable and break up into smaller ones. The waves inside the envelope move with a velocity that differs from the velocity of the soliton, thus, the envelope soliton has an *internal* dynamics. The relative motion of the envelope and carrier wave is responsible for the internal dynamics of the NLS soliton.

The NLS equation is inseparable part of nonlinear optics where the envelope solitons are usually called *dark* and *bright* solitons, and became quasi-three-dimensional. We shall briefly discuss the optical solitons below.

8.4 Solitons in real systems

As a final note to the presentation of the three types of solitons, it is necessary to remark that real systems do not carry exact soliton solutions in the strict mathematical sense (which implies an infinite life-time and an infinity of conservation laws) but *quasi*-solitons which have most of the features of true solitons. In particular, although they do not have an infinite life-time, quasi-solitons are generally so long-lived that their effect on the properties of the system are almost the same as those of true solitons. This is why physicists often use the word soliton in a relaxed way which does not agree with mathematical rigor.

In the following sections, we consider a few examples of solitons in real systems, which are useful for the understanding of the mechanism of high-T_c superconductivity.

9. Solitons in the superconducting state

Solitons are literally everywhere. The superconducting state is not an exception: vortices and *fluxons* are topological solitons. Fluxons are quanta of magnetic flux, which can be studied in *long Josephson junctions*. A long Josephson junction is analogous to the chain of atoms studied by Frenkel and Kontorova. The sine-Gordon equation provides an accurate description of vortices and fluxons.

In a type-II superconductor, if the magnitude of an applied magnetic field is larger than the lower critical field, B_{c1}, the magnetic field begins to penetrates the superconductor in microscopic vortices which form a regular lattice. Vortices in a superconductor are similar to vortices in the ideal liquid, but there is a dramatic difference—they are *quantized.* As shown in Fig. 2.11, inside the vortex tube, there exists a magnetic field. The magnetic flux supported by the vortex is a multiple of the *quantum magnetic flux* $\Phi_0 \equiv hc/2e$, where c is the speed of light; h is the Planck constant, and e is the electron charge. The vortices in type-II superconductors are a pure and beautiful physical realization of the sine-Gordon (topological) solitons. Their extreme stability can be easily understood in terms of the Ginzburg-Landau theory of superconductivity which contains a system of nonlinear coupled differential equations for the vector potential and the wave function of the superconducting condensate. Thus, topological solitons in the form of vortices were found in the superconducting state 20 years earlier than the superconducting state itself was understood.

Let us now consider fluxons in a long Josephson junction. An example of a *long* superconductor-insulator-superconductor (SIS) Josephson junction is schematically shown in Fig. 5.13. The insulator in the junction is thin enough, say 10–20 Å to ensure the overlap of the wave functions. Due to the Meissner effect, the magnetic field may exist only in the insulating layer and in adjacent thin layers of the superconductor.

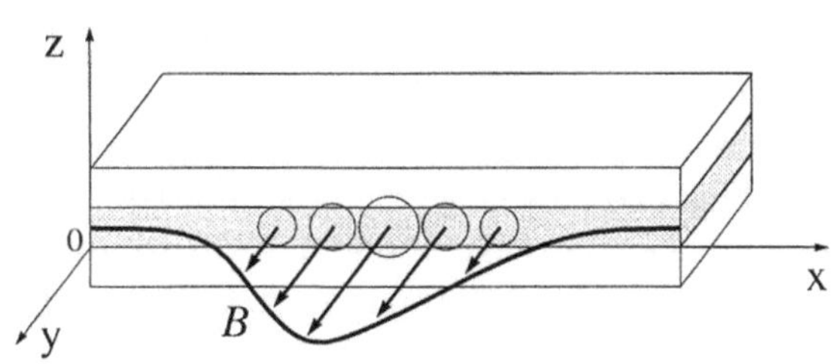

Figure 5.13. Sketch of a soliton in a long Josephson junction.

The *dc* and *ac* Josephson effects were discussed in Chapter 2. The magnitude of the zero-voltage current resulting from the tunneling of Cooper pairs, known as the *dc* Josephson effect, depends on the phase difference between two superconductors as

$$I = I_c \sin\varphi, \tag{5.24}$$

where $\varphi = \varphi_2 - \varphi_1$ is the phase difference, and I_c is the critical Josephson current.

The oscillating current of Cooper pairs that flows when a steady voltage V is maintained across a tunnel barrier is known as the *ac* Josephson effect. The phase difference between the two superconductors in the junction is given by

$$\frac{d\varphi}{dt} = \frac{2e}{\hbar}V. \tag{5.25}$$

Taking into account the inductance and the capacitance of the junction, one can easily get a sine-Gordon equation for $\varphi(x,t)$ forced by a right-hand side term which is associated with the applied bias. For a long Josephson junction, the nonlinear equation exactly coincides with Eq. (5.8). In the long Josephson junction, the sine-Gordon solitons describe quanta of magnetic flux, expelled from the superconductors, that travels back and forth along the junction. Their presence, and the validity of the soliton description, can be easily checked by the microwave emission which is associated with their reflection at the ends of the junction.

For the Josephson junction, the ratio $\lambda_J = c_0/\omega_0$ [see Eq. (5.8)] gives a measure of the typical distance over which the phase (or magnetic flux) changes, and is called the *Josephson penetration length.* This quantity allows one to define precisely a *small* and a *long* junction. A junction is said to be long if its geometric dimensions are large compared with λ_J. Otherwise, the junction is small. The Josephson length is usually much larger than the London penetration depth λ_L.

The moving fluxon has a kink shape, which is accompanied by a *negative* voltage pulse and a *negative* current pulse, corresponding to the space and time derivatives of the fluxon solution. In Fig. 5.13, the arrows in the y direction represent the magnetic field, and the circles show the Josephson currents producing the magnetic field.

A very useful feature of the Josephson solitons is that they are not difficult to operate by applying bias and current to the junction. By using artificially prepared inhomogeneities one can make bound state of solitons. In this way one can store, transform and transmit information. In other words, long Josephson junctions can be used in computers. One of the most useful properties of such devices would be very high performance speed. Indeed, the characteristic time may be as small as 10^{-10} sec, while the size of the soliton may be less than 0.1 mm. The main problem for using the Josephson junctions in electronic devices is the cost of cooling refrigerators. However, I am absolutely sure that they will be commercially used in electronics if room-temperature superconductors become available.

10. Topological solitons in polyacetylene

Let us consider topological solitons and *polarons* in solids. One-dimensional polarons are breathers, consequently, they can be considered as a soliton-antisoliton (kink-antikink) bound state. The term "polaron" should not be confused with the Holtstein (small) polaron which is three-dimensional, and represents a self-trapped state in solids (see Section 12).

Polyacetylene, $(CH)_x$, is the simplest linear conjugated polymer. Polyacetylene exists in two isomerizations: *trans* and *cis*. Let us consider the *trans*-polyacetylene. The structure of *trans*-polyacetylene is schematically shown in Fig. 5.14. The important property of *trans*-polyacetylene is the double degeneracy of its ground state: the energy for the two possible patterns of alternating short (double) and long (single) bonds. The double-well potential is schematically shown in Fig. 5.15. In double-well potential, a kink cannot be

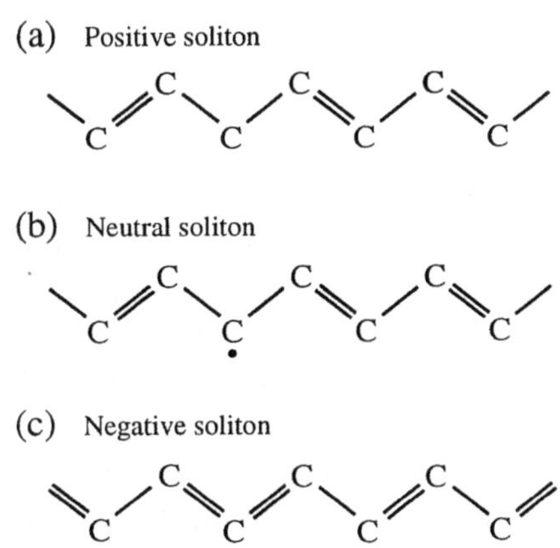

Figure 5.14. Formula of polyacetylene showing the possibility to have a topological soliton: (a) positive; (b) neutral, and (c) negative.

followed by another kink like, for example, in a potential with infinite number of equilibrium states, shown in Fig. 5.9: a kink can be followed only by an antikink, which connect the two potential wells, as shown in Fig. 5.15. In *cis*-polyacetylene, the two ground-state configurations have different energies. Therefore, in *cis*-polyacetylene, the two-well potential is asymmetrical. Further we consider only the *trans*-polyacetylene configuration, sometimes, dropping the prefix "trans".

Undoped polyacetylene which has a half-filled band is an insulator with a charge gap of 1.5 eV. The gap is partially attributed to the so-called Peierls instability of the one-dimensional electron gas. Upon doping, polyacetylene becomes highly conducting. The significant overlap between the orbitals of the neighboring carbon atoms results in a relatively broad band (called π-band) with a width of 10 eV. The polyacetylene chains are only weakly coupled to

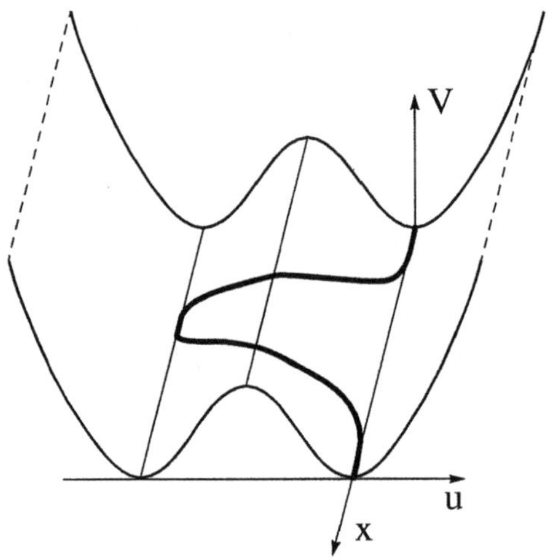

Figure 5.15. Sketch of the double-well potential for *trans*-polyacetylene. The solid curve shows the trajectory of the kink-antikink (breather) solution.

each other: the distance between the nearest carbon atoms belonging to different chains is about 4.2 Å, which is three times larger than the distance between the nearest carbon atoms in the chains. As a consequence, the interchain hopping amplitude is about 30 times smaller than the hopping amplitude along the chain, which makes the intrinsic properties of polyacetylene quasi-one-dimensional.

As a consequence of the two degenerate ground state in *trans*-polyacetylene, it is natural to expect the existence of excitations in polyacetylene in the form of topological solitons, or moving domain walls, separating the two degenerate minima. Depending on doping, there exist 3 different types of topological solitons (kinks) in polyacetylene, having different quantum numbers. As shown in Fig. 5.14a, the removal of one electron from a polyacetylene chain creates a positively charged soliton with charge of $+e$ and spin zero [8]. The addition of one electron to a polyacetylene chain generates a neutral soliton with spin of 1/2, as shown in Fig. 5.14b. Figure 5.14c shows the case when two electrons are added to a polyacetylene chain. In this case, the soliton is negatively charged, having spin zero. Since these three kinds of solitons differ only by the occupation number of the zero energy state, the all have the same creation energy given by

$$E_s = \frac{2}{\pi}\Delta, \tag{5.26}$$

where 2Δ is the width of the charge gap which separates the valence band and the conduction band. All three solitons occupy the midgap state of the charge gap, as schematically shown in Fig. 5.16b. Thus, to create a soliton is more energetically profitable than to add one electron to the conduction band, $E_s < 2\Delta$. Figure 5.16a shows a $tanh$-shaped kink which connects the two electron bands and the electron density of states of the midgap state shown in Fig. 5.16b (compare with Fig. 5.6). In Fig. 5.16a, the spatial density of states is determined by $|\psi(x)|^2$, where

$$\psi(x) = \frac{C}{\cosh(x/d)} = C \times sech(x/d), \tag{5.27}$$

and C and d are constants.

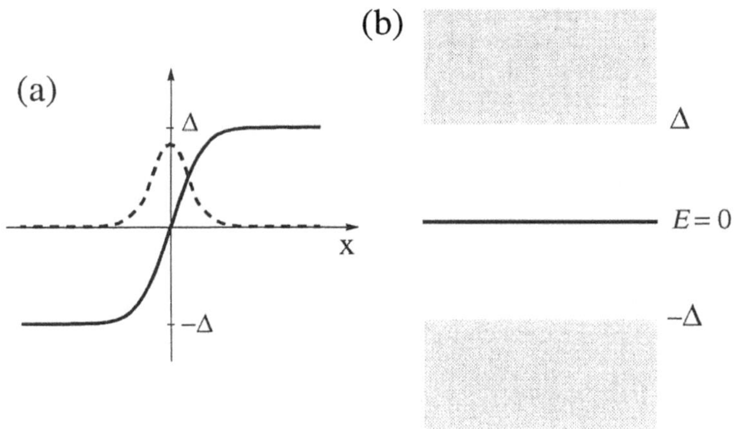

Figure 5.16. (a) A schematic plot of a topological soliton (thick curve) in *trans*-polyacetylene and the electron density of states $|\psi(x)|^2$ for the intragap state (dashed curve). The vertical axis for the density of states is not shown. (b) The spectrum of electron states for the soliton lattice configuration. The lower band corresponds to the valence band, and the upper band to the conduction band.

In Fig. 5.16a, the density of states is presented as a function of soliton position along the main polyacetylene axis. Figure 5.17 depicts the density of states in doped polyacetylene as a function of energy having the origin at the middle of the charge gap. The density of states shown in Fig. 5.17 corresponds to the total number of energy levels per unit volume which are available for possible occupation by electrons. Since the peak in the spatial density of states, shown in Fig. 5.16a, is not the delta function, but has a finite width, then, according to the Fourier transform, the width of corresponding peak in the spectral density of states is also finite. So, the width of the soliton peak shown schematically in Fig. 5.17 is finite. In Fig. 5.17, the height of the soliton peak depends on the density of added or removed electrons: the height increases as the electron density increases [9]. However, one should realize that there exists a maximum density of added or removed electrons, above which the charge gap collapses, and polyacetylene becomes metallic.

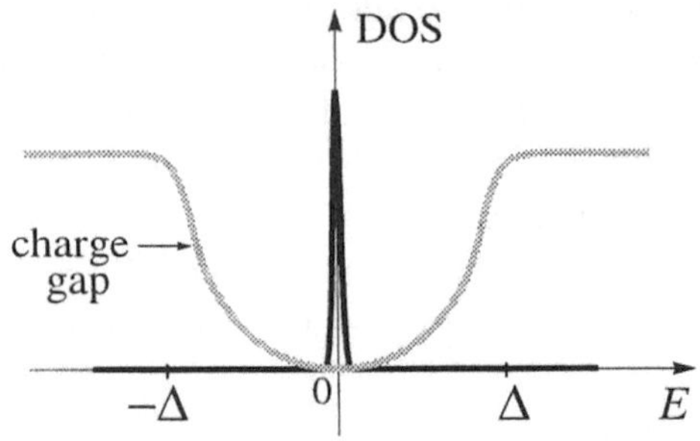

Figure 5.17. Sketch of the electron density of states (DOS) per unit energy interval in *trans*-polyacetylene for the soliton lattice configuration. The height of the soliton peak depends on the density of added or removed electrons [9].

The model was created in order to explain a very unusual behavior of the spin susceptibility in *trans*-polyacetylene. Upon doping, pristine polyacetylene becomes highly conducting. Strangely, the spin susceptibility of *trans*-polyacetylene remains small well into the conducting regime. Using the model, this strange behavior can be well understood: upon doping, the charge carriers in polyacetylene are not conventional electrons and holes, but spinless solitons (kinks).

Following the original analysis [8], the model has been refined and it has been shown that the most probable defects are not the topological solitons but breathers (or polarons) which can be considered as a soliton-antisoliton bound state. In double-well potential, a kink can only be followed by an anti-kink, as shown in Fig. 5.15. Thus, the breather (polaron) solution is a sum of two *tanh* functions:

$$\Delta_{pol} = \Delta - v_F K \left[\tanh\left(K\left(x + \frac{R}{2}\right)\right) - \tanh\left(K\left(x - \frac{R}{2}\right)\right)\right], \quad (5.28)$$

where R is the distance between the soliton and antisoliton; v_F is the velocity on the Fermi surface, and K is determined by

$$v_F K = \Delta \tanh(KR). \quad (5.29)$$

The spectrum of electron states for Δ_{pol} consists of a valence band (with the highest energy -Δ), a conduction band (with the lowest energy +Δ), and two localized intragap states with energies $\pm E_0(R)$, as shown in Fig. 5.18b, where

$$E_0(R) = \frac{\Delta}{\cosh(KR)}. \quad (5.30)$$

The two intragap states are symmetric and antisymmetric superposition of the midgap state localized near the kink and antikink:

$$\psi_{\pm}(x) = \frac{1}{2}\left(\frac{\sqrt{K/2}}{\cosh(K(x-R/2))}|-\rangle \pm \frac{\sqrt{K/2}}{\cosh(K(x+R/2))}|+\rangle\right). \quad (5.31)$$

These results for the electron wave function hold for any distance R between the kink and antikink. However, the configuration Eq. (5.28) is only self-consistent if

$$R_{sc} = \sqrt{2}\ln(1+\sqrt{2})\xi_0, \quad (5.32)$$

where $\xi_0 = v_F/\Delta$ denotes a correlation length. Then, $K\xi_0 = 1/\sqrt{2}$ and $E_0(R_c) = \Delta/\sqrt{2}$.

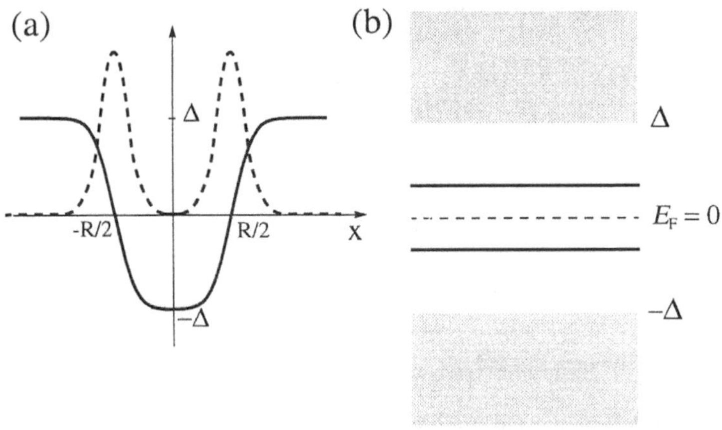

Figure 5.18. A schematic plot of a soliton-antisoliton lattice configuration (thick curve) in *trans*-polyacetylene and the electron density of states $|\psi(x)|^2$ for the two intragap state (dashed curve). The vertical axis for the density of states is not shown. (b) The spectrum of electron states for the soliton-antisoliton lattice configuration. The lower band corresponds to the valence band, and the upper band to the conduction band.

The occupation of the two intragap states $\psi_{\pm}$ cannot be arbitrary: the solution is self-consistent if there is either one electron in the negative energy state ψ_- and no electrons in the positive energy state ψ_+, or if ψ_- is doubly occupied and ψ_+ is occupied by one electron. Only charged polarons are stable, thus they have charge $\pm e$ and spin 1/2. The polaron has the quantum numbers of an electron (a hole). In other words, the polaron is a bound state of a charged spinless soliton and a neutral soliton with spin 1/2. Figure 5.18a shows a bound state of a kink and an antikink separated by distance R, and the electron density of states of the two intragap states shown in Fig. 5.18b.

By contrast, a charged kink and a charged antikink do not form a bound state, as they repel each other, even, in the absence of the Coulomb interaction.

For a given density of added or removed electrons, this repulsion forces the charged kinks and antikinks to form a periodic lattice. At small density of added charges, the soliton lattice has a lower energy than the polaron lattice, since the kink creation energy E_s [see Eq. (5.26)] is smaller than the energy of polaron creation, which is given by

$$E_p = \frac{2\sqrt{2}}{\pi}\Delta = \sqrt{2}E_s. \tag{5.33}$$

In Fig. 5.18a, the density of states is presented as a function of soliton and antisoliton positions along the main polyacetylene axis. Figure 5.19 *schematically* depicts the density of states in doped polyacetylene as a function of energy having the origin at the middle of the charge gap. Since the occupation of the two intragap states cannot be arbitrary, the heights of the two peaks shown in Fig. 5.19 are, in fact, different: the right-hand peak should be lower than the left-hand peak.

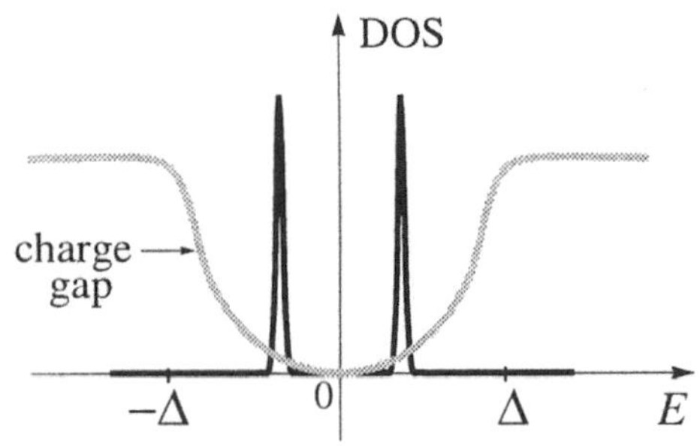

Figure 5.19. Sketch of the electron density of states (DOS) per unit energy interval in *trans*-polyacetylene for the soliton-antisoliton lattice configuration. The height of the soliton-antisoliton peaks depends on the density of added or removed electrons [9].

Finally, it is worth emphasizing two aspects of kink-antikink bound state in any nonlinear system. First, the *distance* between the two peaks in the spectral density of states, shown in Figs. 5.18b and 5.19, depends on the *real distance* between the kink and the antikink, as schematically depicted in Fig. 5.18a. Second, the spatial *and* spectral densities of states of any kink-(anti)kink bound state (if such exists) are not sensitive to the "polarity" of the two kinks: the density of states of a bound state of two (anti)kinks (for example, in the periodic potential shown in Fig. 5.9) coincides with the density of states of a kink-antikink bound state: compare the dashed curves in Figs. 5.20 and 5.18a.

11. Magnetic solitons

Solitons are not restricted to the macroscopic world—they exist on the microscopic scale too.

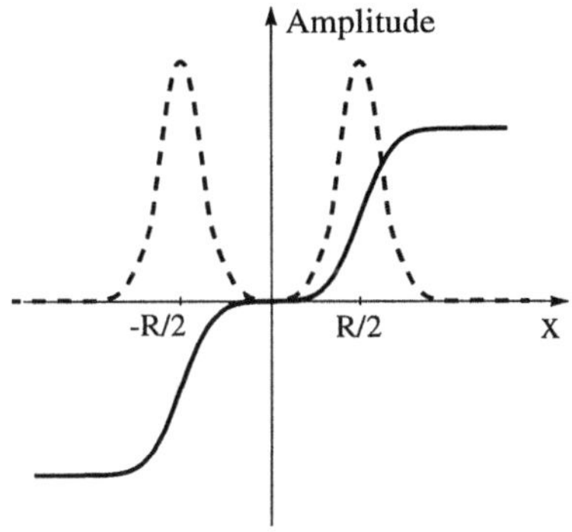

Figure 5.20. A schematic plot of two kinks (solid curve), for example, in infinite-well potential shown in Fig. 5.9, and the corresponding density of states (dashed curve). Compare with Fig. 5.18(a).

Historically, in the theory of solid state magnetism, nonlinearities were considered small and were described in terms of linear theory. In the linear approximation, eigen-excitations in the magnetic medium were considered as an ideal gas of non-interacting spin waves or *magnons*. Taking into account that nonlinearities in real medium always present, the solution of nonlinear equation of motion for magnetic excitations represents a magnetic soliton. The term "magnetic soliton" has a broad meaning: there exist magnetic kinks and envelope solitons which are described by the sine-Gordon and NLS equations, respectively.

Magnetic *kink* solitons clearly show up in the properties of quasi-one-dimensional magnetic materials in which spins interact strongly along one axis of the crystal and very weakly along the other axes. Such spin chains are qualitatively similar to the pendulum chain shown in Fig. 5.7. Let us assume that the pendulums in Fig. 5.7 represents spins in a spin chain. Such a situation when all spins in the chain are oriented in the same direction (down in Fig. 5.7) corresponds to a *ferromagnetic* ordering. And a torsion of the spin lattice, shown in Fig. 5.7, can propagate as a soliton in the crystal. An *antiferromagnetic* ordering represents a staggered spin orientation in the chain, i.e. for each spin in the chain, its nearest neighbors have the opposite orientation. In antiferromagnetically ordered materials, a torsion of the spin lattice also corresponds to a magnetic kink-soliton. Magnetic kink-solitons are domain walls which can be moving or stationary, and described by the sine-Gordon equation. These solitons are usually created thermally and their dynamics can be studied very accurately by using neutron scattering. The soliton concept is again precious because the solitons can be treated as quasiparticles and the results can be obtained by investigating a "soliton gas."

For a qualitative understanding of magnetic solitons, we need not look into details of so-called *exchange* forces between spins. As distinct from quantum exchange forces that really act among the nearest neighbors, these exchange forces are classical and have a *long range*. Accordingly, it becomes energetically favorable for ferromagnetically ordered spins to divide into many groups. In one half of these groups, the spins look up, in the other half, they look down. These groups are also called *domains*, and the borders between domains—domain walls which are similar to the Frenkel-Kontorova dislocations. Therefore, they are solitons, and their evolution is described by the sine-Gordon equation. Like dislocations, the domain walls may move along the crystal if there is no obstruction from the crystal imperfections or from other domain walls.

The latter domain walls are somewhat easier to observe than the dislocations. A typical distance between two magnetic domain walls is about 1 mm, and their thickness is typically 10 μm or so. Covering well-polished surface of a ferromagnetic sample with a thin powder of a magnetic material, it turns out that its particles gather near the domain walls. Thus, *the particles are attracted by the domain walls*.

As mentioned above, there also exist magnetic *envelope* solitons. In the linear approximation, a propagating linear packet of spin waves is spread by dispersion because individual spin waves in the packet are totally independent of each other. The solution of equation of motion for the magnetization, which contains nonlinear effects, represents a magnetic envelope soliton. The latter is a cluster or a bound state of large number of spin waves where the attraction between spin waves compensate the spreading effect of dispersion. The existence of spin wave envelope solitons was predicted theoretically, and later they were experimentally observed in the microwave frequency range. This type of magnetic solitons are described by the NLS equation.

12. Self-trapped states: the Davydov soliton

Solitons represent localized states and, in real systems, there exist *self-localized* states. In other words, there are cases when, in exchange of interaction with a medium, an excitation or a particle locally deforms the medium in a way that it is attracted by the deformation. In the context of condensed matter physics, Landau proposed that an electron in a solid can be considered as being "dressed" by its self-created polarization field, forming a quasiparticle. When the particle-field interaction (electron-phonon coupling) is strong, both the particle wave function and lattice deformation are localized. In the three-dimensional case, this localized entity is known as a (Holstein) polaron and, in one dimension a Davydov soliton or *electrosoliton*. Their integrity is maintained owing to the dynamical balance between the dispersion (exchange inter-site interaction) and the nonlinearity (phonon-electron coupling). Figure

5.21 schematically shows a self-trapped state of a particle (a small polaron or a Davydov soliton). In a self-trapped state, *both* the particle-wave function and lattice deformation are *localized.*

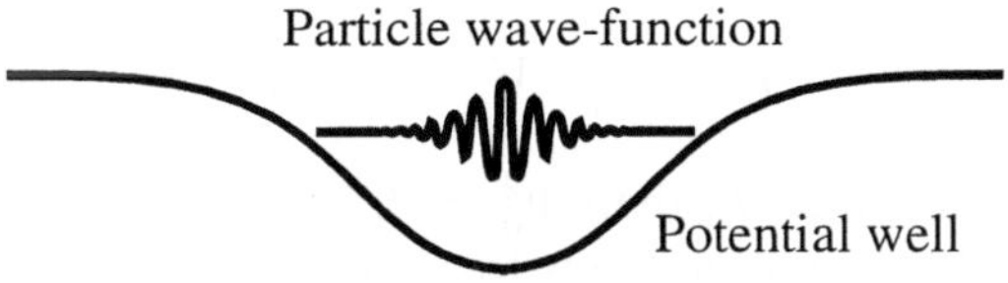

Figure 5.21. Sketch of a self-trapped state: a particle (electron or hole) with a given wave function deforms the lattice inducting a potential well which in turn traps the particle.

Before we discuss the Davydov bisoliton model of high-T_c superconductivity, we consider first the concept of the Davydov soliton [10, 11].

To describe the self-focusing phenomena, the NLS equation is used,

$$\left[i\hbar\frac{\partial}{\partial t} + \frac{\hbar^2}{2m}\frac{\partial^2}{\partial x^2} + G|\psi(x,t)|^2\right]\psi(x,t) = 0. \tag{5.34}$$

The equation is written in the long-wave approximation when the excitation wavelength λ is much larger than the characteristic dimension of discreteness in the system, i.e. under the condition $ka = 2\pi a/\lambda \ll 1$. The equation describes the complex field $\psi(x,t)$ with self-interaction. The function $|\psi|^2$ determines the position of a quasiparticle of mass m. The second term in the NLS equation is responsible for the dispersion, and the third one for nonlinearity. The coefficient G characterizes the intensity of the nonlinearity.

When the nonlinearity is absent ($G = 0$), the NLS equation has solutions in the form of plane waves,

$$\psi(x,t) = \Phi_0 \exp[i(ka - \omega(k)t)], \tag{5.35}$$

with the square dispersion law $\omega(k) = \hbar k^2/2m$. With nonlinearity ($G \neq 0$) in the system having a translational invariance, the excited states move with constant velocity V. Therefore it is convenient to study solutions of the NLS equation in the reference frame

$$\zeta = (x - Vt)/a, \tag{5.36}$$

moving with constant velocity. In this reference frame the NLS equation has solutions in the form of a complex function

$$\psi(x,t) = \Phi(\zeta)\exp[i(kx - \omega t)],\ k = mV/\hbar, \tag{5.37}$$

where the real function $\Phi(\zeta)$ satisfies the nonlinear equation

$$\left[\hbar\omega - \frac{1}{2}mV^2 + J\frac{\partial^2}{\partial x^2} + G_0\Phi^2(\zeta)\right]\Phi(\zeta) = 0, \tag{5.38}$$

where $J = \hbar^2/2ma^2$.

The last equation has two types of solutions: nonlocalized and localized ones. The nonlocalized solution corresponds to a constant value of the amplitude $\Phi(\zeta) = \Phi_0$ and the dispersion law

$$\hbar\omega = \frac{1}{2}mV^2 - G_0\Phi_0^2. \tag{5.39}$$

If a particle is at a distance L, then $\Phi_0^2 = L^{-1}$ and, in the limit $L \to \infty$, the second term in the last equation tends to zero.

The localized solution of the NLS equation normalized by the condition

$$\int \Phi^2(\zeta)d\zeta = 1, \tag{5.40}$$

is represented, as we already know, by the bell-shaped function

$$\Phi(\zeta) = \sqrt{\frac{g}{2}} \times sech(g\zeta), \tag{5.41}$$

with the dimensionless nonlinearity parameter g given by

$$g = G_0/4J. \tag{5.42}$$

The function $\Phi(\zeta)$ is nonzero on a segment $\Delta\zeta \approx 2\pi/g$. The larger the nonlinearity parameter the smaller is the localization region.

The energy of localized excitations is determined by the expression

$$\hbar\omega = \frac{1}{2}mV^2 - Jg^2. \tag{5.43}$$

In the localized solution found above, the nonlinearity is generated by the "self-action" of the field. Let us now study a self-trapping effect when two linear system interact with each other. As an example, we consider the excess electron in a quasi-one-dimensional atomic (molecular) chain. If neutral atoms (molecules) are rigidly fixed in periodically arranged sites na of a one-dimensional chain, then, due to the translational invariance of the system, the lowest energy states of an excess electron are determined by the conduction band. The latter is caused by the electron collectivization. In the continuum approximation, the influence of a periodic potential is taken into account by replacing the electron mass m_e by the effective mass $m = \hbar/2a^2J$ which is inversely proportional to the exchange interaction energy that characterizes the electron jump from one node site into another. In this approximation, the electron motion along an ideal chain corresponds to the free motion of a quasiparticle with effective mass m and electron charge.

Taking into account of small displacements of molecules of mass M ($\gg m$) from their periodic equilibrium positions, there arises the short-range deformation interaction of quasiparticles with these displacements. When the deformation interaction is rather strong, the quasiparticle is self-localized. Local displacement caused by a quasiparticle is manifest as a potential well that contains the particle, as schematically shown in Fig. 5.21. In turn, the quasiparticle deepens the well.

A self-trapped state can be described by two coupled differential equations for the field $\psi(x,t)$ that determines the position of a quasiparticle, and the field $\rho(x,t)$ that characterizes local deformation of the chain and determines the decrease in the relative distance $a \to a - \rho(x,t)$ between molecules of the chain,

$$\left[i\hbar\frac{\partial}{\partial t} + \frac{\hbar^2}{2m}\frac{\partial^2}{\partial x^2} + \sigma\rho(x,t)\right]\psi(x,t) = 0, \tag{5.44}$$

$$\left(\frac{\partial^2}{\partial t^2} - c_0^2\frac{\partial^2}{\partial x^2}\right)\rho(x,t) - \frac{a^2\sigma}{M}\frac{\partial^2}{\partial x^2}|\psi(x,t)|^2 = 0. \tag{5.45}$$

The first equation characterizes the motion of a quasiparticle in the local deformation potential $U = -\sigma\rho(x,t)$. The second equation determines the field of the local deformation caused by a quasiparticle. The system of the two equations are connected through the parameter σ of the interaction between a quasiparticle and local deformation. The quantity $c_0 = a\sqrt{k/M}$ is the longitudinal sound velocity in the chain with elasticity coefficient k. When there is one quasiparticle in the chain, the function $\psi(x,t)$ is normalized by

$$\frac{1}{a}\int_{-\infty}^{\infty}|\psi(x,t)|^2 dx = 1. \tag{5.46}$$

In the reference frame $\zeta = (x - Vt)/a$ moving with constant velocity V, the following equality $\partial\rho(x,t)/\partial t = -V/a \times \partial\rho/\partial\zeta$ holds. Then, the solution for $\rho(x,t)$ has the form

$$\rho(x,t) = \frac{\sigma}{k(1-s^2)}|\psi(x,t)|^2, \; s^2 = V^2/c_0^2 \ll 1. \tag{5.47}$$

Substituting the values $\rho(x,t)$ into Eq. (5.44), we transform it to a nonlinear equation for the function $\psi(x,t)$,

$$\left[i\hbar\frac{\partial}{\partial t} + \frac{\hbar^2}{2m}\frac{\partial^2}{\partial x^2} + 2gJ|\psi(x,t)|^2\right]\psi(x,t) = 0. \tag{5.48}$$

where

$$g \equiv \frac{\sigma^2}{2k(1-s^2)J} \tag{5.49}$$

is the dimensional parameter of the interaction of a quasiparticle with a local deformation. Substituting the function

$$\psi(x,t) = \Phi(\zeta)\exp[i(kx - \omega t)], \; k = mV/\hbar \tag{5.50}$$

into the Eq. (5.48) we get the equation

$$[\hbar\omega - \frac{1}{2}mV^2 - J\frac{\partial^2}{\partial\zeta^2} + 2gJ\Phi^2(\zeta)]\Phi(\zeta) = 0, \tag{5.51}$$

for the amplitude function $\Phi(\zeta)$ normalized by $\int \Phi^2(\zeta)d\zeta = 1$. The solution of this equation is

$$\Phi(\zeta) = \frac{1}{2}\sqrt{g} \times sech(g\zeta/2), \tag{5.52}$$

with the value

$$\hbar\omega = \frac{1}{2}mV^2 - D(s). \tag{5.53}$$

The quantity

$$D(s) = \frac{1}{8}g^2 J \tag{5.54}$$

determines the binding energy of a particle and the chain deformation produced by the particle itself. According to Eq. (5.52), a quasiparticle is localized in moving reference frame

$$\Delta\zeta = 2\pi/g, \tag{5.55}$$

as shown in Fig. 5.22. In this region, the field localization is characterized by the function

$$\rho(\zeta) = \frac{g\sigma}{4k(1-s^2)}sech^2(g\zeta/2). \tag{5.56}$$

The following energy is necessary for deformation:

$$W = \frac{1}{2}k(1+s^2)\int \rho^2(\zeta)d\zeta = \frac{1}{24}g^2J(1+s^2). \tag{5.57}$$

Measured from the bottom of the conduction band of a free quasiparticle, the total energy (including that of deformation) transferred by a soliton moving with velocity V is determined by the expression

$$E_s(V) = W + \hbar\omega = E_s(0) + \frac{1}{2}M_{sol}V^2, \tag{5.58}$$

in which the energy of a soliton at rest $E_s(0)$ and its effective mass M_{sol} are determined, respectively, by the equalities

$$E_s(0) = \frac{1}{12}g^2 J, \tag{5.59}$$

$$M_{sol} = m(1 + \frac{g^2 J}{3a^2 k}). \tag{5.60}$$

The soliton mass M_{sol} exceeds the effective mas of a quasiparticle m, as its motion is accompanied by the motion of a local deformation.

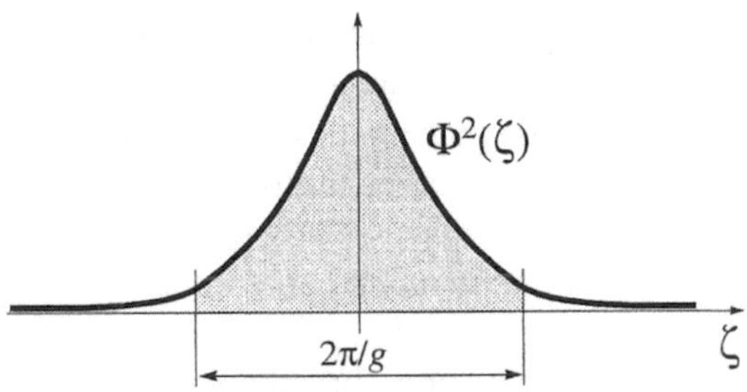

Figure 5.22. A schematic plot of a self-trapped soliton. The soliton is very localized, having a size of $2\pi/g$.

The effective potential well where quasiparticles are placed, is determined, in the reference frame ζ, by the expression

$$U = -\sigma\rho(x,t) = -g^2 J \times sech^2(g\zeta/2). \tag{5.61}$$

The self-trapped soliton is very stable. The soliton moves with the velocity $V < c_0$, otherwise, the local deformation of the chain will be not able to follow the quasiparticle. Alternatively, the soliton can be stationary.

It is worth emphasizing that the Davydov solitons are conceptually different from the small polarons which are three-dimensional. In fact, the small polarons are practically at rest owing to their large mass.

The bisoliton excitations will be discussed separately in Chapter 7.

13. Discrete breathers

In lattices, intrinsic localized modes with internal (envelope) oscillations (see Fig. 5.11c) are often called *discrete breathers* because an envelope mode or nonlinear wavepacket can be considered as the small amplitude limit of a breather-solution for the continuum sine-Gordon equation and also for other systems in the semi-discrete approximation, where the oscillations (discrete carrier wave) vary rapidly inside the slow (continuous) envelope. In this approximation, discrete breathers are linked to envelope solitons. Contrary to continuous breathers which are known to exist only in some particular systems, discrete breathers are structurally stable as soon as nonlinear oscillators are coupled sufficiently weakly and locally.

In purely harmonic lattices, spatially localized modes can occur only when defects or disorder are present, so that the translational invariance of the underlying lattice is broken. Discrete breathers may be created anywhere in a perfect homogenous nonlinear lattice. As an example, let us consider the behavior of a one-dimensional chain of particles with mass m, bound by massless springs. A similar chain was used in the computer simulations by Fermi, Pasta and Ulam (see above). In this chain, a discrete breather is a nonlinear excitation centered at a lattice site, which involves longitudinal displacements of a few masses. The longitudinal displacements of any two neighboring masses are in antiphase, so that the displacement of the central mass is maximal, and two neighboring masses have smaller, antiphase displacements relatively to the central mass etc. This case corresponds to an *odd parity mode*, however, one may also have an *even parity mode* centered at the midpoint between two masses.

The basic properties of the discrete breather or the *intrinsic localized mode* can be summarized as follows [7]:

i. It is a time periodic solution localized in space which may occur in one, two or three dimensions.

ii. It may be centered on any lattice site giving rise to a configurational entropy.

iii. It extends over a few lattice sites.

iv. It has an amplitude dependent frequency.

v. It may be mobile.

Contrary to solitons, discrete breathers do not require integrability for their existence and stability. A particular important condition of breather existence is that all multiples of the fundamental frequency lie outside of the linear excitations spectrum. Discrete breathers are predicted to exist; however, they have not yet been observed experimentally.

14. Structural phase transitions

From Chapter 3 we know that the phase diagram of cuprates is very rich on structural phase transitions. What is a structural phase transition? A structural phase transition corresponds to the shifts of the equilibrium positions of the atoms (molecules). They take place owing to the instability of some lattice displacement pattern, which drives the system from some stable high-temperature phase at $T > T_d$ to a different low-temperature lattice configuration at $T < T_d$, where T_d is the critical temperature. The dynamics of such transitions is frequently characterized by a vibrational or phonon mode whose frequency sharply decreases as the temperature approaches the critical temperature T_d from above. As a result, the restoring force corresponding to that displacement pattern softens, and one calls this particular mode a *soft mode*.

At the critical temperature T_d, the lattice displacements become large and *the dynamics is highly nonlinear*. As a consequence, equations describing the structural phase transition become nonlinear. Simulating structural phase transitions in one-dimensional lattice, a model corresponding to a structural phase transition can be presented by a large number of double-well potentials shown in Fig. 5.15, forming a one-dimensional chain similar to that in Fig. 5.5. Contrary to the case in Fig. 5.5, in the latter model, the balls (atoms) cannot jump from one double-well to another, however, being able to change the minima in each double-well potential. Writing down the Hamiltonian, the solution of the obtained nonlinear equation strongly depends on the coupling (spring) strength between nearest-neighbor balls (atoms) and the potential-barrier height inside each double well. In the case of strong coupling between balls (atoms), the equation takes a form of the sine-Gordon equation. The solution of the equation coincides with Eq. (9), having a $tanh$ shape and corresponding to a localized kink soliton.

So, the main point of this section is that a structural phase transition can be presented as a localized soliton (domain wall). Owing to preparation, impurity and defects content, the crystal may be in a state of coexistence of two or more structural phases, forming *domains* separated by *walls*.

15. Tunneling and the soliton theory

What can we expect from tunneling measurements in a system having solitons?

As discussed in Chapter 2, a tunneling conductance $dI(V)/dV$ obtained in a superconductor-insulator-normal metal (SIN) junction directly relates to the electron density of states per unit energy interval since voltage V multiplied by e represents energy. All the soliton solutions considered above are real-space functions, i.e. functions of x and t. Therefore, before we discuss tunneling measurements in terms of the solitons solutions, we need to transform the solutions from real space into momentum realm.

First, let us *qualitatively* determine the shape of tunneling characteristics caused by solitons. Taking into account that $V = E/e$, Figure 5.17 is represented in conductance units, dI/dV vs V, in Fig. 5.23a. Bearing in mind that tunneling current is the sum under the conductance curve, $I(V) = \int \frac{dI(V)}{dV} dV + C$, where C is the constant, the $I(V)$ characteristic of the soliton peak shown in Fig. 5.23a at zero bias is schematically depicted in Fig. 5.23b. The constant C is defined by the condition $I(V = 0) = 0$.

The same procedure has been done for two polaron peaks shown in Fig. 5.19: the $dI(V)/dV$ and $I(V)$ characteristics of a polaron are shown in Figs. 5.24a and 5.24b, respectively.

Let us now *quantitatively* determine the shape of tunneling conductance peak, obtained in a system having topological solitons. In general, the total energy of a soliton consists of the energy of a static soliton and its kinetic energy. For simplicity, consider a static soliton, thus, a soliton with a velocity $v = 0$. The spatial density of states of a topological soliton is determined by $|\psi(x)|^2$, where $\psi(x) = C \times sech(x/d)$ given by Eq. (5.27). Then, the spectral

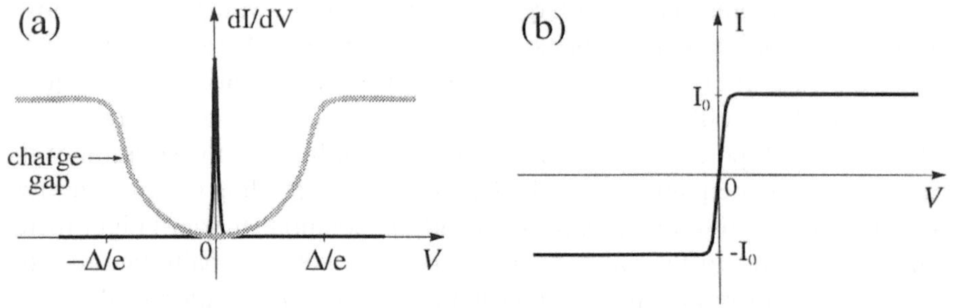

Figure 5.23. (a) Same as Fig. 5.17 but in conductance units: dI/dV vs V (for an SIN junction); and (b) the $I(V)$ characteristic corresponding to the soliton peak in plot (a). The $I(V)$ characteristic of the charge gap is not shown.

function $\varphi(k)$ can be found by applying the Fourier transform to $\psi(x)$, where k is the wave number:

$$\varphi(k) = \int_{-\infty}^{\infty} \psi(x) e^{-ikx} dx. \tag{5.62}$$

Bearing in mind that the Fourier transform of the $sech(\pi x)$ function gives $sech(\pi k)$ [12], then the energy spectrum of a topological soliton is also enveloped by the $sech$ hyperbolic function, centered at some k_0. Then, for $\Delta k = (k - k_0)$, we have $\varphi(\Delta k) = C_1 \times sech(d_1 \Delta k)$, where C_1 and d_1 are constants.

Since $E = \frac{\hbar^2 k^2}{2m}$, where m is the mass, then

$$E(k) - E(k_0) = \Delta E = \frac{\hbar^2 k^2}{2m} - \frac{\hbar^2 k_0^2}{2m} \simeq \frac{\hbar^2 k_0}{m} \Delta k \tag{5.63}$$

valid in the vicinity of k_0, i.e. for $\Delta k \ll k_0$. Substituting $\Delta k \to \Delta E$, one obtains

$$\varphi(\Delta E) = C_1 \times sech(d_1 \frac{m}{\hbar^2 k_0} \Delta E). \tag{5.64}$$

Assuming that the soliton energy remains constant, i.e. k_0 = constant, then, the spectral density of states is determined by

$$|\varphi(\Delta E)|^2 = C_1^2 \times sech^2(\Delta E / E_0), \tag{5.65}$$

where $E_0 = \frac{\hbar^2 k_0}{m d_1}$. Thus, in the vicinity of peak energy, the energy density of states is proportional to $sech^2(\Delta E / E_0)$.

Finally, in SIN tunneling measurements performed in a system having topological solitons, for example, in polyacetylene, one can expect that, near peak bias,

$$\frac{dI(V)}{dV} \simeq A \times sech^2(V/V_0) \tag{5.66}$$

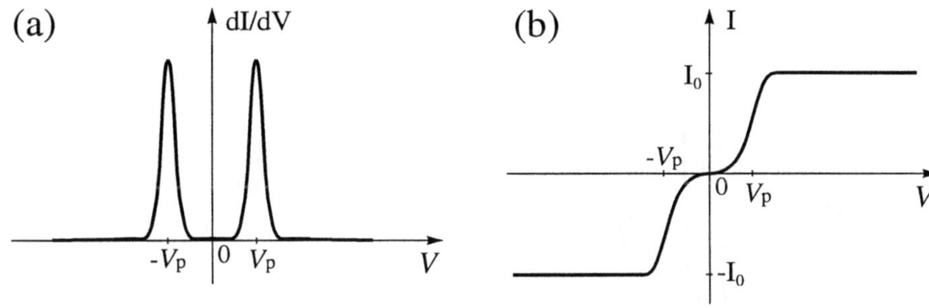

Figure 5.24. (a) Same as Fig. 5.19 but in conductance units: dI/dV vs V (for an SIN junction). The charge gap is not shown, and the width of the conductance peaks is slightly enlarged in comparison with that in Fig. 5.19. (b) The corresponding $I(V)$ characteristic.

and

$$I(V) \simeq B \times \tanh(V/V_0), \tag{5.67}$$

where V is the applied bias, and A, B and V_0 are the constant. In the equations, $\Delta V \to V$ since the peak is centered at zero bias, as shown in Fig. 5.23a.

By the same token, for the polaron characteristics shown in Fig. 5.24, one can obtain that, in an SIN junction,

$$\frac{dI(V)}{dV} \simeq A \times \left[sech^2 \left(\frac{V + V_p}{V_0} \right) + sech^2 \left(\frac{V - V_p}{V_0} \right) \right] \tag{5.68}$$

and

$$I(V) \simeq B \times \left[\tanh \left(\frac{V + V_p}{V_0} \right) + \tanh \left(\frac{V - V_p}{V_0} \right) \right], \tag{5.69}$$

where V_p is the peak bias, as shown in Fig. 5.24. The latter equations are valid in the vicinities of $\pm V_p$.

What is interesting is that the results obtained above for a topological soliton are also valid for a static envelope soliton which is schematically shown in Fig. 5.25a. As discussed above, in real space coordinates, the enveloping curve of an envelope soliton has the $sech$-function shape (the dashed curve in Fig. 5.25a). The energy spectrum of the envelope soliton is schematically presented in Fig. 5.25b. From the Fourier transform, one can easily obtain that the enveloping curve of energy spectrum of an envelope soliton, shown in Fig. 5.25b by the dashed line, is also the $sech$ hyperbolic function. Hence, for a static envelope soliton, the density-of-state peak of energy spectrum has the $sech^2$-function shape, as for a topological soliton.

Thus, quasiparticle peaks in a conductance obtained in an SIN tunneling junction in a system having either envelope or topological solitons can be fitted by the $sech^2$ hyperbolic function, and the corresponding tunneling $I(V)$ characteristic by the $tanh$ hyperbolic function. As a consequence, quasiparticle

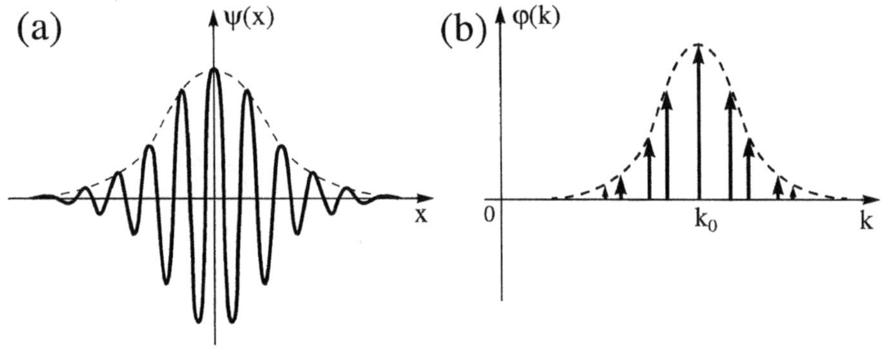

Figure 5.25. (a) An envelope soliton in real space, and (b) its energy spectrum, shown schematically.

peaks in a conductance obtained in a superconductor-insulator-superconductor (SIS) tunneling junction have to be fitted by the convolution of the $sech^2$ function with itself [see Eq. (2.18)]. The convolution cannot be resolved analytically. In Chapter 12, we shall see that $sech^2$ - and $tanh$ -function fits can be applied also to tunneling spectra obtained in an SIS junction. Therefore, independently from the type of tunneling junction, SIN or SIS, the $sech^2$ function can be used to fit quasiparticle peaks in conductances, and the $tanh$ function to fit $I(V)$ characteristics, caused by solitons. However, one has to realize that quasiparticle conductance peaks caused by solitonic states will appear in the "background" originated from other electronic states present in the system.

For a moving soliton, i.e. for $v \neq 0$, the results obtained above are also valid if its velocity v is small.

16. Modern solitons

The modern soliton theory is a very broad area of research not only in different branches of physics but in different branches of science, such as chemistry, biology, medicine, astronomy etc. Practical devices based on the soliton concept, for example, in optics are now a multimillion-dollar industry.

In 1962, that is even before Kruskal and Zabusky's work, Skyrme suggested that elementary particles, for example, protons can be regarded as solitons (in modern usage). He used kink solitons being the exact solutions of the sine-Gordon equation to describe the dynamics of elementary particles. He demonstrated that his solitons behave like fermions. At that time, his ideas were very unpopular. However, rigorous proofs of the equivalence between the sine-Gordon theory and the theory of fermions were given only 15 years later by others.

Attempts to use optical pulses for transmitting information started as soon as good enough quality optical fibers became available. Since even superb quality fibers have dispersion, it is a more or less evident idea to use optical solitons for the transmission of information. The nonlinearity which is normally very small in optical phenomena results in the self focusing of the laser beam. As we already know, the self-focusing phenomena are described by the NLS equation. Thus, the optical soliton is in fact an envelope soliton. In optics, the envelope soliton shown in Fig. 5.11c is called a *bright soliton* because it corresponds to a pulse of light. There also exists another type of envelope solitons which are called in optics *dark solitons*, because they correspond to a hole in the continuous light (carrier) wave. Using solitons for transmitting the information, the transmission speed of optical fibers is enormous, and can be about 100 Gigabits per second. In addition, soliton communication is more reliable and less expensive.

Solitons are encountered in biological systems in which the nonlinear effects are often the predominant ones [13]. For example, many biological reactions would not occur without large conformational changes which cannot be described, even approximately, as a superposition of the normal modes of the linear theory.

The shape of a nerve pulse was determined more than 100 years ago. The nerve pulse has a bell-like shape and propagates with the velocity of about 100 km/h. The diameter of nerves in mammals is less than 20 microns and, in first approximation, can be considered as one-dimensional. For almost a century, nobody realized that the nerve pulse is the soliton. So, all living creatures including humans are literally stuffed by solitons. Living organisms are mainly organic and, in principle, should be insulants—solitons are what keeps us alive.

The last statement is true in every sense: the blood-pressure pulse seems to be some kind of solitary wave [7]. The muscle contractions are stimulated by solitons [11].

The so-called *Raman effect* is closely related to supplying solitons with additional energy. The essence of the Raman effect is that the spectrum of the scattered (diffused) light is changed by its interaction with the molecules of the medium. Crudely speaking, the incoming wave is modulated by vibrating molecules. These vibrations are excited by the wave, but the frequency of the vibrations depends only on properties of the molecules.

So, I could continue to enumerate different cases and different systems where solitons exist. Nevertheless, I think that it is already enough information for the reader to understand the concept of the soliton, and to perceive them as real objects because we deal with them every day of our lives.

17. Neither a wave nor a particle

At the end, I have a proposal. Solitons are often considered as "new objects of Nature" [5]. It is not true in the global sense. Solitons exist since the existence of the universe. It is true that solitons are "new objects of Nature" *for humans*. However, they existed even before the life began.

Solitons are waves, however, very strange waves. They are not particles, however, they have particle-like properties. They transfer energy, and do it very efficiently. Since the formulation of the relativistic theory, we know that the mass is an equivalent of energy, and relates to it by

$$E = mc^2.$$

From this expression, it is not important if an object has a mass or not: it is an object if it has an ability to transfer (localized) energy.

In my opinion, solitons do not have to be associated neither with waves nor with particles. They have to remain as *solitons*. It is simply a *separate* (and very old) form of the existence of matter. And, they should remain as they are.

I am sure that, in the future, we shall add to the group of *waves, solitons*, and *particles* new names—new forms of the existence of matter. However, it will happens in the future and, right now, let us return to our problem—the mechanism of high-T_c superconductivity in cuprates.

APPENDIX: Books recommended for further reading

- M. Remoissenet, *Waves Called Solitons* (Springer-Verlag, Berlin, 1999).
- A. S. Davydov, *Solitons in Molecular Systems* (Kluwer Academic, Dordrecht, 1991).
- A. T. Filippov, *The Versatile Soliton* (Birkhäuser, Boston, 2000).
- G. Eilenberger, *Solitons* (Springer, New-York, 1981).
- F. K. Kneubühl, *Oscillations and Waves* (Springer, Berlin, 1997).
- *Solitons and Condensed Matter Physics*, A. R. Bishop and T. Schneider, (ed.) (Springer, Berlin, 1978).
- *Nonlinearity in Condensed Matter*, A. R. Bishop, D. K. Cambell, P. Kumar, and S. E. Trullinger, (ed.) (Springer, Berlin, 1987).
- M. Lakshmanan, *Solitons* (Springer, Berlin, 1988).

Chapter 6

EVIDENCE FOR SOLITON-LIKE EXCITATIONS IN CUPRATES

Imagination is more important than knowledge.

—Albert Einstein

The superconducting condensate in cuprates consists of pairs of quasiparticles as in conventional superconductors (see Chapter 3). The identification of Cooper-pair constituents is crucial for understanding the mechanism of high-T_c superconductivity. In classical superconductors, the Cooper pairs are composed of two electrons. The components of the Cooper pairs in cuprates remained unknown until 2001. The main purpose of this Chapter is to present evidence for the presence of soliton-like excitations in cuprates, which form the superconducting condensate. Key evidence for solitons in cuprates is provided by tunneling and acoustic measurements.

As emphasized in the previous Chapter, physicists use the word "soliton" in a relaxed way which does not always agree with mathematical rigor. In this Chapter and in the following ones, we shall often use the term "soliton" in its broader sense as a soliton-like excitation (or a quasi-soliton).

1. Tunneling measurements in Bi2212

For a reader who has attentively read Chapters 4 and 5, one piece of evidence for soliton-like excitations in Bi2212 should be more or less evident: indeed a visual inspection of Figs. 4.6 and 5.24 shows the striking similarity between the two graphs. The tunneling data presented in Figs. 4.5 and 4.6 correspond to direct contributions to $I(V)$ and $dI(V)/dV$ characteristics from the superconducting condensate, indicating that the condensate consists of pairs of soliton-like excitations, not electrons (holes). New tunneling data presented here confirm this scenario.

In Chapter 4, the contribution to the $I(V)$ curves from the superconducting condensate was estimated by extracting it from measured tunneling characteristics. If contributions to tunneling spectra from the superconducting condensate and from the pseudogap are independent of each other, then they can in principle be measured separately. In this Section, we shall discuss tunneling data obtained in three sets of Bi2212 single crystals having different doping levels: underdoped, overdoped and Ni-doped (overdoped in oxygen). The data, presented also elsewhere [1, 2], indeed, show that the two components can be observed separately (or almost separately) and the "superconducting" component is in good agreement with the soliton theory.

Experimental details of the growth and characterization of the three sets of Bi2212 single crystals can be found in Appendix A of Chapter 12. Tunneling measurements were performed by break junctions, and they provide superconductor-insulator-superconductor (SIS) junction data. The description of break-junction setup can be found in Appendix B of Chapter 12. We start with measurements run in underdoped Bi2212.

1.1 Underdoped Bi2212

Figure 6.1a shows the SIS tunneling $dI(V)/dV$ and $I(V)$ characteristics obtained in an underdoped Bi2212 single crystal having a critical temperature of 51 K. Using the empirical relation $T_c/T_{c,max} = 1 - 82.6(p - 0.16)^2$ [see Eq. (3.1)] and $T_{c,max}$ = 95 K for Bi2212, one can easily obtain that the critical temperature T_c = 51 K corresponds approximately to a doping level of $p \simeq 0.085$. The characteristics in Fig. 6.1a resemble the usual tunneling spectra obtained in Bi2212 (see, for example, Fig. 4.2a). The gap magnitude Δ_{sc} = 64 meV is in good agreement with other tunneling measurements performed in underdoped Bi2212 single crystals [3, 4]. In Fig. 6.1a, the Josephson $I_c R_n$ product is estimated to be 13.4 mV, where I_c is the maximum (critical) Josephson current, and R_n the resistance of tunneling junction in the normal state.

The $dI(V)/dV$ and $I(V)$ characteristics shown in Fig. 6.1b are obtained within the *same* underdoped single crystal as those of Fig. 6.1a; however, they look different. In Fig. 6.1b, the gap between the two humps, having a magnitude Δ = 130 meV, is too large to be a superconducting gap. Moreover, the absence of the Josephson current at zero bias in the spectra shown in Fig. 6.1b indicates that these conductance humps are incoherent, and suggests that the spectra shown in Fig. 6.1b are related to the normal-state pseudogap. The differences observed between the spectra shown in Fig. 6.1a and Fig. 6.1b therefore correspond to a contribution from the superconducting condensate.

The differences between the spectra shown in Figs. 6.1a and 6.1b are presented in Fig. 6.2. The tunneling characteristics were normalized before subtraction as those in Fig. 4.4, i.e. (i) the conductances were normalized at high bias, and (ii) the $I(V)$ curve from Fig. 6.1b was normalized by its value at

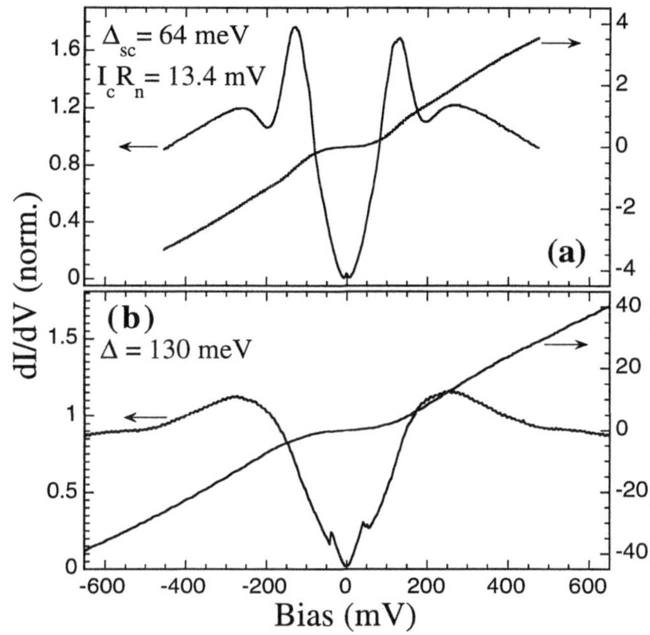

Figure 6.1. (a) and (b) SIS tunneling $dI(V)/dV$ and $I(V)$ characteristics measured at 14 K within the same underdoped Bi2212 single crystal having T_c = 51 K. The $dI(V)/dV$ in both plots are normalized at -400 mV. In plot (a), I_cR_n denotes the Josephson product.

maximum positive bias, and the $I(V)$ curve from Fig. 6.1a is adjusted to be parallel at high bias to the first normalized curve. Comparing Figs. 4.6 and 6.2, one can conclude that the spectra shown in Fig. 6.2 are, indeed, caused by the superconducting condensate. Some parts of the $dI(V)/dV$ curve in Fig. 6.2 are slightly below zero because the spectra depicted in Figs. 6.1a and 6.1b are not taken under the same conditions. In Fig. 6.2, the small humps in conductance at high bias will be discussed below.

Thus, we find that the tunneling characteristics shown in Fig. 6.1a consist of two contributions: one from the pseudogap (the conductance humps at high bias) and the other from the superconducting condensate (the conductance quasiparticles peaks). The tunneling characteristics of the superconducting condensate in Fig. 6.2 are similar to the characteristics of bisoliton-like excitations which were discussed in Chapter 5 (see Fig. 5.24).

The pseudogap will be discussed below, as well as the tunneling spectra of the superconducting condensate shown in Fig. 6.2. First, we shall discuss tunneling data obtained in slightly overdoped Bi2212.

(In fact, the conductance humps in Fig. 6.1b do not correspond *directly* to the pseudogap—it is a product of peak-to-pseudogap tunneling, as shown in

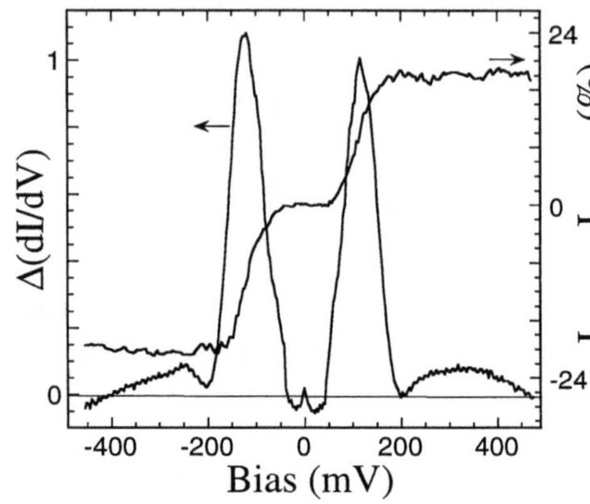

Figure 6.2. Differences $(dI/dV)_a - (dI/dV)_b$ and $I_{a,norm} - I_{b,norm}$ between the two sets of tunneling spectra shown in Figs. 6.1a and 6.1b, respectively. The $I(V)$ curves were normalized before subtraction as those in Fig. 4.4b.

Fig. 12.14. Nevertheless, in this Chapter, we shall continue to associate these humps exclusively with the pseudogap.)

1.2 Overdoped Bi2212

Figure 6.3 shows two sets of SIS tunneling $dI(V)/dV$ and $I(V)$ characteristics obtained within the *same* overdoped Bi2212 single crystal with T_c = 88 K. Using the empirical relation for doping level, a critical temperature T_c = 88 K corresponds to a doping level of $p \simeq 0.19$. In both plots of Fig. 6.3, the gap magnitude Δ_{sc} = 23 meV is in good agreement with other tunneling measurements performed in overdoped Bi2212 single crystals [3, 4]. In Figs. 6.3a and 6.3b, the Josephson I_cR_n product is estimated to be 6 and 7.5 mV, respectively. The tunneling characteristics presented in Fig. 6.3a resemble the usual tunneling spectra obtained in Bi2212.

The $dI(V)/dV$ and $I(V)$ curves in Fig. 6.3b are different from those of Fig. 6.3a: they resemble the tunneling spectra shown in Fig. 6.2. This means that the contribution from the pseudogap in the spectra in Fig. 6.3b is small in comparison with the contribution from the superconducting condensate, *at least*, at low bias. However, at high bias, the contribution from the pseudogap in tunneling characteristics will be always predominant, even, if the pseudogap is weak. The tunneling spectra presented in Fig. 6.3b will be discussed below.

In Chapter 4, the tunneling measurements performed in the same set of overdoped Bi2212 single crystals by point contacts, thus in superconductor-insulator-normal metal (SIN) junctions, show that the "abnormality" in the tunneling spectra is not an SIS-junction effect but intrinsic to Bi2212.

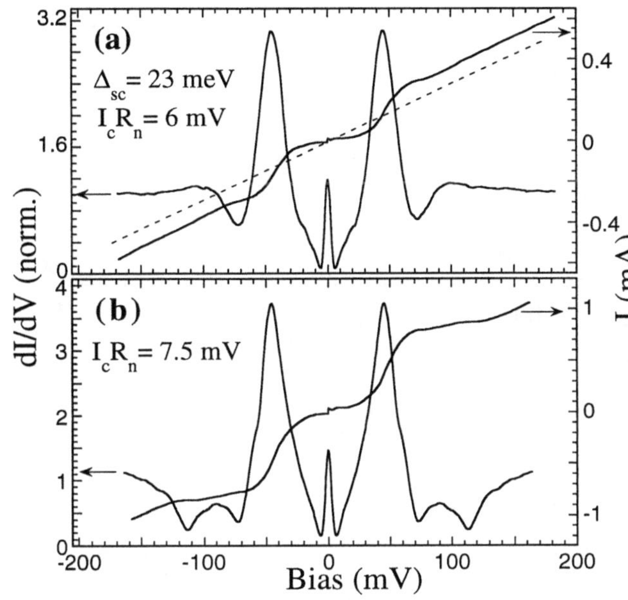

Figure 6.3. (a) and (b) SIS tunneling $dI(V)/dV$ and $I(V)$ characteristics measured at 14 K within the same overdoped Bi2212 single crystal having T_c = 88 K. In both plots, $\Delta_{sc} \simeq 23$ meV, and I_cR_n denotes the Josephson product. The dashed line in plot (a) is parallel at high bias to the $I(V)$ curve.

1.3 Ni-doped Bi2212

Tunneling measurements were also performed in Ni-doped Bi2212 single crystals which are overdoped in oxygen. The Ni content with respect to Cu is approximately 1.5% (see Appendix A of Chapter 12). Figure 6.4 presents two sets of SIS tunneling $dI(V)/dV$ and $I(V)$ characteristics obtained within the *same* Ni-doped Bi2212 single crystal having T_c = 75 K. Using the same empirical relation for doping and $T_{c,max}$ = 87 K for Ni-doped Bi2212, one can estimate the doping level of the Ni-doped Bi2212 single crystal, $p \simeq 0.2$. Thus, the hole concentration in the Ni-doped Bi2212 sample is higher than in the overdoped Bi2212 single crystal (p = 0.19).

In both plots of Figs. 6.4, the gap magnitude is about Δ_{sc} = 17.5 meV. In Figs. 6.4a and 6.4b, the Josephson I_cR_n product is estimated to be 3.1 and 24.5 mV, respectively. The SIS $dI(V)/dV$ and $I(V)$ curves presented in Fig. 6.4a resemble the usual tunneling spectra obtained in Bi2212. The tunneling $dI(V)/dV$ and $I(V)$ characteristics in Fig. 6.4b are reminiscent of the spectra shown in Figs. 6.2 and 6.3b. The effect of the absence of a contribution from the pseudogap in the tunneling spectra in Fig. 6.4b is even stronger than that in Fig. 6.3b.

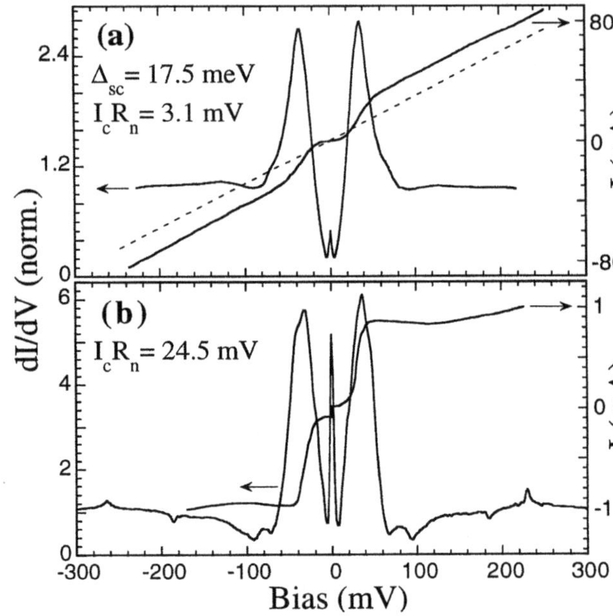

Figure 6.4. (a) and (b) SIS tunneling $dI(V)/dV$ and $I(V)$ characteristics measured at 15 K within the same Ni-doped Bi2212 single crystal having T_c = 75 K. In both plots, $\Delta_{sc} \simeq 17.5$ meV, and I_cR_n denotes the Josephson product. The dashed line in plot (a) is parallel at high bias to the $I(V)$ curve.

Figure 6.5a displays the temperature dependence of the conductance shown in Fig. 6.4b. The $dI(V)/dV$ and $I(V)$ characteristics obtained at 70.3 K are presented separately in Fig. 6.5b, and they resemble the characteristics of a single soliton shown in Fig. 5.23. As one can see in Fig. 6.5a, the quasiparticle peaks disappear *below* $T_c = 75$ K. This fact can be easily understood by taking into account that tunneling spectroscopy probes the *local* density of states, while the critical temperature T_c is a macroscopic characteristic. Thus, in Fig. 6.5a, $T_{c,local} \simeq 70$–71 K. Apparently, the data shown in Fig. 6.5 were measured near an impurity (i.e. near a Ni atom).

1.4 Two components in tunneling spectra

In Figs. 6.3 and 6.4, the "usual" tunneling spectra are depicted in the upper plots in order to show that these Bi2212 single crystals are "usual". The "usual" tunneling spectra consist of two contributions: one from the superconducting condensate and the other from the normal-state pseudogap. In Figs. 6.1a, 6.3a and 6.4a, both contributions are present. The absence (see Fig. 6.1b)

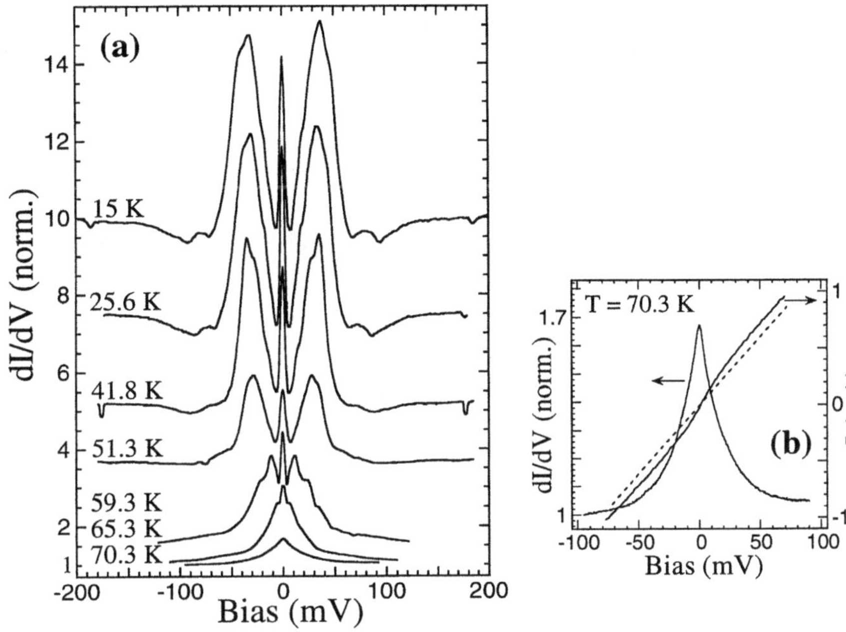

Figure 6.5. (a) Temperature dependence of the conductance in Fig. 6.4b. The conductance scale corresponds to the 70.3 K spectrum, the other spectra are offset vertically for clarity. (b) The $dI(V)/dV$ curve from plot (a), obtained at 70.3 K, and the corresponding $I(V)$ characteristic. The dashed line is parallel to the $I(V)$ curve at high bias.

or weak contribution of one of the components (see Figs. 6.3b and 6.4b) makes the tunneling spectra look "unusual".

Why do the two contributions behave differently at different dopings? As one can see in Fig. 6.1b, in the underdoped region, the contribution from the superconducting condensate is absent or, at least, very weak. At the same time, in the overdoped region the contribution from the superconducting condensate is predominant, as shown in Figs. 6.3b and 6.4b. For the pseudogap component, the opposite is true: it is strong in the underdoped region and weak in the overdoped region.

This effect is due to the fact that superconductivity in Bi2212 is weak in the heavily underdoped region, while for slightly overdoped Bi2212, superconductivity is the strongest (see Fig. 3.21b). As to the pseudogap component, its "strength" is important in the underdoped region but weakens as the hole concentration increases. Therefore, these two contributions to tunneling spectra have opposite trends. It is worth noting that tunneling spectroscopy probes the *local* density of states, therefore, the absence or weak contribution of one component does not mean its absence on a macroscopic scale.

1.5 "Second-harmonic" humps

Figure 6.6 shows three sets of "unusual" tunneling spectra discussed above. In Fig. 6.6, the horizontal scales are chosen to align the conductance peaks.

In Fig. 6.6, one can see first that the three sets of spectra are similar, indicating that the data are all intrinsic to Bi2212. Second, in Figs. 6.6b and 6.6c, one can observe the presence of small humps in the conductances at bias equal approximately to *double* peak bias. These small humps can be perceived as some

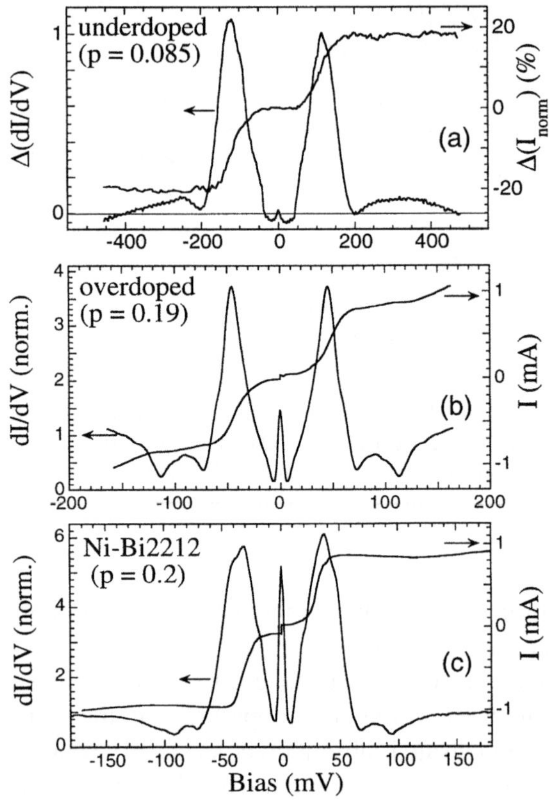

Figure 6.6. (a) The tunneling spectra from Fig. 6.2. (b) The spectra from Fig. 6.3b, and (c) from Fig. 6.4b. In all plots, the horizontal scales are chosen to align the conductance peaks.

sort of "second harmonic" of the superconducting condensate. In Fig. 6.6a, the conductance has a hump at negative bias, which is reminiscent of those in Figs. 6.6b and 6.6c. The temperature dependence of the humps shown in Fig. 6.5a clearly indicates that these humps relate to the quasiparticle peaks, thus to the superconducting condensate. The origin of these humps will be explained in Chapter 12—they originate from tunneling between electronic states of the superconducting condensate and the normal-state pseudogap (see Fig. 12.14).

1.6 Bisoliton-solution fits

Here we compare the tunneling data with the predictions of the soliton theory. The first question is what type of solitons are involved in high-T_c superconductivity? From the previous Chapter, there are three types of true solitons: the Korteweg-de Vries soliton, the topological soliton and the envelope soliton. Using common sense, it is clear that solitons in cuprates must have an amplitude which does not depend on the soliton velocity. This condition leaves us with two choices: the topological soliton and the envelope soliton.

In Chapter 5, it is shown that, independently of the type of tunneling junction—SIN or SIS—the $sech^2$ hyperbolic function can be used to fit the quasiparticle peaks in conductances, while the $tanh$ hyperbolic function the $I(V)$ characteristics, if they are caused by either envelope or topological solitons. In this case, the conductance peaks appear in the "background" originated from other electronic states present in the system. In our case, the background is caused by the electronic states of the pseudogap.

Taking into account that $V_p = \Delta/e$, we rewrite Eq. (5.68) as

$$q_n(V) = A_q \times \left[sech^2\left(\frac{eV + \Delta_{sc}}{eV_0}\right) + sech^2\left(\frac{eV - \Delta_{sc}}{eV_0}\right)\right], \quad (6.1)$$

where e is the electron charge; V is the bias; Δ_{sc} is the maximum superconducting gap, and A_q and V_0 are constants ("q" denotes quasiparticle peaks, and "n" an SIN junction). The $q_n(V)$ function is to fit coherent quasiparticle peaks in SIN-junction conductances. Then, quasiparticle peaks in SIS-junction conductances can be fitted by

$$q_s(V) = A_q \times \left[sech^2\left(\frac{eV + 2\Delta_{sc}}{eV_0}\right) + sech^2\left(\frac{eV - 2\Delta_{sc}}{eV_0}\right)\right], \quad (6.2)$$

where "s" denotes an SIS junction.

Figure 6.7 shows the SIS-junction conductances from Fig. 6.6 and the SIN-junction conductance from Fig. 4.2c. In Fig. 6.7, the SIS-junction data are fitted by the $q_s(V)$ function, and the SIN-junction conductance presented in Fig. 6.7d by the $q_n(V)$ function. The fits are applicable only to the coherent quasiparticle peaks: The humps in conductances at high bias originate from the pseudogap, and the zero-bias peak is caused by a *dc* Josephson current. In Fig. 6.7d, the $q_n(V)$ fit had to be shifted up in order to fit the SIN-junction data at low bias. This means that a "leakage", i.e. side current, was present in this SIN junction. In Fig. 6.7, one can see that the conductance data and the bisoliton-solution fits are in good agreement.

Figure 6.8 depicts the $I(V)$ characteristics which were discussed above. In Fig. 6.8, for simplicity, we analyze the $I(V)$ data only at positive bias. Figure 6.8a shows the two $I(V)$ curves from Figs. 6.2 and 6.1b. The $I(V)$ characteristic from Fig. 6.2 is caused by the superconducting condensate, and

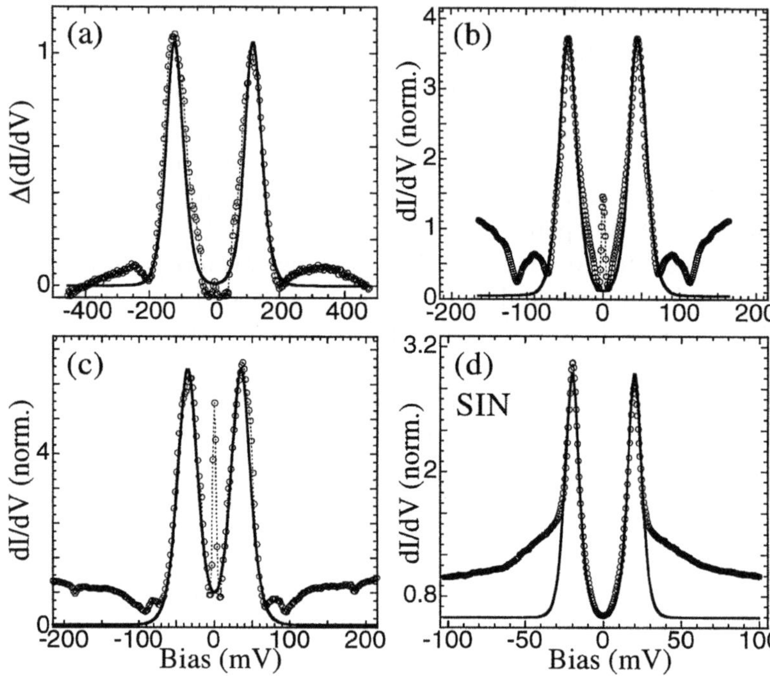

Figure 6.7. $dI(V)/dV$ curves (circles) obtained in Bi2212 and the $q(V)$ fits (see text). The conductances are from (a) Fig. 6.2, (b) Fig. 6.3b, and (c) Fig. 6.4b. In plots (a)–(c), the conductances are obtained in SIS junctions, and are fitted by the $q_s(V)$ (see text). (d) The SIN conductance from Fig. 4.2c and the $q_n(V)$ fit (see text).

the $I(V)$ characteristic from Fig. 6.1b by the normal-state pseudogap. As shown in Fig. 6.8a, the $I(V)$ data from Fig. 6.2 can be fitted quite well by the hyperbolic function [see Eq. (5.69)]

$$c_s(V) = A_c \times \left[\tanh\left(\frac{eV + 2\Delta_{sc}}{eV_0}\right) + \tanh\left(\frac{eV - 2\Delta_{sc}}{eV_0}\right)\right], \qquad (6.3)$$

where A_c is a constant ("c" denotes current, and "s" an SIS junction).

As we already know, any tunneling characteristic obtained either in an SIN or SIS junction consists of the two contributions issued from the superconducting condensate and the pseudogap. These two contributions to $I(V)$ characteristics are shown separately in Fig. 6.8a. The other $I(V)$ characteristics depicted in Fig. 6.8 are not resolved into these two components. So, we apply the $c_s(V)$ [or $c_n(V)$] fit to the $I(V)$ data expecting that their difference will resemble the $I(V)$ characteristic of the pseudogap shown in Fig. 6.8a.

Figure 6.8b depicts the $I(V)$ data from Fig. 4.1a, the $c_s(V)$ fit, and their difference. The data are obtained in an underdoped Bi2212 single crystal with

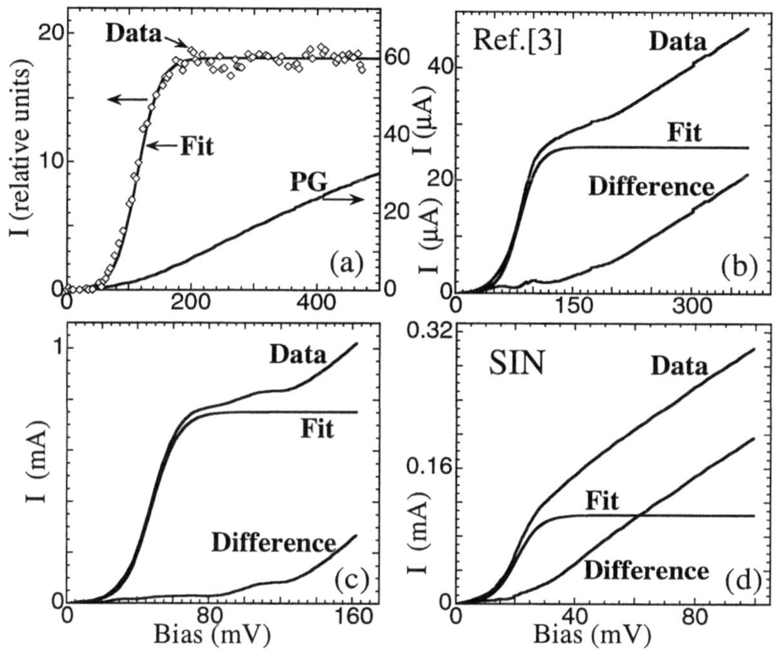

Figure 6.8. $I(V)$ curves obtained in Bi2212 and the $c(V)$ fits (see text). (a) The SIS $I(V)$ curve (diamonds) from Fig. 6.2, the $I(V)$ characteristic of pseudogap from Fig. 6.1b, and the $c_s(V)$ fit. (b) The SIS $I(V)$ curve from [3], shown in Fig. 4.1a, the $c_s(V)$ fit, and their difference. (c) The SIS $I(V)$ curve from Fig. 6.3b, the $c_s(V)$ fit, and their difference. (d) The SIN $I(V)$ curve from Fig. 4.2c, the $c_n(V)$ fit, and their difference. In all plots, for simplicity, the $I(V)$ curves are presented only at positive bias. The spectra in plots (a) and (b) are measured in underdoped Bi2212, and in plots (c) and (d) in overdoped Bi2212. In plots (b) and (c), the Josephson currents at zero bias are not shown.

$T_c = 83$ K [3]. Figure 6.8c depicts the two components in the $I(V)$ curve from Fig. 6.3b (and Fig. 6.6b). The SIN-junction $I(V)$ characteristic from Fig. 4.2c is shown in Fig. 6.8d, as well as the hyperbolic fit

$$c_n(V) = A_c \times \left[\tanh\left(\frac{eV + \Delta_{sc}}{eV_0}\right) + \tanh\left(\frac{eV - \Delta_{sc}}{eV_0}\right)\right], \tag{6.4}$$

and their difference ("n" denotes an SIN junction). In Fig. 6.8, one can see that all plots are similar. In Figs. 6.8b–6.8d, the amplitude of the fit, A_c, can be changed—this will only affect the scale but not the shape of the differences corresponding to the pseudogap. Thus, any $I(V)$ characteristic measured in Bi2212 can be easily resolved into the two components if the "superconducting" component is interpolated by the bisoliton-solution fit obtained in Chapter 5.

The $I(V)$ characteristic shown in Fig. 6.4b (and Fig. 6.6c), however, consists of three components. Figure 6.9a shows the $I(V)$ curve from Fig. 6.4b, the $c_s(V)$ fit and their difference. As seen in Fig. 6.9a, the difference itself consists of two components: At high bias, the pseudogap component is predominant as that of Fig. 6.8a, while a new component at 35 mV dominates at low bias. From analysis of neutron-scattering data, we shall see in Chapter 9 that this peak corresponds to electron tunneling assisted by spin excitations. In other words, this new component is caused by the magnetic resonance peak which was discussed in Chapter 3.

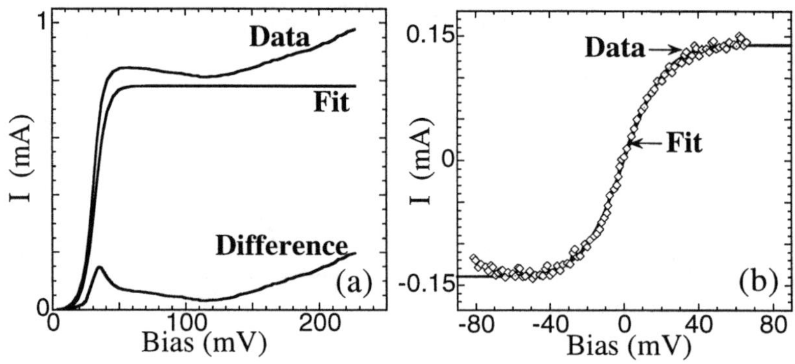

Figure 6.9. SIS $I(V)$ curves obtained in a Ni-doped Bi2212 single crystal and the $c_s(V)$ and $k(V)$ fits (see text). (a) The $I(V)$ curve from Fig. 6.4b, the $c_s(V)$ fit, and their difference. (b) The $I(V)$ curve from Fig. 6.5b (diamonds) (a linear background is subtracted) and the $k(V)$ fit. In plot (a), the Josephson current at zero bias is not shown.

1.7 Single-soliton fit

Figure 6.9b shows the $I(V)$ curve from Fig. 6.5b without the linear contribution (the dashed line in Fig. 6.5b) and the hyperbolic function [see Eq. (5.67)]

$$k(V) = A_k \times \tanh\left(\frac{V}{V_0}\right) \tag{6.5}$$

which is used to fit the data. Physically, this means that, at this temperature, the soliton-like excitations are decoupled. It is worth noting that such a decoupling occurring below the bulk critical temperature is due to the presence of an impurity close to the spot where the density of states is tested. The conductance shown in Fig. 6.5b can be fitted by

$$p(V) = A_p \times sech^2\left(\frac{V}{V_0}\right), \tag{6.6}$$

given by Eq. (5.66). Thus, there is good agreement between the data below T_c and near (*local*) T_c.

As a final note to this Section, the tunneling characteristics obtained in Bi2212 with different dopings are in good agreement with the soliton theory. Moreover, even, "usual" tunneling conductances measured in Bi2212 are much better fitted by the bisoliton-solution rather by a d-wave BCS interpolation [5]. All above facts provide undisputed evidence for the presence of solitons in Bi2212, which form the superconducting condensate.

1.8 Tunneling pseudogap

It is not the purpose of this Chapter to discuss the origin of a pseudogap, nevertheless, any additional piece of evidence is useful.

In Chapter 3, Figures 3.33 and 3.34 show two normal-state pseudogaps in cuprates: a charge gap on charge stripes and a spin gap in local antiferromagnetic domains. What pseudogap is tested in tunneling measurements?

In Fig. 6.4a, one can see that the humps in the conductance obtained in the Ni-doped Bi2212 single crystal are much weaker than those in Figs 6.1a and 6.3a. Since the humps are caused by a pseudogap, this means that Ni destroys the pseudogap observed in tunneling measurements. This statement is true from another point of view: many attempts to observe the pseudogap in tunneling measurements above T_c in Ni-doped Bi2212 single crystals have failed. As discussed in Chapter 3, a Ni atom doped into a CuO_2 plane is magnetic. If the pseudogap observed in tunneling measurements has a *magnetic* origin, then Ni would not destroy it. Therefore, tunneling measurements in cuprates probe the charge gap located on charge stripes.

2. Tunneling measurements in YBCO

We discuss here tunneling measurements performed on chain layers in near-optimally doped YBCO, which show the presence of solitons on the chains.

Chains in YBCO are one-dimensional. As discussed in Chapter 3, at low temperature, chains in YBCO are locally insulant, having a well-defined $2k_F$-modulated charge-density wave (CDW) order, where k_F is the momentum at the Fermi surface. The CDW order is clearly observed in tunneling [6] and nuclear quadrupole resonance [7] measurements. However, at a macroscopic scale, the chains have conducting properties as observed by infrared and microwave measurements.

The tunneling spectroscopic studies of the surface CuO chains in YBCO, performed at 4.2 K, show sharp resonances ("spikes") inside the CDW gap [8]. Figure 6.10 schematically depicts an averaged conductance curve obtained on a chain in YBCO. The conductance is averaged along a few nanometers. The gap magnitude varies in space between 15 and 20 meV. The spikes appear inside the gap at different bias (if they appear at all); however, the averaged spectrum has three well defined peaks, as shown in Fig. 6.10.

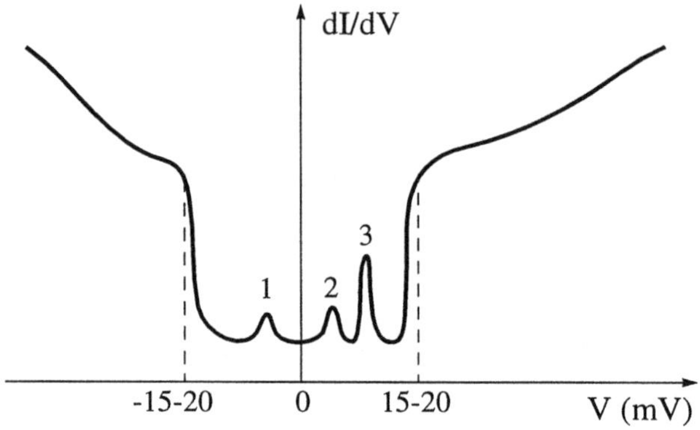

Figure 6.10. An averaged conductance obtained at 4.2 K by STM in near optimally doped YBCO on the chain layer [9]. The conductance is averaged along a few nanometers on a chain (see text for more details).

It is important that the spectrum is averaged in space because, in any one-dimensional system, the density of solitons is small: the soliton density cannot be large, as the system would then be metallic. In order to observe solitons by tunneling measurements in *one-dimensional system*, the solitons have either to be pinned by an inhomogeneity or be very heavy (and, consequently, slow). Any inhomogeneity on a chain is undesirable (remember, Zn located on chains in YBCO destroy superconductivity very efficiently). Another solution is to average the spectra at thermal equilibrium.

In Fig. 6.10, the symmetrical soliton peaks 1 and 2 are centered at about ±5–6 mV. The peaks correspond to an induced superconducting gap on the chain. The origin of peak 3 is also solitonic; however, it is not clear what role it plays in the superconducting state of YBCO. As proposed elsewhere [8], it may be connected to the presence of charge stripes in the neighboring CuO_2 layer. In any case, the most important conclusion which can be drawn from the data is that, in YBCO, superconductivity on the chains is induced and has a solitonic origin.

3. Acoustic measurements in LSCO

We discuss here ultrasound measurements in non-superconducting La_2CuO_4 (LaCO) single crystals.

Below 38 K, acoustic measurements of elastic coefficients in non-superconducting untwinned LaCO single crystals show an activation of domain-wall motion [9]. Specific-heat measurements performed on the same set of LaCO single crystals confirm this fact [9]. The acoustic measurements are performed by a small-sample resonant ultrasound technique. Figure 6.11 shows the temperature dependence of the resonance frequency which primarily depends on

c_{11} and c_{22} longitudinal coefficients. These data were not deconvoluted to obtain the variation in elastic coefficients because the deconvolution process produces additional scatter making the trends harder to see.

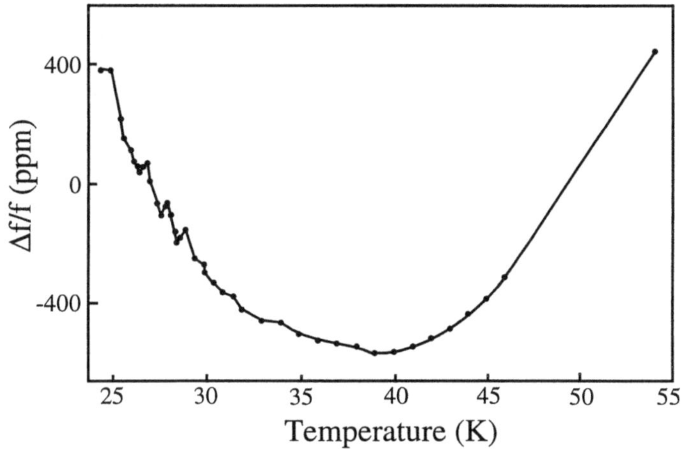

Figure 6.11. Temperature dependence of a resonant frequency (in ultrasound region) in a LaCO single crystal, which primarily depends on c_{11} and c_{22} elastic coefficients.

In Fig. 6.11, one can observe that, around 38 K, the resonant frequency has a broad minimum. Such a minimum in the resonance frequency (thus in the longitudinal sound velocity) is not typical of crystalline solids. It is common, however, in systems having low-frequency relaxation processes, such as ferromagnets in which the activation of domain-walls motion influences the temperature dependence of the elastic properties [9]. In the same set of LaCO single crystals, specific-heat measurements show the presence of a broad peak near 38 K, which was also associated with domain motion [9]. Some other acoustic measurements performed in LSCO ($0.05 \leq x \leq 0.3$) were also interpreted in terms of domain-wall motion [10, 11].

Unfortunately, acoustic measurements alone are not apt to distinguish between two possible types of domain walls present in cuprates: a domain wall moving along a charge stripe and the side motion of a charge stripe as a whole.

Finally, it is worth mentioning that the temperature of 38 K corresponds to the maximum superconducting transition temperature $T_{c,max}$ in La compounds. Thus, it appears that the observed thermodynamic and conductivity features present near $T_{c,max}$ do not require the presence of superconductivity. This silent feature of cuprates such as LSCO, YBCO, Bi2212 and Tl2212 was already emphasized in Chapter 3.

As discussed in Chapter 3, transport measurements in LBCO suggest that the LTO→LTT structural phase transition induces an intra-gap electronic states in the middle of a pseudogap [12], meaning the appearance of solitonic states.

4. Nickelates and manganites

Stripe phases in cuprates and some other doped Mott insulators such as nickelates and manganites are very similar. In nickelates and manganites, stripe phases are understood much better than in cuprates because charge stripes in the former are *quasi*-static. In cuprates, the stripes fluctuate too fast to be able to study them in detail. Therefore, knowledge of stripe excitations in nickelates and manganites is very helpful for understanding the nature of quasiparticles in cuprates.

4.1 NMR measurements in $La_2NiO_{4.17}$

In nickelates, NiO_2 planes are analogous with CuO_2 planes in cuprates. Quasi-one-dimensional charge stripes in NiO_2 planes are assembled by Ni^{3+} ions carrying spin $S = 1/2$. They separate antiferromagnetic domains formed by Ni^{2+} ions with spin $S = 1$.

Nuclear magnetic resonance (NMR) measurements performed on $La_2NiO_{4.17}$ show that the structure of charge stripes is strongly *solitonic* and that the stripes are *site* centered [13]. The latter means that the charge stripes in $La_2NiO_{4.17}$ are not bond centered. This is very useful information for identifying quasiparticle excitations and the stripe structure in cuprates.

4.2 Tunneling measurements in $La_{1.4}Sr_{1.6}Mn_2O_7$

We discuss here remarkable tunneling data obtained in $La_{1.4}Sr_{1.6}Mn_2O_7$ (LSMO) perovskite.

Manganites are Mott insulators, never exhibiting superconductivity. They are known for the effect of colossal magnetoresistance which occurs near the metal-insulator transition. In manganites, MnO_2 planes are analogous with CuO_2 planes in cuprates. In contrast to the spin pairing in cuprates and nickelates, the pairing of the magnetic moments on the Mn sites in a MnO_2 plane is always parallel. Depending on temperature and doping level in its crystal structure, the interlayer coupling can be either tilted ferromagnetic or canted antiferromagnetic, leading to the establishment of long-range three-dimensional magnetic order. Quasi-one-dimensional charge stripes in doped MnO_2 planes are assembled by Mn^{4+} ions, separating ferromagnetic domains formed by Mn^{3+} ions.

The simplest view of the double-layer LSMO manganite is a stuck of ferromagnetic, metallic sheets consisting of the MnO_2 bilayers that are respectively separated by nonmagnetic insulating $(La,Sr)_2O_2$ layers. Thus, the layered LSMO crystal forms a virtually infinite array of ferromagnetic/metallic—insulator—ferromagnetic/metallic intrinsic junctions. In LSMO, the three-dimensional ferromagnetic order occurs at a Curie temperature $T_C \sim 90$ K [14]. The metal-insulator transition appears somewhat above T_C. Two-dimensional in-plane ferromagnetic correlations survive up to about 340 K.

In LSMO single crystals, intrinsic tunneling measurements performed at low temperature on micron-size mesas [15] reveal striking similarity with tunneling data obtained in *superconducting* Bi2212. However, LSMO never exhibits superconductivity. What is common to LSMO and Bi2212 is the presence of charge stripes.

Figure 6.12 shows the two $I(V)$ characteristics obtained at 4.2 K in a LSMO mesa having a size of 5×5 μm^2 and thickness of about 200 Å. The measurements were performed in magnetic fields $B = 0$ and $B = 0.7$ T. In the latter case, the magnetic field is oriented perpendicular to the layers. The $I(V)$ curves, recorded in magnetic fields $0 < B < 0.7$ T with a step of 0.1 T, lie between the two $I(V)$ characteristics shown in Fig. 6.12 [15]. Thus, the resistance of interlayer tunneling is lowered with increasing magnetic field. This is due to the fact that, in the intrinsic junctions of LSMO, electron tunneling is spin polarized [15]. Owing to the same effect, the $I(V)$ characteristic in a magnetic filed of 0.7 T exhibits a small hysteresis, as can be seen in Fig. 6.12. The inset of Fig. 6.12 shows the corresponding $dI(V)/dV$ curves. In Fig. 6.12, one can observe that the $I(V)$ and $dI(V)/dV$ characteristics are slightly asymmetric relatively to zero bias.

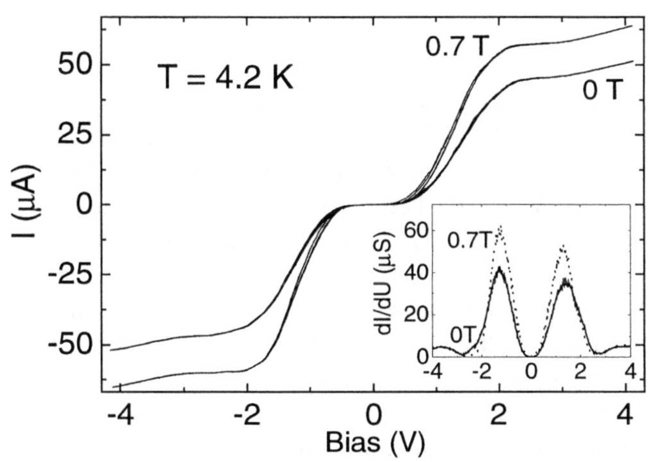

Figure 6.12. Tunneling $I(V)$ characteristics obtained at 4.2 K in a micron-size LSMO mesa in magnetic fields $B = 0$ and $B = 0.7$ T [15]. The magnetic field is oriented perpendicular to the layers. The inset shows the corresponding $dI(V)/dV$ curves.

Any intrinsic junction is analogous with an SIS junction. Therefore, in conductances obtained in intrinsic junctions, a half of the distance between conductance peaks corresponds to the double magnitude of a binding energy $2\Delta_b$. In the inset of Fig. 6.12, quasiparticle peaks in conductance, recorded in zero field, are located approximately at ± 1.4 V. Taking into account that the crystal constant of LSMO in the c-axis direction is about 20 Å [14], one obtains that the value 1.4 V corresponds *approximately* to $\Delta_b \approx 70$ meV per junction. In $B = 0.7$ T, the conductance peaks are shifted to ± 1.25 V. In zero magnetic field, the temperature dependence of quasiparticle peaks is shown in Fig. 6.13. Above 150 K, the quasiparticle peaks are too weak to be distinguished from the background, so, it was not possible to trace them at higher temperatures [16].

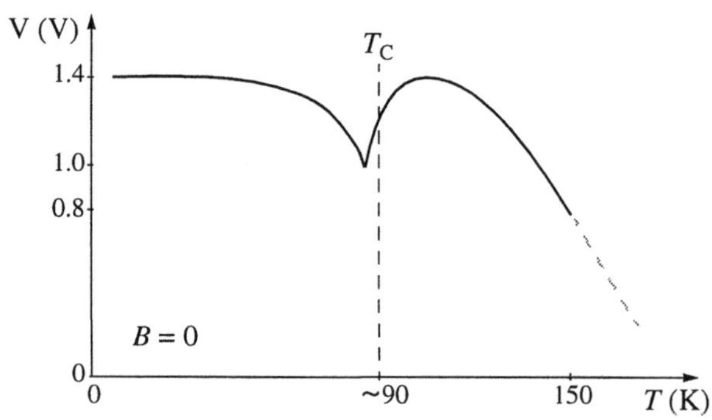

Figure 6.13. Temperature dependence of the conductance peaks shown in the inset of Fig. 6.12 by the solid line ($B = 0$) [16].

The $c_s(V)$ and $q_s(V)$ fits (see above) applied respectively to the $I(V)$ and $dI(V)/dV$ characteristics obtained in LSMO describe the data quite well. As an example, Figure 6.14a shows the $I(V)$ characteristic from Fig. 6.12, taken in $B = 0.7$ T, and the $c_s(V)$ fit (dashed line). Since the $I(V)$ curve is asymmetrical, the $c_s(V)$ fit was applied only to the $I(V)$ curve at negative bias. In Fig. 6.14a, one can see that there is good agreement between data and fit. Below -3 V, the data deviate from the fit because of a pseudogap contribution. By changing the parameters of the $c_s(V)$ fit, the $I(V)$ curve at positive bias can be also described by the fit; however, in this case the fit and the $I(V)$ curve at negative bias will not match.

Figure 6.14b shows the corresponding $dI(V)/dV$ characteristic and the $q_s(V)$ fit which was applied to fit only the quasiparticle peak at negative bias. In Fig. 6.14b, one can observe that there is good agreement between the data and $q_s(V)$ fit. The other $I(V)$ and $dI(V)/dV$ characteristics shown in Fig. 6.12 can be also described by the $c_s(V)$ and $q_s(V)$ fits, respectively.

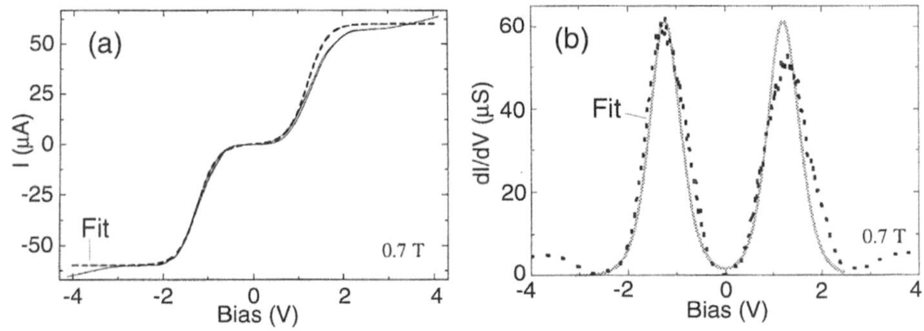

Figure 6.14. (a) The $I(V)$ characteristic from Fig. 6.12 (solid curve), obtained in LSMO in a magnetic field $B = 0.7$ T, and the $c_s(V)$ fit (dashed curve). (b) The corresponding $dI(V)/dV$ characteristic from the inset of Fig. 6.12 (dashed curve) and the $q_s(V)$ fit (grey curve). The $c_s(V)$ and $q_s(V)$ fits were applied to fit the data exclusively at negative bias.

From a simple visual inspection of Figs. 6.6 and 6.12, one can conclude that the data obtained in *superconducting* Bi2212 and in *non*-superconducting LSMO show a *striking* similarity. The main difference between the two sets of tunneling spectra in Figs. 6.6 and 6.12 is the presence of a *dc* Josephson current in the Bi2212 characteristics at zero bias. As discussed in Chapter 2, the Josephson current in tunneling spectra is caused by the long-range phase coherence. This means that the presence/absence of phase coherence does not affect paired quasiparticles in Bi2212/LSMO. Thus, in Bi2212, quasiparticles can remain coupled without phase coherence, as they exist in LSMO. Since, in LSMO, quasiparticles (charge carriers) and paired quasiparticles reside on charge stripes, by analogy, (paired) quasiparticles in Bi2212 also reside on charge stripes.

Secondly, since the presence/absence of phase coherence does not affect paired quasiparticles in Bi2212, this indicates that the pairing mechanism and the mechanism for establishing phase coherence do not correlate with each other in Bi2212. Thus, these two mechanisms have most likely different origins. In Fig. 6.13, the temperature dependence of LSMO quasiparticle peaks looks almost identical to the temperature dependence of Bi2212 quasiparticle peaks (see Fig. 12.20), if we adjust the two temperatures $T_C = T_c$. Since the Curie temperature in LSMO is analogous with the critical temperature T_c in Bi2212, this indicates that the origin of phase-coherence mechanism in Bi2212 is most likely *magnetic*. As a consequence, the pairing mechanism in Bi2212 *and* LSMO has a non-magnetic origin.

One may then wonder why manganites never become superconducting, if the Cooper pairs exist? Or nickelates? This is a good question—the answer will be given in Chapter 10.

Chapter 7

THE BISOLITON MODEL OF HIGH-T_C SUPERCONDUCTIVITY

What Nature created at the Big Bang—the spin of the electron—she later tried to "get rid" of in living matter.

—From the text

We discuss here a model of high-T_c superconductivity proposed by Davydov [1, 2], which is based on the so-called bisoliton model. Initially, the bisoliton model did not have any relation with superconductivity, and was created by Davydov and co-workers in order to explain electron transfer in living tissues. From the physical point of view, the general understanding of many biological processes is still very limited. However, it is known that (i) in redox reactions occurring in living organisms, electrons are transferred from one molecule to another in pairs with opposite spins; and (ii) electron transport in the synthesis process of ATP (adenosine triphosphate) molecules in conjugate membranes of mitochondria and chloroplasts is realized by pairs, but not individually [1, 2]. Apparently, in living tissues, electron transfer is preferable in pairs in which two electrons are in a singlet state. In organic compounds which are normally insulants, such a composite boson can move much easier than a single electron.

At the Big Bang, spin was attached to what are now called the fermions in order to create diversity of possible forms of the existence of matter. Seemingly, in living tissues which appeared later, the spin of the electron became rather of an obstacle in the evolution of living matter. In many biological processes, in order to get rid of electron spin, two electrons with opposite spins are coupled, forming a composite boson with $2e$ charge and zero spin. It happened that in inorganic solids two electrons, at some circumstances, can be paired too. This state of matter, which is in fact an instability in solids, is now called the superconducting state. Thus, the understanding of some biological processes can lead to better understanding of the phenomenon of superconductivity in solids. This is particularly true in the case of high-T_c superconductivity. Superconductivity does not occur in living tissues because it requires not only the electron pairing but also the phase coherence among the pairs.

In Chapter 5, we considered the concept of a self-localized state in one-dimensional systems. In one-dimensional systems, a self-trapped quasiparticle is called *the Davydov soliton, an electrosoliton* or *a solitonic polaron.* We are now going to discuss the bisoliton model of electron transfer in a molecular chain, which was later utilized for explaining the phenomenon of superconductivity in quasi-one-dimensional organic compounds and in cuprates. The bisoliton theory is based on the concept of *bisolitons*, or electron (or hole) pairs coupled in a singlet state due to local deformations of the lattice. Thus, the bisolitons are formed owing to electron-phonon interactions which are moderately strong and nonlinear [1, 2].

1. The bisoliton model

If a quasi-one-dimensional soft chain is able to keep some excess electrons with charge e and spin 1/2, they can be paired in a singlet state due to the interaction with local chain deformation created by them. The potential well formed by a short-range deformation interaction of one electron attracts another electron which, in turn, deepens the well.

In a simple model of pairing of two quasiparticles in a soft quasi-one-dimensional chain, developed by Brizhik and Davydov in 1984 [3], the relative motion of quasiparticles in a paired state was not taken into account. Here we consider the pairing which does take into account their relative motion [1, 2].

Assume that along a molecular chain, the elementary cells of mass M are separated by a distance a from one other. Within the continuum approach, we characterize their position by a continuous variable $x = na$. The equation of motion of two quasiparticles with effective mass m in the potential field

$$U(x,t) = -\sigma\rho(x,t), \tag{7.1}$$

created by a local deformation $\rho(x,t)$ of the infinite chain, takes the form

$$\left[i\hbar\frac{\partial}{\partial t} + \frac{\hbar^2}{2m}\frac{\partial^2}{\partial x_i^2} + U(x_i,t)\right]\psi_j(x_i,t) = 0, \quad \text{where} \quad i,j = 1,2, \tag{7.2}$$

and $\psi_j(x_i,t)$ is the coordinate function of quasiparticle i in the spin state j.

The local deformation $\rho(x,t)$ is caused by two quasiparticles due to their interaction with displacements from equilibrium positions. The function $\rho(x,t)$ characterizing this local deformation of the chain is determined by the equation

$$\left(\frac{\partial^2}{\partial t^2} - c_0^2\frac{\partial^2}{\partial x^2}\right)\rho(x,t) + \frac{\sigma a^2}{M}\frac{\partial^2}{\partial x^2}\left(|\psi_1(x,t)|^2 + |\psi_2(x,t)|^2\right) = 0, \tag{7.3}$$

where $c_0 = a\sqrt{k/M}$ is the longitudinal sound velocity in the chain, $a\sigma$ is the energy of deformation interaction of quasiparticles with the chain, and k is the coefficient of longitudinal elasticity.

Owing to the translational symmetry of an infinite chain, it is possible to study the excitations propagating along the chain with a constant velocity $V < c_0$. In this case, the forced solution of Eq. (7.3) which satisfies the condition $\rho(x,t) \neq 0$ for all x and t under which $|\psi_j(x,t)|^2 \neq 0$ relative to the reference frame $\zeta = (x - Vt)/a$ moving with velocity V, has the form

$$\rho(\zeta) = \frac{\sigma}{k(1-s^2)}\left[|\psi_1(\zeta)|^2 + |\psi_2(\zeta)|^2\right], \quad \text{where} \quad s^2 \equiv V^2/c_0^2. \tag{7.4}$$

The total energy of the local deformation of the chain is determined by

$$W = \frac{1}{2}k(1+s^2)\int \rho^2(\zeta)d\zeta. \tag{7.5}$$

Substituting first $\rho(\zeta)$ into Eq. (7.1) and the latter into Eq. (7.2), one obtains the equation for the function $\Psi(x_1, x_2, t)$ which determines the motion of a pair of quasiparticles in the potential field (7.1), disregarding the Coulomb repulsion of electrons:

$$\left[i\hbar\frac{\partial}{\partial t} + J\left(\frac{\partial^2}{\partial\zeta_1^2} + \frac{\partial^2}{\partial\zeta_2^2}\right) + G(|\psi_1(\zeta_1)|^2 + |\psi_1(\zeta_2)|^2 \right.$$
$$\left. + |\psi_2(\zeta_1)|^2 + |\psi_2(\zeta_2)|^2)\right]\Psi(x_1, x_2, t) = 0. \tag{7.6}$$

Here, we introduce the following notations

$$J \equiv \frac{\hbar^2}{2ma^2} \quad \text{and} \quad G \equiv \frac{\sigma^2}{k(1-s^2)}. \tag{7.7}$$

We consider further the states with a small velocity of motion, $s^2 \ll 1$. In this case, the parameter G can be replaced by the constant

$$G_0 \simeq \frac{\sigma^2}{k}. \tag{7.8}$$

The coordinate function of a pair of quasiparticles in a singlet spin state is symmetric and can be written in the form

$$\Psi(x_1, x_2, t) = \frac{1}{\sqrt{2}}\left[\psi_1(\zeta_1)\psi_2(\zeta_2) + \psi_1(\zeta_2)\psi_2(\zeta_1)\right]e^{-iE_pt/\hbar}, \tag{7.9}$$

where E_p is the energy of two paired quasiparticles in the potential field $U(\zeta)$.

In a chain consisting of a large number N of elementary cells and containing N_1 pairs of quasiparticles, the pairing is realized only from those states of free quasiparticles which have a wave number close to the wave number of the Fermi surface $k_F = \pi N_1/2aN$. Due to the conservation law of quasi-momentum, a pair moving with a velocity $V = \hbar k/m$ can be formed from two quasiparticles with the wave numbers

$$k_1 = 2k - k_F \quad \text{and} \quad k_2 = k_F. \tag{7.10}$$

If the wave functions of quasiparticles in the paired state are represented by the modulated plane waves

$$\psi_j(\zeta_i) = \Phi(\zeta_i)\exp(ik_j\zeta_i), \quad \text{where} \quad i, j = 1, 2, \tag{7.11}$$

the coordinate function of paired quasiparticles transforms into the following form

$$\Psi(x_1, x_2, t) = \sqrt{2}\Phi(\zeta_1)\Phi(\zeta_2)\cos[(k - k_F)(\zeta_1 - \zeta_2)] \times e^{i[k(\zeta_1+\zeta_2)-E_pt/\hbar]}. \tag{7.12}$$

The appearance here of the cosine function results from the conditions of symmetry imposed.

Substituting the function $\Psi(x_1, x_2, t)$ into Eq. (7.6), one obtains the equation for the amplitude functions $\Phi(\zeta_i)$

$$\left[\frac{\partial^2}{\partial\zeta_i^2} + 4\mathrm{g}\Phi^2(\zeta_i) + \Lambda\right]\Phi(\zeta_i) = 0, \quad \text{where} \quad i = 1, 2, \tag{7.13}$$

with the dimensionless parameter

$$\mathrm{g} \equiv \frac{G}{2J} = \frac{\sigma^2}{2kJ(1 - s^2)} \approx \frac{G_0}{2J} = \frac{\sigma^2}{2kJ} \quad (\text{ for} \quad s^2 \ll 1), \tag{7.14}$$

that characterizes the coupling of a quasiparticle with the deformation field. The energy $E(V)$ of a pair of quasiparticles in this field is expressed in terms of the eigenvalue Λ of Eq. (7.13) via the relation

$$E(V) = E_p(0) + 2\frac{mV^2}{2} - \hbar V k_F, \tag{7.15}$$

where

$$E_p(0) = \Lambda J + E_F, \tag{7.16}$$

characterizes the position of the energy level of a static pair of quasiparticles beneath their Fermi level

$$E_F = \frac{\hbar^2 k_F^2}{m}. \tag{7.17}$$

The deformation field $\rho(\zeta)$ is expressed in terms of the function $\Phi(\zeta)$ as follows

$$\rho(\zeta) = -\frac{2\sigma}{k(1-s^2)}\Phi^2(\zeta). \tag{7.18}$$

Therefore, the energy of the local chain deformation in Eq. (7.5) is defined by

$$W = \frac{2G(1+s^2)}{1-s^2}\int \Phi^4(\zeta)d\zeta. \tag{7.19}$$

Equation (7.13) admits periodic solutions corresponding to the uniform distribution of a pair of quasiparticles over the chain. Real functions $\Phi(\zeta_i)$ of these solutions must satisfy the conditions of periodicity

$$\Phi(\zeta_i) = \Phi(\zeta_i + L), \quad \text{where} \quad L = N/N_1, \tag{7.20}$$

and normalization

$$\int_0^L \Phi^2(\zeta)d\zeta = 1. \tag{7.21}$$

The latter requires that each pair of quasiparticles should be within each period. The exact periodic solutions of Eq. (7.13) is expressed in terms of the Jacobian elliptic function $\mathrm{dn}(u,q)$ via the relation

$$\Phi_q(\zeta_i) = \sqrt{\frac{\mathrm{g}}{2}}E^{-1}(q) \times dn(u,q), \quad \text{where} \quad u = \mathrm{g}\zeta/E(q). \tag{7.22}$$

An explicit form of the Jacobian function $\mathrm{dn}(u,q)$ depends on a specific value of the modulus q taking continuous values in the interval [0, 1]. The eigenvalue Λ of Eq. (7.13) is also given in terms of the modulus q by the relation

$$\Lambda_q = -\mathrm{g}^2q^2/E(q). \tag{7.23}$$

The function $E(q)$ is a complete elliptic integral of the second kind which is defined by

$$E(q) \equiv \int_0^{K(q)} dn^2(u,q)du, \tag{7.24}$$

where $K(q)$ is the complete elliptic integral of the first kind:

$$K(q) = \int_0^1 \left[(1-t^2)(1-q^2t^2)\right]^{-1/2} dt. \tag{7.25}$$

The elliptic integrals $E(q)$ and $K(q)$ are shown in Fig. 7.1 as functions of the modulus q. The integrals are given below explicitly for two cases: $q \approx 1$ and

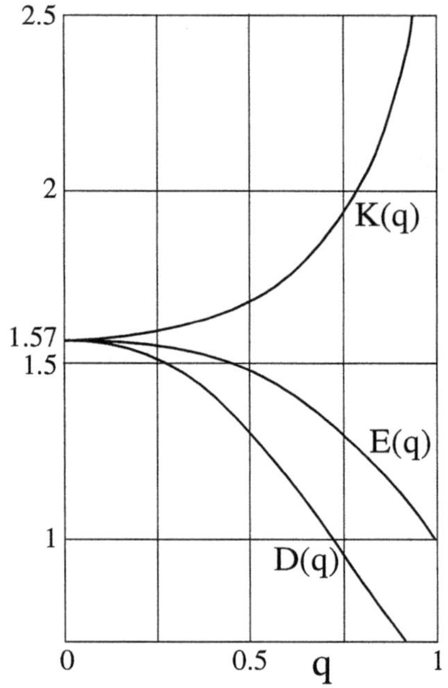

Figure 7.1 Elliptic integrals $|K(q)|$ and $|E(q)|$ and the $D(q)$ function (see text).

$q \ll 1$.

The value of the modulus q is determined by the product of the period L with the dimensionless coupling parameter g of the electron-phonon interaction, and the elliptic integrals as follows

$$\mathrm{g}L = 2E(q)K(q). \tag{7.26}$$

The spatial distributions of both quasiparticles within a bisoliton is characterized by $|\Phi_q(\zeta_i)|^2$. According to Eq. (7.18), the field of the local chain deformation is also the periodic function

$$\rho_q(\zeta) = \frac{\sigma \mathrm{g}}{k(1-s^2)E^2(q)} dn^2 \left(\frac{\mathrm{g}\zeta}{E(q)}, q \right). \tag{7.27}$$

Substituting $\Phi(\zeta)$ into Eq. (7.19) one obtains the energy of the local chain deformation per period

$$W_q(V) = \mathrm{g}^2 J \frac{(1+s^2)D(q)}{2(1-s^2)E(q)}, \tag{7.28}$$

where

$$D(q) \equiv \frac{1}{3}[2(2-q^2)E(q) - (1-q^2)K(q)]. \tag{7.29}$$

The function $D(q)$ ($q < 1$) is depicted in Fig. 7.1.

According to Eqs. (7.15), (7.16) and (7.23), the total energy $\ni_q (V) = E_p(V) + W_q(V)$, in terms of the Fermi energy (7.17), takes at small velocities the following form

$$\ni_q (V) = -\Pi_0(q) + \frac{1}{2}M_{bs}V^2 - \hbar V k_F, \tag{7.30}$$

where

$$\Pi_0(q) = \mathrm{g}^2 q^2 J/E(q) - W_q(V) = \mathrm{g}^2 JF(q), \tag{7.31}$$

characterizes the energy of pairing of quasiparticles including the energy of formation of the local chain deformation. The function $F(q)$ is determined by

$$F(q) \equiv \frac{2}{E^2(q)}[2 - q^2 - D(q)/E(q)]. \tag{7.32}$$

The quantity

$$\Delta = \frac{1}{2}\Pi_0(q), \tag{7.33}$$

determines the energy gap in the quasiparticle spectrum resulting from a pairing. Thus, the pairing gap is a half of the energy of the formation of a static bisoliton. The effective bisoliton mass within each period is

$$M_{bs} = 2m + \frac{4\mathrm{g}^2 J}{c_0^2}\frac{D(q)}{E^2(q)}, \tag{7.34}$$

where m is the effective mass of a quasiparticle.

The energy gap Δ is obtained regardless of the screened Coulomb repulsion between quasiparticles. Such an approximation is justified if the spatial extension of each bisoliton, equal to

$$d \simeq 2\pi a/\mathrm{g}, \tag{7.35}$$

exceeds the lattice constant a.

The energy gap of a static bisoliton is independent of the ion mass M, i.e. there is no isotope effect, although the electron-phonon interaction underlies the pairing. The isotope effect manifests itself only through the bisoliton kinetic energy and, at small velocities, $s^2 = V^2/c_0^2 \ll 1$, the isotope effect is also small.

Within each period L of an infinite chain there exists a correspondence between the energy $\ni_q (V)$ and the coordinate function $\Psi_q(x_1, x_2, t)$. The latter characterizes the bisoliton distribution given for a fixed value of k and the modulus q which is determined by Eq. (7.26) with the parameters g and L characterizing the chain. This becomes

$$\Psi_q(x_1, x_2, t) \quad = \quad \frac{\sqrt{2}}{E^2(q)} \cos[(k - k_F)(\zeta_1 - \zeta_2)]$$

$$\times dn\left(\frac{g\zeta_1}{E(q)}, q\right) dn\left(\frac{g\zeta_2}{E(q)}, q\right) e^{i\left[\delta(k)+\frac{E_p t}{\hbar}\right]}. \quad (7.36)$$

Here, $\delta(k)$ is the phase given by

$$\delta(k) = 2k[\frac{1}{2}(x_1 + x_2) - Vt], \quad (7.37)$$

which smoothly varies when the coordinate of the bisoliton center of gravity changes.

1.1 Small density of doped charge carries

At rather small values of the density of quasiparticles, when the inequality

$$gL \gg 1, \quad (7.38)$$

is valid, the complete elliptic integrals have the following asymptotics

$$E(q) \approx 1 \quad \text{and} \quad K(q) \approx \ln(4/\sqrt{1-q^2}). \quad (7.39)$$

In this case, Equation (7.26), which determines the modulus q of the Jacobian elliptic function, becomes

$$\sqrt{1-q^2} = \frac{1}{4}\exp(-gL) \ll 1. \quad (7.40)$$

This equation is satisfied if $q \approx 1$. Within this approximation, the function $D(q)$ takes the value

$$D(q) \approx \frac{2}{3}[1 - 4gL\exp(-gL)]. \quad (7.41)$$

Then the function $F(q)$ has the form

$$F(q) \approx \frac{2}{3}[1 + 4gL\exp(-gL)], \quad (7.42)$$

and the energy of formation of a bisoliton is

$$\Pi_0 = \frac{2}{3}g^2 J[1 + 4gL\exp(-gL)] \simeq \frac{2}{3}g^2 J. \quad (7.43)$$

If the condition $gL \gg 1$ holds, the effective mass of a bisoliton

$$M_{bs} \simeq 2m + \frac{8g^2 J}{3c_0^2}, \quad (7.44)$$

exceeds two effective masses of quasiparticles forming the bisoliton. The mass of a static bisoliton is independent of the mass of the heavy molecules M and, at small velocities, depends weakly on M. From Eqs. (7.33) and (7.43), the energy gap takes the value

$$\Delta = \frac{1}{3}\mathrm{g}^2 J. \tag{7.45}$$

The enveloping wave functions $\Phi(\zeta)$ of quasiparticles within each period are approximated by the hyperbolic functions

$$\Phi_q(\zeta) = \sqrt{\frac{\mathrm{g}}{2}}(1 + e^{-\frac{1}{2\mathrm{g}L}}[\mathrm{g}\zeta \tanh(\mathrm{g}\zeta) + \sinh^2(\mathrm{g}\zeta)]) \times \frac{1}{\cosh(\mathrm{g}\zeta)}, \tag{7.46}$$

having an asymptotic value

$$\Phi(\zeta) = \sqrt{\frac{\mathrm{g}}{2}} \times \frac{1}{\cosh(\mathrm{g}\zeta)} = \sqrt{\frac{\mathrm{g}}{2}} \times sech(\mathrm{g}\zeta). \tag{7.47}$$

Within the coordinate frame ζ, the functions $\Phi(\zeta)$ presents bell-like elevations with width

$$\Delta\zeta \approx 2\pi/\mathrm{g}, \tag{7.48}$$

separated by a distance L from each other. For a static bisoliton (i.e. $\zeta = x$), Equation (7.47) transforms into

$$\Phi(x) = \sqrt{\frac{\mathrm{g}}{2}} \times sech(\mathrm{g}x). \tag{7.49}$$

In a bisoliton, the value

$$\Delta x = 2\pi a/\mathrm{g} \tag{7.50}$$

determines the correlation length.

When there is only one bisoliton in the chain, two quasiparticles in the bisoliton move in the combined effective potential well

$$U_{\uparrow\downarrow}(\zeta) = -2\mathrm{g}^2 J \times sech^2(\mathrm{g}\zeta), \tag{7.51}$$

The radius of this well is twice as small as that of an isolated soliton, and its depth twice as large as that of a soliton (see Eq. (3.61) in Chapter 5).

1.2 Large density of doped charge carries

At rather large density of quasiparticles in the chain, when $\mathrm{g}L < 5$, the complete elliptic integrals $E(q)$ and $K(q)$ have the following asymptotics

$$E(q) \approx \frac{\pi}{2}\left[1 - \frac{1}{4}q^2 - \frac{3}{64}q^4 + \cdots\right] \quad \text{and} \tag{7.52}$$

$$K(q) \approx \frac{\pi}{2}\left[1 + \frac{1}{4}q^2 + \frac{9}{64}q^4 + \cdots\right] \quad \text{for} \quad q^2 \ll 1. \tag{7.53}$$

In this case, from Eq. (7.26), one obtains that

$$q^2 = 4\sqrt{2}\left[\frac{2gL}{\pi^2} - 1\right]^{1/2} \ll 1, \tag{7.54}$$

which determines the modulus q of the Jacobian elliptic function. By virtue of Eq. (7.54), the periodic solutions of Eq. (7.36) in the form of Jacobian elliptic functions exist under the condition $gL < \pi^2/2$. Therefore, the maximum density of quasiparticles on a chain in the form of the bisoliton condensate corresponds to the minimum period

$$L_{min} = \frac{\pi^2}{2g}. \tag{7.55}$$

If the inequality

$$\pi^2/2 < gL < 5, \tag{7.56}$$

is obeyed, the Jacobian functions are then approximated as

$$\Phi_q(\zeta) \approx \frac{\sqrt{2g}}{\pi}\left[1 - \frac{q^2}{2}\sin^2(2g\zeta/\pi)\right], \quad \text{where} \quad q^2 \ll 1. \tag{7.57}$$

At $q^2 \ll 1$, the function $D(q)$ becomes

$$D(q) \approx \frac{\pi^2}{2}\left[1 - \frac{3}{4}q^2 + \frac{9}{64}q^4\right]. \tag{7.58}$$

Then, the energy gap takes the value

$$\Delta \simeq g^2 J\left[\frac{2}{\pi}\right]^3, \tag{7.59}$$

which is smaller than that in Eq. (7.45). Thus, the magnitude of the pairing gap decreases as the density of quasiparticles increases. Moreover, at a large density of bisolitons, their Coulomb interaction reduces the magnitude of the energy gap in Eq. (7.59).

1.3 The Coulomb repulsion

All the results obtained above did not take into account the Coulomb repulsion of quasiparticles. If we take into account the Coulomb repulsion as a perturbation, then, at small velocities, a pairing is still energetically profitable if the dimensionless coupling constant g is greater than some critical value,

$$g_{cr} \approx \left[\frac{e_{eff}^2}{4a\pi^2 J}\right]^{1/2}, \tag{7.60}$$

where e_{eff} is the effective screened charge. The critical value is estimated from the condition that the displacement of quasiparticles caused by the Coulomb repulsion is less than the bisoliton "dimensions" $2\pi a/g$.

1.4 Stability of the bisolitons

Bisolitons *do not* interact with acoustic phonons since this interaction is completely taken into account in the coupling of quasiparticles with a local deformation. Therefore, they do not radiate phonons. At low temperature, bisolitons are stable if the gain in the binding energy under their coupling exceeds the screened Coulomb repulsion of their charges.

The velocity of bisoliton motion should not exceed whichever is the smallest of the following velocities:

i. the longitudinal sound velocity c_0 (in cuprates, $c_0 \sim 10^5$ cm/c);

ii. the maximum group velocity V_g of free quasiparticles within their conduction band, which is determined by the exchange integral J characterizing the width of this band,

$$V_g = \frac{2aJ}{\hbar} = \frac{\hbar}{am}, \tag{7.61}$$

where m is the effective mass of a quasiparticle (in cuprates, $V_g \sim 10^6$ cm/c).

The bisolitons do not undergo a self-decay if their velocity is smaller than the critical one

$$V_{cr} = \frac{2\Delta}{\hbar k_F}, \tag{7.62}$$

where k_F is the momentum at the Fermi surface. The last condition is obtained from Eq. (7.30).

2. Bisoliton superconductivity

The bisoliton model proposed by Brizhik and Davydov [3] was first utilized in order to explain superconductivity in quasi-one-dimensional organic conductors, which was discovered in 1979.

In quasi-one-dimensional organic conductors, the bonds between plane molecules in stacks arise from weak van der Waals forces. Hence, due to the deformation interaction, a quasiparticle (electron or hole) causes a local deformation of a stack of molecules, which also induces intramolecular atomic displacements in molecules. The deformation interaction results in the nonlinear equations which were considered above.

2.1 The critical temperature

In the framework of the bisoliton model of superconductivity, Davydov defined the critical temperature T_c as the temperature at which the energy gap vanishes.

If the density of excess pairs of quasiparticles is small, and the condition g < 3 holds, an organic or a perovskite crystal becomes superconducting as the temperature decreases. By analogy with the BCS theory of superconductivity, the critical temperature can be estimated from the value of pairing energy gap as

$$k_B T_c \sim 2\Delta, \tag{7.63}$$

where k_B is the Boltzmann constant.

Below the critical temperature, the phase $\delta(k)$ in Eq. (7.37) is coherent for the whole chain.

2.2 Superconductivity in cuprates

When Davydov proposed the bisoliton model as the mechanism for high-T_c superconductivity in cuprates, he did not know about the existence of quasi-one-dimensional charge stripes in CuO_2 planes. Hence, he needed to locate one dimensionality in the CuO_2 planes. Since the CuO_2 planes in cuprates consist of quasi-infinite parallel chains of alternating ions of copper and oxygen, Davydov assumed that each -Cu-O-Cu-O- chain in a CuO_2 plane can be considered as a quasi-one-dimensional system. Therefore the current flows along these parallel chains. In the framework of the bisoliton model, he studied the charge migration in one of these chains and all the results obtained above were directly applied to the superconducting condensate in cuprates.

According to Davydov, the doping level in superconducting cuprates is small; thus, the obtained results for small density of doped charge carriers are valid for cuprates, i.e. when $\mathrm{g}L \gg 1$. Davydov emphasized that the energy of pairing Δ in Eq. (7.43) does not depend on the mass M of an elementary cell. This mass only appears in the kinetic energy of the bisolitons. Therefore the isotope effect is very small, notwithstanding the fact that the basis of pairing is the electron-phonon interaction.

In superconductors described by the BCS theory, the dimensionless coupling parameter g is the order of 10^{-3}–10^{-4}, while it is around 1 in cuprates.

2.3 A concluding remark

The main ideas of the bisoliton model are formulated above as they were presented by Davydov [1, 2]. In the next Chapter, we shall compare the predictions of the bisoliton model with real data obtained in cuprates; therefore, we do not discuss here the estimates made by Davydov in his works [1, 2]. We shall see that the main idea of the bisoliton model is correct and, in first approximation, the model can describe *pairing* characteristics of hole-doped cuprates; however the Davydov model of high-T_c superconductivity is not complete: it lacks the mechanism for establishing phase coherence.

Chapter 8

THE BISOLITON MODEL AND DATA OBTAINED IN CUPRATES

When a physicist doesn't understand a problem he writes a lot of formula; when the main point is understood, the number of formula remaining is maximum two.

—Niels Bohr

We compare here predictions of the bisoliton model of high-T_c superconductivity, considered in detail in the previous Chapter, and experimental data obtained in cuprates.

The bisoliton model of high-T_c superconductivity was proposed by Davydov [1, 2]. Davydov's genius is that, from only three experimental facts given below, he immediately understood that high-T_c superconductivity is caused by soliton-like excitations. These three facts were: (i) the isotope effect in optimally doped cuprates is absent (see Fig. 3.17); (ii) the coherence length in hole-doped cuprates is very short, $\xi \simeq 15$ Å, and (iii) cuprates become superconducting only when they are slightly doped. He concluded that the sharp decrease of the coherence length in cuprates in comparison with metallic superconductors indicates a rather strong interaction of quasiparticles with the acoustic branch of the lattice vibrations inherent in cuprates. However, Davydov was not a specialist in the field of superconductivity. Therefore, many of his estimations [1, 2] were simply unrealistic. It is the purpose of this Chapter to compare data obtained in cuprates with predictions of the bisoliton model.

We shall see below that the main idea of the model is correct and, in first approximation, the model can describe *pairing* characteristics of hole-doped cuprates; however, it lacks a mechanism for establishing phase coherence.

1. Main results of the bisoliton model

Let me briefly recall the main results of the bisoliton model, which we shall use below.

In a molecular one-dimensional chain, the energy spectrum of quasiparticle excitations has a gap which separates the filled valence band and the empty conduction band. Owing to the moderately strong phonon-electron (hole) interaction, the most energetically profitable transfer of quasiparticles in the chain is by means of self-trapped states: *electrosolitons* and *bisolitons.* The system of an electron surrounded by local chain deformation is called an *electrosoliton.* A *bisoliton* is the electron (or hole) pair coupled in a singlet state due to local deformation of the lattice. The energy levels of electrosolitons and bisolitons are located below the conduction band, thus inside the gap. According to the model, acoustic phonons (see Figs. 2.4 and 2.6) are responsible for the formation of electrosolitons and bisolitons.

A bisoliton can be formed either from two electrosolitons or directly from two quasiparticles. The decay of a bisoliton can lead to formation of one or two electrosolitons. The energy level of the bisolitons is below the Fermi level of quasiparticles. The position of the Fermi level of quasiparticles relative to the valence and conduction bands was not specified by Davydov. Figure 8.1 schematically shows the energy levels of electrosolitons and bisolitons inside the gap according to Davydov's description [1, 2].

In the framework of the bisoliton model, superconducting cuprates have small density of doped quasiparticles, i.e. when the inequality $gL \gg 1$ holds,

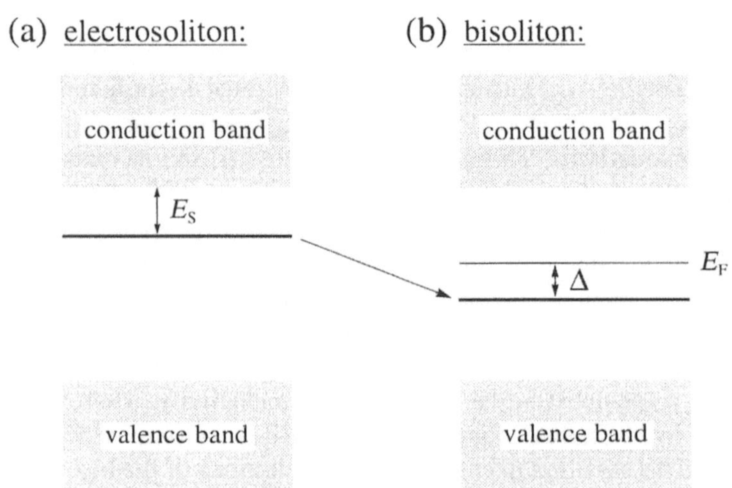

Figure 8.1. Energy levels of (a) an electrosoliton in the framework of the Davydov-soliton model (see Chapter 5), and (b) a bisoliton in the framework of the bisoliton model, shown at low temperature ($T \to 0$).

where g is the dimensionless parameter which characterizes the coupling of a quasiparticle with the chain, and L is the dimensionless distance between two bisolitons. Two quasiparticles having the mass m form a static bisoliton with the mass

$$M_{bs} \simeq 2m + \frac{8\text{g}^2 J}{3c_0^2} \tag{8.1}$$

which exceeds two effective masses of quasiparticles, where J is the exchange interaction energy and c_0 is the longitudinal sound velocity in the chain. The energy gap in the quasiparticle spectrum resulting from a pairing is determined by

$$\Delta \simeq \frac{1}{3}\text{g}^2 J. \tag{8.2}$$

The energy gap is half of the energy of formation of a static bisoliton. The bisoliton-formation energy includes not only the energy of pairing of quasiparticles but also the energy of formation of the local chain deformation. The magnitude of the energy gap decreases as the density of quasiparticles increases.

The energy gap Δ and the bisoliton mass M_{bs} do not depend on the mass M of an elementary cell. This mass only appears in the kinetic energy of the bisolitons. Therefore the isotope effect is very small, notwithstanding the fact that the basis of pairing is the electron-phonon interaction.

The correlation length in a bisoliton (the size of a bisoliton) is given by

$$d = \frac{2\pi a}{\text{g}}, \tag{8.3}$$

where a is the lattice constant. The enveloping wave functions $\Phi_i(x)$ of quasiparticles within each period are approximated by the hyperbolic function

$$\Phi(x) = \sqrt{\frac{\text{g}}{2}} \times sech(\text{g}x), \tag{8.4}$$

where x is the axis along the chain.

In the framework of the bisoliton model of superconductivity, the critical temperature T_c is defined as the temperature at which the energy gap vanishes.

2. Phase coherence in cuprates

In the bisoliton model, the mechanism for establishing phase coherence among bisolitons is not specified. Therefore, we first discuss the mechanism for the establishment of phase coherence in superconducting cuprates in the framework of this model.

In the framework of the bisoliton model, phonons cannot mediate the phase coherence because, as emphasized by Davydov, bisolitons do not interact with acoustic phonons. This interaction is completely taken into account in the coupling of quasiparticles with a local lattice deformation. Implicitly, Davydov assumed that the phase coherence among bisolitons is established due to the overlap of their wave functions (*the wave-function coupling*) as in the BCS theory for conventional superconductors (see Chapter 2). The problem is that, in the framework of the bisoliton model, the wave-function coupling cannot mediate the phase coherence among bisolitons because the wave function of any self-trapped state is by definition very localized.

In the BCS theory, the phase coherence among the Cooper pairs can be established by the wave-function coupling since the average distance between the Cooper pairs is much smaller than the coherence length (the size of a Cooper pair). In cuprates, the distance between the Cooper pairs (bisolitons) is similar to the size of a bisoliton.

Secondly, Davydov defined the critical temperature T_c as the temperature at which the energy gap vanishes. With such a definition for T_c, the wave-function coupling can formally be considered the mediator of the phase coherence. In reality, however, the wave-function coupling cannot be responsible for mediating the phase coherence among the bisolitons. This was the reason why Davydov avoided discussing the bell-like shape of the $T_c(p)$ dependence in cuprates (see Fig. 3.13), where p is the hole concentration in CuO_2 planes—he could not explain it in the framework of the model. The bisoliton theory predicts that by increasing the hole (electron) concentration the magnitude of pairing gap decreases. Therefore, if the wave-function coupling mediates the phase coherence among bisolitons, then, from $T_c \sim \Delta$, one obtains that, by increasing the hole concentration, T_c will monotonically decrease, contrary to experiment. Thus, in cuprates, *the phase coherence among bisolitons is established due to a non-phonon mechanism which is different from the wave-function coupling*.

In the next Chapter, we shall see that in cuprates spin fluctuations mediate the phase coherence. Thus, the bisoliton model is the theory of quasiparticle pairing, but it lacks the mechanism for establishing phase coherence. Therefore, we discuss below solely the pairing characteristics.

3. Pairing characteristics of cuprates

In this Section, we discuss experimental data obtained in hole-doped cuprates, such as Bi2212, YBCO and LSCO, and in electron-doped NCCO.

3.1 Polaron and bisoliton energy levels

According to the Davydov-soliton model (see Chapter 5) (which has no relation with high-T_c superconductivity), the energy level of an electrosoliton inside the gap is measured from the lowest level of the conduction band, as

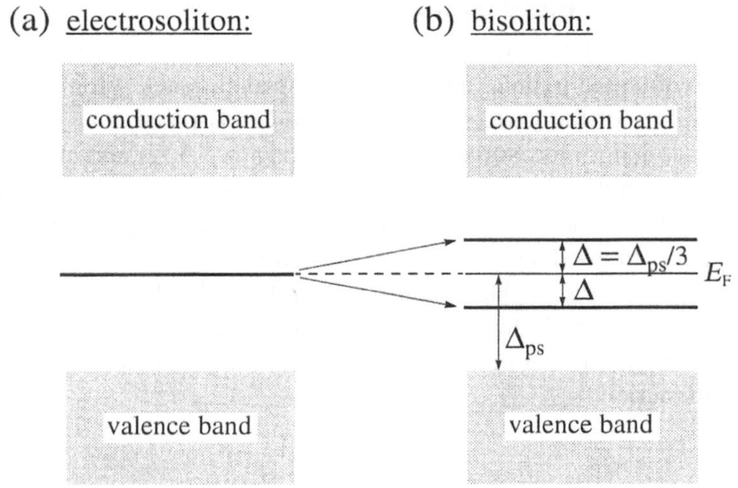

Figure 8.2. Energy levels in Bi2212 of (a) an electrosoliton (polaron), and (b) a bisoliton (bipolaron), shown at low temperature ($T \rightarrow 0$). In frame (b), the magnitude of the pairing gap is about one third of the magnitude of the charge gap (pseudogap), $\Delta \simeq \Delta_{ps}/3$. In frame (b), the lower energy level corresponds to the bonding bisoliton level, and the upper energy level to the antibonding bisoliton level.

schematically shown in Fig. 8.1a. In the framework of the bisoliton model of high-T_c superconductivity, the energy level of bisolitons is determined by the Fermi level of quasiparticles, as shown in Fig. 8.1b. In his works, Davydov did not specify the location of the Fermi level of quasiparticles relative to the conduction and valence bands.

From the data discussed in Chapter 6, the electrosoliton and bisoliton energy levels in Bi2212 are schematically depicted in Fig. 8.2. The Fermi level in Bi2212 is also shown in Fig. 8.2, which lies in the middle of the charge gap (the tunneling pseudogap). In Fig. 8.2a, electrosolitons (polarons) propagate at the Fermi level and, on cooling, give rise to bisolitons. The energy level of bisolitons is defined by the magnitude of the pairing gap analogous to that in Eq. (8.2).

In the framework of the bisoliton model, Davydov did not indicate that, in addition to the "bonding" bisoliton energy level, there always exists the corresponding "antibonding" energy level, as shown in Fig. 8.2b. The antibonding bisoliton energy level is in fact empty but potentially available for possible occupation, for example, by tunneling quasiparticles. Therefore, there are two quasiparticle peaks in tunneling conductances, as shown in Fig. 6.7. From tunneling and angle-resolved photoemission measurements, the magnitude of the pairing gap in Bi2212 is one third of the charge gap (pseudogap), $\Delta = \Delta_{ps}/3$, as indicated in Fig. 8.2b.

3.2 The coupling parameter g

Let us estimate the coupling parameter g from known values of the coherence length obtained in hole- and electron-doped cuprates. From Eq. (8.3), by using the values of the lattice constant of the CuO_2 planes ($a \simeq 3.85$ Å) and the coherence length measured in hole-doped ($\xi \approx 15$ Å) and electron-doped ($\xi \approx 80$ Å) cuprates, one obtains $g_h \simeq 1.6$ and $g_e \simeq 0.3$, where the letters "h" and "e" denote hole- and electron-doped cuprates, respectively. The ratio $g_h/g_e \simeq 5.3$ is in good agreement with similar ratio estimated for organic C_{60} where $g_h/g_e \approx 5$–6 [3]. This means that, independently of the material, the maximum T_c value will always be higher in hole-doped superconductors.

Consider an additional meaning of the coupling parameter of "the electron-phonon interaction"

$$g \equiv \frac{G}{2J}. \tag{8.5}$$

In a system having solitons, the balance between nonlinearity and dispersion is responsible for the existence of the solitons. The bisoliton model is based on the nonlinear Schrödinger equation. As discussed in Chapter 5, in the nonlinear Schrödinger equation the second term is responsible for the dispersion and the third one for the nonlinearity. The coefficient in the second term, the energy of the exchange interaction, $2J$, characterizes the "strength" of the dispersion and the coefficient in the third term, the nonlinear coefficient of the electron-phonon interaction, G, characterizes the "strength" of the nonlinearity. The parameter g represents the ratio between the two coefficients G and $2J$. Therefore, in a sense, the coupling parameter g reflects the balance between the nonlinear and dispersion forces. As a consequence, it cannot be very small $g \ll 1$, or very large $g \gg 1$. If $g \ll 1$, dispersion will prevail, and the bisolitons will gradually diffuse, giving rise to "bare" quasiparticles. If $g \gg 1$, nonlinearity effects prevails, and the bisolitons will become immobile, localized. Thus, $g \sim 1$.

Discussing the critical temperature in the framework of the bisoliton model, Davydov underlined that the parameter g should be $g < 3$; however, he did not explain why. Apparently, it followed from the condition that the electron-phonon interaction is *moderately* strong. Therefore, the value $g \sim 1$ is in a sense "the optimal value." Indeed, as mentioned by Davydov, in superconductors described by the BCS theory, the parameter g is the order of 10^{-3}–10^{-4}, meaning that, in conventional superconductors, the electron-phonon interaction is very weak and superconductivity is "linear."

In cuprates, the "optimal" distance d between quasiparticles in a bisoliton, i.e. when $g \sim 1$, can be estimated from Eq. (8.3): the variation of the parameter $1/3 < g < 3$ corresponds to the variation of the distance 8 Å $< d <$ 70 Å, and $d \simeq 24$ Å if $g = 1$, as schematically shown in Fig. 8.3. The doping dependence of the parameter g in Bi2212 will be considered below.

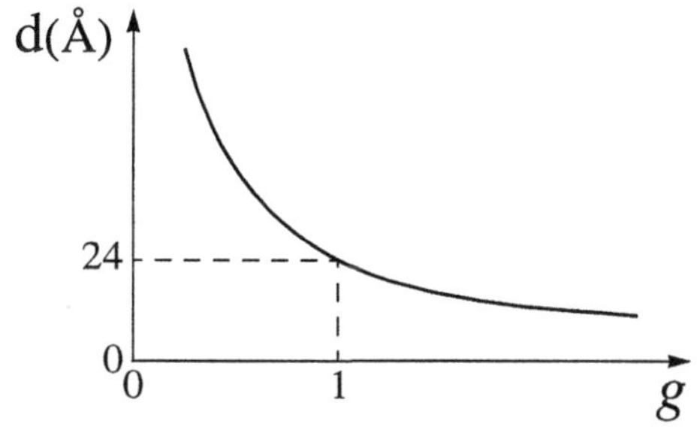

Figure 8.3. The size of a bisoliton (the distance between quasiparticles in a bisoliton) in cuprates as a function of the coupling parameter g at low temperature ($T \longrightarrow 0$).

In electron-doped NCCO, the estimate of the parameter g is about g $\simeq 0.3$. Such a value of g may signify that the bisolitons in NCCO have a short lifetime.

3.3 Doping dependence of g and the energy gap in Bi2212

What is the doping dependence of the parameter g? From common sense, the electron-phonon interaction weakens as the doping p increases, so the parameter G decreases with doping. At the same time, the exchange energy arising from the overlap of electron wave functions on the nearest sites, J, also decreases as p increases. So, in general, the parameter g can increase or decrease with increase in doping. Let us estimate the doping dependence g(p) in Bi2212. First, we estimate the value of J. By using Eq. (7.7) given in Chapter 7, one obtains that $J = \frac{\hbar^2}{2ma^2} \simeq$ 50–60 meV, where we have used the value of the lattice constant in CuO_2 planes, a = 3.85 Å and the effective quasiparticle mass in cuprates, $m \simeq$ (4–5)m_e where m_e is the electron mass.

Secondly, we assume that the exchange energy J has a doping dependence similar to the doping dependence of the *magnetic* exchange energy J_m. In first approximation, this assumption is reasonable since both the exchange energies reflect the electron-electron interactions on the nearest sites. Thus, we assume that $J(p) \approx \alpha J_m(p)$, where α is the coefficient of proportionality. Then, from Eq. (8.2), one can obtain

$$J(p) = \frac{3\Delta(p)}{g^2(p)} \simeq \alpha J_m(p). \tag{8.6}$$

Figure 8.4 shows for Bi2212 the doping dependence $J_m(p)$ measured in Raman scattering measurements [4], together with the dependence $\Delta_p(p)$ obtained in tunneling and angle-resolved photoemission measurements (see Fig. 3.22). In Fig. 8.4 in optimally doped Bi2212, thus near $p_{opt} = 0.16$, $\Delta_p(0.16) \simeq$ 40 meV and $J_m(0.16) \simeq$ 75 meV. Assuming that the estimated value g $\simeq 1.6$

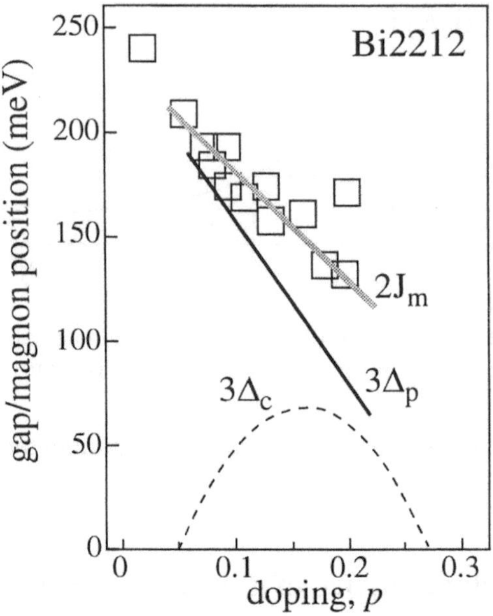

Figure 8.4. Rescaled two-magnon data, $2J_m$, (squares) as a function of doping level, measured in Bi2212 and YBCO at low temperature [4]. The solid and dashed lines represent $3\Delta_p$ and $3\Delta_c$ energy gaps from Fig. 3.22, respectively.

in hole-doped cuprates corresponds to the value near optimal doping, one can obtain that $\alpha \simeq 0.61$ in Bi2212 from Eq. (8.6). This value $\alpha \simeq 0.61$ is in good agreement with the value $J \simeq$ 50–60 meV estimated above. Then, from Eq. (8.2), an estimate of the doping dependence $\mathrm{g}(p)$ is shown in Fig. 8.5. One can see that the parameter g decreases as the doping increases, varying between 1.8 and 1.3 over the interval $0.05 < p < 0.23$.

3.4 Bisoliton mass

The effective mass of a static bisoliton in Eq. (8.1) can be represented as

$$M_{bs} \simeq 2m + \frac{8\Delta}{c_0^2}, \tag{8.7}$$

where Δ is the energy gap in Eq. (8.2). Since the magnitude of the energy gap decreases as the doping level p increases, the bisoliton mass also decreases with increase of p.

In superconducting LSCO, acoustic measurements [5] show that the longitudinal sound velocity c_0 in Eq. (8.7) depends moderately on doping level, and its doping dependence has a bell-like shape similar to that in Fig. 3.1. Therefore, the dependence $M_{bs}(p)$ is not linear: upon increasing the doping level, the value of the bisoliton mass decreases fast in the underdoped region

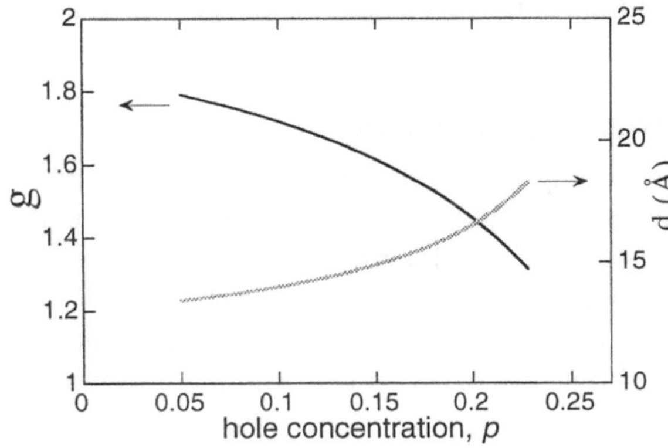

Figure 8.5. The doping dependence of the coupling parameter g (solid curve), obtained from the data shown in Fig. 8.4 and the corresponding average size of bisolitons (grey curve) in Bi2212 at low temperature ($T \to 0$).

and then much slower in the overdoped region. It is important to note that this bisoliton mass is the in-plane mass.

3.5 Coherence length

In conventional superconductors, the average distance between the Cooper pairs is much smaller than the size of a Cooper pair. Therefore, the notion of *coherence length* denotes the size of a Cooper pair, i.e. the average distance between two electrons in a Cooper pair. In cuprates, there are three characteristic distances: the out-of-plane coherence length and two in-plane characteristic lengths. The in-plane characteristics are the average distance between quasiparticles in a bisoliton and the phase-coherence length which is determined by magnetic fluctuations into CuO_2 planes. We shall discuss the phase-coherence length in Chapter 10.

In Bi2212, the doping dependence of the distance between holes in a bisoliton estimated from the $\Delta(p)$ and $J_m(p)$ dependences is shown in Fig. 8.5. As one can see, the *estimated* size of a bisoliton in Bi2212 is about 13 Å in the underdoped region, and increases as the doping level increases.

In conventional superconductors, the coherence length directly relates to the energy gap as $\xi \sim 1/\Delta$. In cuprates, the situation is different. The distance d between two holes (electrons) in a bisoliton does not relate directly to the pairing energy gap. However, they are correlated through the dimensionless parameter g [see Eqs. (8.2) and (8.3)]. One of the reasons for the absence of a direct correlation between Δ and d is that, in the framework of the bisoliton model, the pairing energy includes not only the energy of pairing of quasiparticles but also the energy of formation of the local chain deformation. At the same time, the coherence length obtained, for example, from penetration-depth measurements, and the coherence energy gap (see Fig. 3.22) should correlate with each other.

3.6 Tunneling characteristics

The spatial enveloping wave functions of quasiparticles in a bisoliton, given by Eq. (8.4), are approximated by the $sech$ hyperbolic function. Then, as discussed in Chapter 5, the spectral density of states of a bisoliton is represented by the $sech^2$ function. In Chapter 6, we used the $sech^2$ and $tanh$ functions to fit tunneling $dI(V)/dV$ and $I(V)$ characteristics in Bi2212, respectively. Thus, the tunneling data are in agreement with the bisoliton model of high-T_c superconductivity.

Let us estimate in the framework of the model the peak width in the spectral density of states of a bisoliton. In so doing, we shall estimate the width of coherent quasiparticle peaks in tunneling conductances. The size of a bisoliton d is given by Eq. (8.3). Then, in momentum space, the approximate spread in wave number associated with the bisoliton is

$$\Delta k \gtrsim \frac{1}{d}. \tag{8.8}$$

The corresponding spread in energy can be estimated from

$$\Delta E = \frac{\hbar^2 (\Delta k)^2}{2m}. \tag{8.9}$$

By using the values of the lattice constant of the CuO_2 planes $a \simeq 3.85$ Å, the distance $d \simeq 15$ Å and the quasiparticle mass in cuprates $m \sim (4 \div 5)m_e$, where m_e is the electron mass, one can obtain that $\Delta E > 4$ meV. In near optimally doped Bi2212, the width of coherent quasiparticle peaks is equal to or larger than 15 meV. Thus, the bisoliton model of high-T_c superconductivity gives a good estimate for the width of quasiparticle conductance peaks.

3.7 Phonon spectrum in Bi2212

In the bisoliton model, the acoustic branch of phonons (see Fig. 2.4) is responsible for the formation of electrosolitons and bisolitons. Figure 8.6 shows the phonon spectrum $F(\omega)$ obtained in slightly overdoped Bi2212 by inelastic neutron scattering measurements [6]. In The phonon spectrum $F(\omega)$ consists of the acoustic branch ($\omega < 50$ meV) and the optical branch ($\omega > 50$ meV). Figure 8.6 also shows two spectral functions $\alpha^2 F(\omega)$ obtained in slightly overdoped Bi2212 in two independent tunneling measurements [7, 8]. The spectral function $\alpha^2 F(\omega)$ is the parameter of the electron-phonon interaction in the Eliashberg equations, which characterizes the coupling strength between charge carriers and phonon vibrations (see Chapter 2).

In Fig. 8.6, one can observe that in slightly overdoped Bi2212 charge carriers are strongly coupled to the 20 meV acoustic mode and to the 73 meV opti-

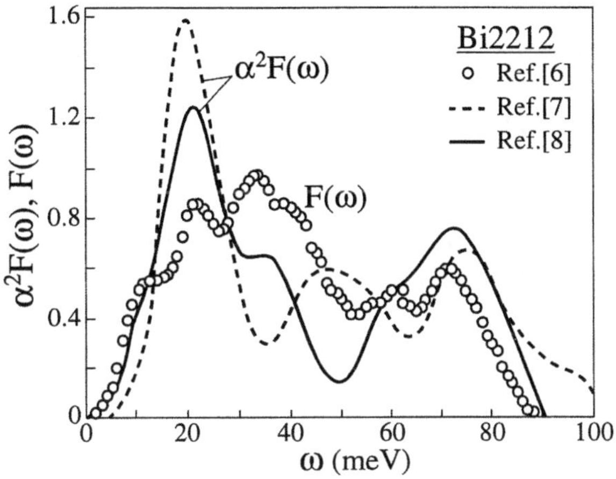

Figure 8.6. Phonon spectrum $F(\omega)$ (circles) obtained in inelastic neutron scattering measurements in Bi2212 [6] and two spectral functions $\alpha^2F(\omega)$ (solid and dashed curves) obtained in tunneling measurements by break-junctions [7, 8]. All the data are measured in slightly overdoped Bi2212 at low temperature ($T \ll T_c$).

cal mode. The latter is caused by the Cu–O bond-stretching mode that propagates in the CuO_2 plane. The role of phonons at 50 meV is controversial: one spectral function $\alpha^2F(\omega)$ shows the peak at 50 meV (the dashed curve), while the other spectral function $\alpha^2F(\omega)$ exhibits the dip. Leaving aside the question of the 50 meV phonons, it is clear that the optical phonons with $\omega = 73$ meV are coupled to charge carriers in Bi2212. Angle-resolved photoemission measurements performed in LSCO, YBCO, Bi2212 and Bi2201 show a kink in the dispersion at 55–75 meV, thus the optical phonons are coupled to charge carriers in hole-doped cuprates. Therefore, there is a clear disagreement between the bisoliton model and the data.

On the other hand, it is possible that the optical phonons are responsible exclusively for the formation of charge stripes, while the acoustic phonons (20 meV mode in Bi2212) for the bisoliton formation.

3.8 Electron-doped NCCO

In electron-doped NCCO, an estimate of the parameter g, obtained from the value of the coherence length ξ = 80 Å, is g $\approx$ 0.3 (see above). Such a value for the parameter g seems too small. Let us estimate the parameter g from the energy gap.

In underdoped NCCO, the magnitude of the pairing gap is $\Delta_p \simeq 13$ meV [9]. In antiferromagnetic NCCO, the energy of the magnetic exchange interactions is $J_m \simeq 155$ meV [10]. Then, on increasing the doping level, the magnitude of J_m decreases, thus, in underdoped NCCO, $J_m < 155$ meV. Assuming that in NCCO $J \simeq \alpha J_m$, where α = 0.61 obtained for Bi2212, from Eq. (8.2), one obtains that in underdoped NCCO the parameter g is in fact g > 0.64. Consequently, the distance between two electrons in a bisoliton is $d <$ 38 Å, which is much smaller than ξ = 80 Å.

What is interesting is that the value $d < 38$ Å is in good agreement with the value $\xi_{ab}(0) \simeq 30$ Å obtained recently from tunneling measurements in high magnetic fields [11]. Thus, it is possible that the value of the coherence length in NCCO $\xi \simeq 80$ Å has to be reconsidered.

3.9 Concluding remarks

A comparison of the main characteristics of the bisoliton model and the data obtained in cuprates shows that the bisoliton model is not a theory for high-T_c superconductivity. First, it lacks the mechanism for establishing phase coherence. Secondly, even though the bisoliton model can describe *some* pairing characteristics, it is only *in first approximation.* This is probably because in the bisoliton model the Coulomb repulsion between quasiparticles in a bisoliton is not taken into account.

However, the main idea of the bisoliton model is correct: the moderately strong and nonlinear electron-phonon interaction mediates the pairing in cuprates. The main result of the model is that in the presence of strong electron-phonon interaction the BCS isotope effect can be absent or small.

In spite of the fact that the Cooper pairs in cuprates are not Davydov's bisolitons in a *direct* sense, we shall continue to use the term "bisolitons" in the remainder of the book. The Cooper pairs in cuprates consist of two quasi-one-dimensional soliton-like excitations, therefore the name "bisolitons" is probably the best choice. For example, the name "bipolarons" does not reflect the presence of one dimensionality in the system.

This Chapter and Chapter 7 should serve as a starting point for the future theory of high-T_c superconductivity.

4. Key experiments for bisoliton superconductivity

In Chapters 10 and 11, we shall see that bisoliton superconductivity exists not only in cuprates but in some other materials as well. This raises the question: what key experimental techniques can be used for identifying bisoliton superconductivity?

The first indication of bisoliton superconductivity is the magnitude of the coherence length (the distance between quasiparticles in a bisoliton). A short coherence length of, say, less than 50 Å can be a good sign of bisoliton superconductivity. Second, tunneling measurements always remain key experiments for any type of superconductivity. It is worth remembering that tunneling measurements provided evidence for bisoliton superconductivity in Bi2212 (see Chapter 6).

Third, by analogy with the isotope effect in conventional superconductors, the study of *elastic* properties of a superconducting material can serve as a key experiment for revealing the bisoliton superconductivity, i.e. acoustic measurements of elastic coefficients.

As an example, let us consider elastic properties of high-T_c superconductors, which are remarkably different from those of conventional superconductors. In cuprates, longitudinal and transverse elastic coefficients reveal not only anisotropic lattice properties in the normal state, but also anisotropic coupling between superconductivity and lattice deformation.

Figure 3.31 in Chapter 3 shows the elastic transverse coefficient $(C_{11} - C_{12})/2$ obtained in LSCO ($x = 0.14$) as a function of temperature and of applied magnetic field. Other elastic coefficients for various modes in LSCO also show anisotropic lattice properties in the normal state and anisotropic coupling between superconductivity and lattice deformation below the critical temperature [5]. All these unusual elastic properties can naturally be understood in the framework of the bisoliton model of high-T_c superconductivity: in contrast to characteristics of conventional superconductors, all bisoliton characteristics depend on the longitudinal elastic coefficients and the longitudinal sound velocity (see Chapter 7). Second, since the bisolitons are aligned along -Cu-O-Cu-O- bonds in the CuO_2 planes, it is *expected* that the elastic coefficients for different modes exhibit anisotropic lattice properties below and somewhat above T_c.

What is even more striking in Fig. 3.31 is the effect of magnetic field applied along various axes on the $(C_{11} - C_{12})/2$ coefficient: this coefficient exhibits a strong and anisotropic dependence on the direction and magnitude of the applied magnetic field for temperatures *below* T_c. This mysterious behavior of the $(C_{11} - C_{12})/2$ coefficient in applied magnetic field below T_c takes place because in cuprates spin fluctuations (excitations) mediate the phase coherence, and are coupled below T_c to the bisolitons and thus to the lattice. Consequently, the lattice is affected by the applied magnetic field through the intermediary of spin fluctuations. Finally, in Fig. 3.31, a hardening of the $(C_{11} - C_{12})/2$ elastic coefficient at very low temperature occurs due to the formation of spin glass.

Chapter 9

THE MECHANISM OF *C*-AXIS PHASE COHERENCE IN CUPRATES

And dearest to me are analogies, my most reliable teachers.

—Johannes Kepler

In Chapters 6 and 8, we discussed the pairing mechanism in cuprates, which is based on the moderately strong, nonlinear electron-phonon interaction. The purpose of this Chapter is to consider the mechanism of c-axis phase coherence in superconducting cuprates, i.e. the mechanism of the interlayer coupling.

Superconductivity requires not only electron pairing but also long-range phase coherence. In conventional superconductors described by the BCS theory, the Cooper pairs are formed at T_c owing to the electron-phonon interaction. The phase coherence among the Cooper pairs also occurs at T_c due to the overlap of Copper-pair wave functions. In conventional superconductors, the wave-function coupling is possible because the average distance between the Cooper pairs is much smaller than the coherence length (the size of a Cooper pair). In cuprates, the distance between the Cooper pairs (bisolitons) is similar to the size of bisolitons. Therefore, the Cooper pairs in cuprates exist well above T_c, especially, in the underdoped region, and the long-range phase coherence appears only at T_c.

It was already established a year after the discovery of high-T_c superconductors that superconductivity in cuprates is quasi-two-dimensional, occurring in CuO_2 planes. The mechanism of long-range phase coherence was not studied in detail, since it was always assumed that the long-range phase coherence appearing at T_c originates from the Josephson coupling between superconducting CuO_2 layers, i.e. due to an overlap of in-plane wave functions. In Chapter 8, it was shown that, in cuprates, the phase coherence among bisolitons is established due to a non-phonon mechanism which is *different* from the wave-function coupling. Experimental results also indicate that the Josephson coupling (the wave-function coupling) between CuO_2 planes cannot be responsible for the c-axis phase coherence in cuprates. Indeed, measurements of out-of-plane (ρ_c) resistivity in Tl2212 as a function of applied pressure show that the temperature dependence $\rho_c(T)$ shifts smoothly down as the pressure increases, at the same time, the critical temperature T_c first increases and then *decreases* [1]. This result can not be explained by the interlayer Josephson-coupling mechanism. The authors conclude [1]: "Any model that associates high-T_c with the interplane Josephson coupling should therefore be revisited."

In this Chapter, we shall see that the long-range phase coherence in cuprates occurs due to magnetic fluctuations along the c axis.

1. Superconductivity and magnetism

Superconductivity and magnetism were earlier regarded as mutually exclusive phenomena. A considerable amount of experimental and theoretical research on the interplay of superconductivity and magnetism has followed the discovery of the superconducting compounds $R\text{Mo}_6\text{X}_8$ (R = Gd, Tb, Dy and Er, and X = S, Se) and $R\text{Rh}_4\text{B}_4$ (R = Nd, Sm and Tm) [2]. They are usually called *the Chevrel phases* which will be discussed in Chapter 11. The rapid progress made in this field during the last few years revealed a rich variety of extraordinary superconducting and magnetic phenomena in novel materials that are due to the interaction between superconductivity and magnetism [3].

1.1 Superconductivity and antiferromagnetism

Many of the $R\text{Mo}_6\text{X}_8$ and $R\text{Rh}_4\text{B}_4$ compounds exhibit the coexistence of superconductivity and long-range antiferromagnetic order [2]. For conventional superconductors which are characterized by a large-size coherence length, this may not seem too surprising since, over the scale of the coherence length, the exchange field of an antiferromagnet averages to zero. Nevertheless, antiferromagnetic order affects superconductivity, and does it by means of several mechanisms. For example, in $ErMo_6S_8$ and $SmRh_4B_4$, the magnitude of second critical magnetic field B_{c2} increases below the Néel temperature T_N, while in others compounds, such as $R\text{Mo}_6\text{S}_8$ (R = Tb, Dy and Gd), as well as $NdRh_4B_4$, B_{c2} decreases rather abruptly below T_N [2]. Theoretical analysis of the temperature dependence of B_{c2} for the $R\text{Mo}_6\text{S}_8$ antiferromagnetic superconductors revealed an additional pairbreaking parameter whose temperature dependence is similar to that of the antiferromagnetic order parameter (sublattice magnetization) [2].

Coexistence of superconductivity and antiferromagnetic order was later found in many other superconducting materials, such as U-based heavy fermions (UPt_3, URu_2Si_2, UNi_2Al_3, UPd_2Al_3, U_6Co, and U_6Fe), heavy fermions $R\text{Rh}_2\text{Si}_2$

(R = La and Y), $Cr_{1-x}Re_x$, $CeRu_2$, borocarbides RNi_2B_2C (R = Tm, Er, Ho, Dy), most organic superconductors [3, 4], and the new heavy fermion $CeRh_{1-x}Ir_xIn_5$ [5]. In $CeRh_{0.5}Ir_{0.5}In_5$, the bulk superconductivity coexists *microscopically* with small-moment magnetism ($\leq 0.1\mu_B$) [5]. In heavy fermion $CeIrIn_5$, the onset of a small magnetic field ($\sim$ 0.4 Gauss) sets exactly at T_c [5]. In other heavy fermions, there are strong antiferromagnetic correlations present in the superconducting state [3, 4]. Another class of materials in which superconductivity and antiferromagnetism coexist are cuprates, such as YBCO and LSCO compounds [3]. Magnetic properties of cuprates will be discussed separately.

Below a certain coherence temperature ($\sim$ 20–100 K), heavy-fermion metals, which contain magnetic ions with $4f$ or $5f$ electrons, are characterized by a large quasiparticle mass. The magnetic ions are responsible for the magnetic properties of heavy fermion materials. In heavy fermions UPt_3, URu_2Si_2, UCu_5 and $CeRhIn_5$, magnetic correlations lead to an itinerant spin-density-wave order, while, in UPd_2Al_3 and $CeCu_2Si_2$, to a localized antiferromagnetic order. In the latter heavy fermions, the antiferromagnetic order appears first, followed by the onset of superconductivity.

As discussed in Chapter 2, superconductivity is destroyed by a sufficiently high magnetic field. However, this is not always true. In the quasi-two-dimensional organic conductor λ-$(BETS)_2FeCl_4$, where BETS stands for bis-(ethylenedithio)tetraselenafulvalene, the superconducting phase is induced by a magnetic field exceeding 18 Tesla [6, 7]. Crystalline λ-$(BETS)_2FeCl_4$ consists of layers of highly conducting BETS sandwiched between insulating layers of iron chloride $FeCl_4$. The field is applied parallel to the conducting layers. This is particularly remarkable since this compound, at zero field, is an antiferromagnetic insulator below 8.5 K. Field-induced superconductivity was earlier reported for $Eu_xSn_{1-x}Mo_6S_8$ but this compound is paramagnetic above T_c = 3.8 K [3]. In λ-$(BETS)_2FeCl_4$, the Fe^{3+} ions within the $FeCl_4$ molecules are responsible for a long-range antiferromagnetic order.

Field-induced superconductivity in magnetic materials is usually discussed in terms of the Jaccarino-Peter compensation, in which the applied field compensates the internal magnetic field provided by the magnetic ions. By increasing the magnitude of applied magnetic field above $B = 41$ Tesla, the compound λ-$(BETS)_2FeCl_4$ becomes metallic at 0.8 K [7]. The dependence $T_c(B)$ has a bell-like shape with a maximum $T_c \simeq$ 4.2 K near 33 Tesla. Interestingly, a magnetic field of only 0.1 Tesla, applied perpendicular to the conducting BETS planes, destroys the superconducting state. This indicates that, in this organic conductor, superconductivity is robust along the c axis and, at the same time, week in the planes (in cuprates, it is the other way round). We shall discuss the symmetry of superconducting order parameter (energy gap) in λ-$(BETS)_2FeCl_4$ in the following Chapter.

1.2 Superconductivity and ferromagnetism

Can superconductivity and *ferromagnetism* coexist? At first glance, ferromagnetism and superconductivity appear mutually hostile, and all experimental facts known before 2000 supported this point of view. In fact, the answer to the question is "yes."

In erbium rhodium boride ($ErRh_4B_4$) and holmium molybdenum sulphide ($HoMo_6S_8$), superconductivity is destroyed by the onset of a first-order ferromagnetic phase transition [2]. For example, in $ErRh_4B_4$, the critical temperature is 8.7 K. On cooling, the superconductivity is destroyed at 0.8 K by the onset of a first-order ferromagnetic phase transition which starts at 1 K. Between 0.8 and 1 K, superconductivity in $ErRh_4B_4$ coexists with a *local* ferromagnetic order which becomes long-ranged at 0.8 K. In all other known singlet ferromagnetic superconductors, like $ErRh_4B_4$ and $HoMo_6S_8$, a non-uniform magnetic phase appears in the superconducting state rather than a ferromagnetic phase. The opposite is also true: it is very unlikely for singlet superconductivity to appear in the ferromagnetic state because the exchange interaction forbids the formation of Cooper pairs.

So, it was a surprise when the coexistence of superconductivity and ferromagnetism was discovered in an alloy of uranium and germanium, UGe_2 [8]. At ambient pressure, UGe_2 is known as a metallic ferromagnet with a Curie temperature of $T_C = 53$ K. However, as increasing pressure is applied to the ferromagnet, T_C falls monotonically, and appears to vanish at a critical pressure of $P_c \simeq 16$–17 kbars. In a narrow range of pressure below P_c and thus *within* the ferromagnetic state, the superconducting phase appears in the millikelvin temperature range below the critical temperature (see Fig. 9.1b). Above P_c, UGe_2 is paramagnetic.

As a matter of fact, magnetic fluctuations are strongest when magnetic order is about to form or disappear, a point known as *the quantum critical point.* Quantum critical points have attracted a great deal of attention because the large slow spin fluctuations that occur near the critical pressure (critical density) play a key role in the making and breaking of Cooper pairs. In UGe_2, the critical pressure $P_c \simeq 17$ kbars is a quantum critical point.

Soon after the discovery of superconductivity in itinerant ferromagnet UGe_2, two new itinerant ferromagnetic superconductors were discovered—zirconium zinc $ZrZn_2$ and uranium rhodium germanium URhGe [9, 10]. $ZrZn_2$ superconducts only when it is a ferromagnet, i.e. below the critical pressure $P_c \simeq 21$ kbars. Above P_c, it is a paramagnet showing no trace of superconductivity. In $ZrZn_2$, the maximum critical temperature is slightly less than 3 K at ambient pressure, and decreases with increasing pressure. URhGe is also a superconductor at ambient pressure, and has many similar properties of high-pressure UGe_2—it loses its resistance below 9.5 K, exhibits the Meissner effect and has a large specific-heat anomaly at the superconducting critical temperature.

The archetypal ferromagnet, iron, is found to superconduct at high pressure between 15 and 30 kbars [11]. Albeit, at such pressures, iron ceases to be ferromagnetic, and there is evidence that, at low temperature, it is weakly antiferromagnetic.

The recent discovery of superconductivity, antiferromagnetism and weak ferromagnetism in ruthenocuprate $RuSr_2RCu_2O_8$ (R = Eu, Gd and Y) was interpreted as the coexistence of superconductivity and magnetism with a ferromagnetic component at the same temperature and pressure but located in different layers of the layered perovskite structure: Superconductivity and antiferromagnetism occur in the CuO_2 planes, as in cuprates, but the ferromagnetic order arises from Ru moments in the RuO_2 layers. The crystal structure of $RuSr_2RCu_2O_8$ is similar to that of YBCO except for the replacement of one-dimensional CuO chains by two-dimensional RuO_2 layers. Indeed, measurements show that the ferromagnetic component arises in the RuO_2 layers. However, superconductivity and ferromagnetism do occur in the same location: The superconducting gap coexists with the ferromagnetic component in the RuO_2 planes on a microscopic scale [12, 13]. We shall discuss these remarkable data in the next Chapter.

Interestingly, besides the presence of antiferromagnetic spin-density-wave order at low temperatures, the *microscopic* coexistence of spontaneous weak ferromagnetism with superconductivity is found in borocarbide $ErNi_2B_2C$ [14]. At the same time, there is evidence that, in borocarbides, superconductivity is singlet and mediated by phonons. An explanation of these discordant facts will be given in Chapter 10.

1.3 Magnetically-mediated superconductivity

In conventional superconductors, the Cooper pairs are formed via interactions between electrons and lattice vibrations (phonons). In superconducting heavy fermions, spin fluctuation (electron-electron interactions) are believed to mediate the electron pairing that leads to superconductivity [15]. The spin-fluctuation theory describes how electrons are influenced by the fields produced by others in "the Fermi sea."

For the heavy fermions $CeRh_{0.5}Ir_{0.5}In_5$, $CeIrIn_5$ [5], UGe_2 [8], URhGe [9], $ZrZn_2$ [10], $CeIn_3$, $CePd_2Si_2$ [16] and UPd_2Al_3 [17–19], there is indirect evidence for spin-fluctuation mechanism of superconductivity. In organic superconductors based on charge-transfer salts of the ion BEDT-TTF (or -ET), there is also indirect evidence for magnetically-mediated superconductivity, where BEDT stands for bis(ethylenedithio)tetrathiafulvalene [20]. This intimate relationship between superconductivity and magnetism also appears to be central in cuprates which inherited their magnetic properties from the parent compounds—antiferromagnetic Mott insulators. Many theoretical studies suggest that superconductivity in cuprates is mediated via the exchange of antiferromagnetic spin fluctuations. The purpose of this Chapter is to show that the long-range phase coherence in cuprates occurs owing to magnetic fluctuations.

1.4 Characteristic features of magnetically-mediated superconductivity

In conventional superconductors, superconductivity described by the BCS theory has its specific features, for example, the isotope effect, the s-wave symmetry of the order parameter, etc. What features are inherent to magnetically-mediated superconductivity?

Quantum critical point. As discussed above, magnetically-mediated superconductivity occurs near a quantum critical point. Figure 9.1 schematically shows two phase diagrams: the first diagram is typical for antiferromagnetic heavy-fermions compounds, and the second is the phase diagram of the ferromagnetic heavy fermion UGe_2. In both phase diagrams, the density is changed by applying a pressure. The "doping" by pressure is equivalent to chemical doping or to gate doping in a field-effect transistor geometry. Near the quantum critical density n_c, where the magnetic transition temperature vanishes, magnetic interactions become strong and long-range. Near n_c, magnetic interactions overwhelm other channels. Magnetically-mediated superconductivity may exist only in samples in which the carrier mean free path exceeds the superconducting coherence length. In most cases this requires samples of very high purity.

As one can see in Fig. 9.1, independently of the nature of magnetic interactions—antiferromagnetic or ferromagnetic—the superconducting phase occurs

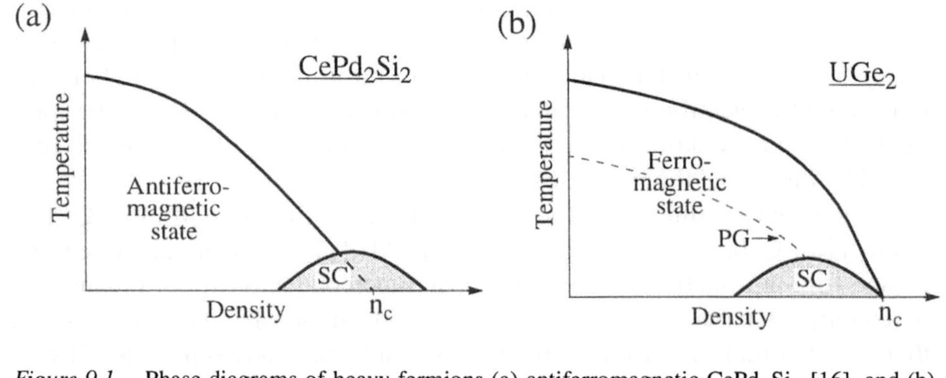

Figure 9.1. Phase diagrams of heavy fermions (a) antiferromagnetic $CePd_2Si_2$ [16], and (b) ferromagnetic UGe_2 [8], shown schematically (SC = superconductivity). In both plots, the density is varied by pressure, and n_c is the quantum critical point. The phase diagram in plot (a) is typical for antiferromagnetic superconductors. In plot (b), the dashed line shows a pseudogap (PG) found in resistivity and magnetization measurements in UGe_2, which may indicate the presence of another critical point.

near a quantum critical point, where magnetic fluctuations are strongest and, consequently, where superconductivity has the best chance to "survive." In a sense, the superconducting phase is "attracted" by a quantum critical point.

Superconducting phase. In Fig. 9.1, one can see that the superconducting phase, as a function of doping, has a bell-like shape. Such a shape of the superconducting phase is typical for magnetically-mediated superconductivity.

Symmetry of the order parameter. Theoretically, the order parameter in antiferromagnetic superconducting compounds, has a d-wave symmetry. The d-wave symmetry of the order parameter was indeed observed in a few magnetic compounds, including cuprates.

For ferromagnetic superconducting materials, the situation is still not clear. Theoretically, the triplet (p-wave) electron pairing is favorable in superconductors with ferromagnetic correlations. However, there is no experimental confirmation of this.

Temperature dependence. In magnetic superconductors, the temperature dependence of *coherence* superconducting characteristics is specific, and differs from the s-wave BCS dependence. Figures 9.2a and 9.2b show respectively the temperature dependence of superfluid density in heavy fermion $CeIrIn_5$, measured by muon-spin relaxation (μSR) [5], and the temperature dependence of Andreev-reflection gap in heavy fermion UPt_3 [21]. In Fig. 9.2, one can observe that the two temperature dependences are similar, and lie below the BCS temperature dependence. Such a temperature dependence can be considered as *typical* and, in first approximation, represents the squared BCS temperature dependence. Magnetic electron-electron interactions being responsible for the occurrence of magnetically-mediated superconductivity result in the squared BCS temperature dependence.

Enhancement of spin fluctuations and the magnetic resonance peak. In antiferromagnetic superconductors, magnetic fluctuations which often, but not always, exist above the superconducting critical temperature, T_c, are enhanced on passing below T_c. Second, below T_c, there often, but not always, occurs a specific magnon-like magnetic excitation called *the magnetic resonance peak* (see Chapter 3). The magnetic resonance peak is observed in inelastic neutron scattering (INS) measurements. In all known cases, the energy position of the magnetic resonance peak, E_r, is temperature independent, and $E_r/k_BT_c \simeq$ 5–10. However, the *intensity* of the resonance mode depends on temperature, and increases as the temperature decreases. Since its width is very narrow, the magnetic resonance peak cannot be caused by magnons. Alternatively, it was proposed that the resonance mode corresponds to magnetic *excitons* [17]. We shall discuss the origin of the magnetic resonance peak in the next Chapter.

For example, in heavy fermion UPd_2Al_3, the antiferromagnetic ordering temperature is about $T_N \simeq 14.3$ K, and the superconducting critical temperature $T_c \simeq 2$K. On passing into the superconducting phase, an abrupt enhance-

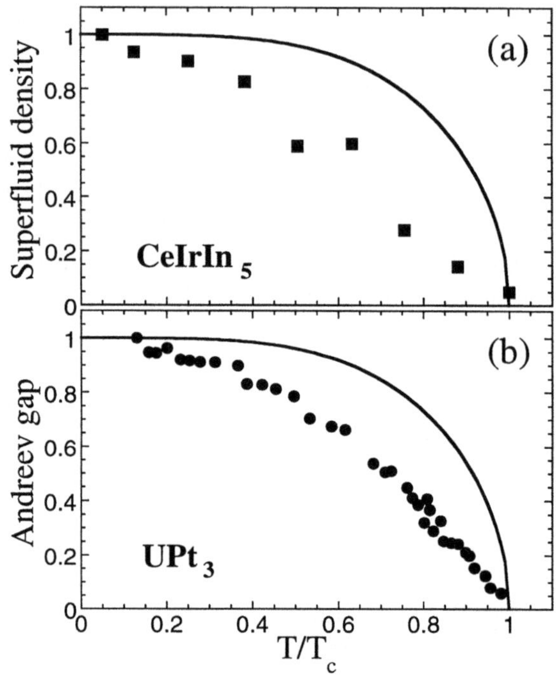

Figure 9.2. Temperature dependences of (a) superfluid density (squares) in heavy fermion $CeIrIn_5$ (T_c = 0.4 K) [5] and (b) Andreev gap (dots) in heavy fermion UPt_3 ($T_c \sim$ 440 mK) [21]. In both plots, the BCS temperature dependence is shown by the solid line. In plot (a), the superfluid density is obtained by μSR measurements [5].

ment of magnetic fluctuations is observed in INS measurements [22]. In addition to this enhancement, the magnetic resonance peak appears at energy $E_r/k_BT_c \simeq 9.2$ [23].

2. Layered compounds with magnetic correlations

In Chapter 3, we briefly discussed how a long-range antiferromagnetic or ferromagnetic order arises in layered compounds: in all known layered magnetic compounds, including undoped cuprates, the long-range antiferromagnetic (ferromagnetic) order develops at Néel temperature T_N (Curie temperature T_C) along the c axis [24]. At the same time, in-plane magnetic correlations exist above T_N (T_C). Thus, in quasi-two-dimensional magnetic materials, the coupling along the c axis represents the last step in establishing a long-range magnetic order.

For example, in layered heavy fermion URu_2Si_2 (T_c = 1.2 K), an antiferromagnetic order develops at a Néel temperature of T_N = 17.5 K along the c axis [25]. A few other examples can be found elsewhere [24].

3. Phase coherence in cuprates

In this Section, we shall see that the long-range phase coherence in superconducting cuprates is mediated by spin fluctuations, and the critical temperature T_c is determined by the magnetic interlayer coupling.

3.1 Cuprates: two energy scales

In cuprates, there are two energy scales (gaps): the pairing energy scale, Δ_p, and the phase-coherence scale, Δ_c, depicted in Fig. 3.22. They have different doping dependences: Δ_p increases almost linearly as the doping level p decreases, whereas Δ_c has approximately a parabolic dependence on p, scaling with T_c. As discussed in Chapter 8, the electron-phonon interaction being moderately strong and nonlinear is responsible for pairing in cuprates, thus for the pairing energy scale. At the same time, the phase coherence which occurs/disappears at T_c is not mediated by phonons (or the wave-function coupling). As a consequence, the two energy scales do not correlate with each other, as shown in Fig. 3.22, and have different temperature dependences.

Figure 9.3 shows two temperature dependences of quasiparticle peaks in tunneling conductances, obtained in slightly overdoped Bi2212 single crystals: in-plane and out-of-plane. Both the tunneling conductances are measured in superconductor-insulator-superconductor (SIS) tunneling junctions. The in-plane tunneling measurements will be considered in detail in Chapter 12. Tunneling measurements along the c-axis were performed on micron-size mesas [26] and, represent *intrinsic* properties of Bi2212. In Fig. 9.3, one can see that the interlayer coherence disappears/appears at T_c, while quasiparticles remain paired in the CuO_2 planes.

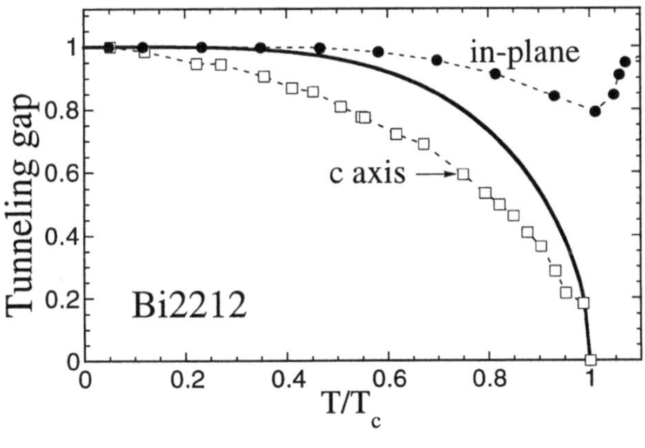

Figure 9.3. Temperature dependences of in-plane (dots) and c-axis (squares) tunneling conductance peaks in slightly overdoped Bi2212 single crystals, $\Delta(T)/\Delta(T_{min})$. The BCS temperature dependence is shown by the solid line. The c-axis data are obtained by micron-size mesas [26]. The in-plane tunneling data are discussed in Chapter 12.

3.2 Magnetic properties

The magnetic properties of cuprates have already been presented in Chapter 3. Here and in the following two Subsections, evidence for the presence of magnetically-mediated superconductivity in cuprates are considered.

In LSCO, there exists direct evidence that superconductivity is intimately related to the establishment of antiferromagnetic order along the c axis: μSR measurements performed in non-superconducting Eu-doped LSCO [27] show that, at different doping levels, the superconducting phase of pure LSCO is replaced in Eu-doped LSCO by a second antiferromagnetic phase, as depicted in Fig. 9.4. Thus, in LSCO, it is possible to switch the entire doping-dependent phase diagram from superconducting to antiferromagnetic. Since LSCO is a layered compound, the main antiferromagnetic phase of Eu-doped LSCO and its second antiferromagnetic phase develop along the c axis. Hence, the superconducting phase of pure LSCO is replaced in Eu-doped LSCO by the antiferromagnetic phase which arises along the c axis. This clearly indicates that superconductivity in LSCO is intimately related to the onset of long-range antiferromagnetic order along the c axis.

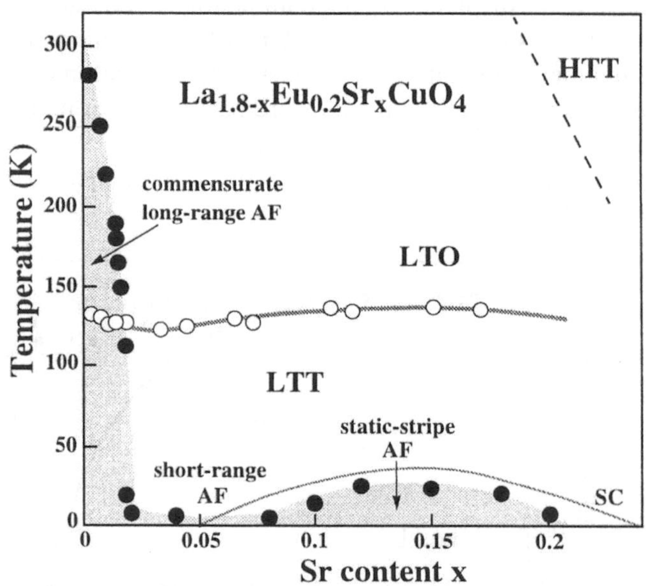

Figure 9.4. Phase diagram of $La_{1.8-x}Eu_{0.2}Sr_xCuO_4$ obtained by μSR measurements [27]. Full and open circles denote the magnetic and the structural transition temperatures, respectively. The superconducting phase in pure LSCO is marked by "SC". The structural transition from the high-temperature tetragonal (HTT) to the low-temperature orthorhombic (LTO) is indicated by the dashed line. The low-temperature tetragonal phase (LTT) appears below 120–130 K (AF = antiferromagnetic).

Infrared reflectivity measurements performed in high-quality single crystals of LSCO show that the superconducting transition is accompanied by the onset of coherent charge transport along the c axis, which is blocked above T_c [28]. Thus, in LSCO, the long-range phase coherence occurs at T_c along the c axis.

In superconducting YBCO, the antiferromagnetic ordering starts to develop above 300 K [29–31]. An antiferromagnetic commensurate ordering with a small moment was observed in underdoped [29,30] and optimally doped [31] YBCO. The magnetic-moment intensity *increases* in strength as the temperature is reduced below T_c, as shown in Fig. 9.5. The magnetic-moment direction is found to be, in one study, along the c axis [29] and in-plane in the other [30]. In heavy-fermion URu_2Si_2, the magnetic moment found below a Néel temperature of T_N = 17.5 K is oriented along the c axis [25].

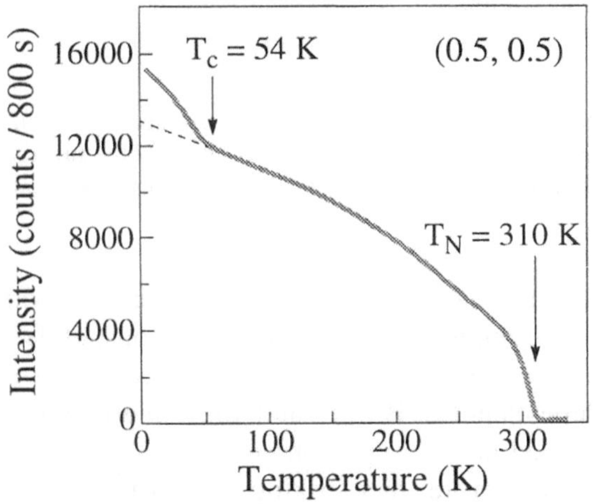

Figure 9.5. Temperature dependence of the magnetic intensity in underdoped YBCO (T_c = 54–55 K), measured at the wave vector **Q** = (0.5, 0.5) by polarized and unpolarized neutron beams [30]. The antiferromagnetic order appears at a Néel temperature of $T_N \simeq 310$ K. The dashed line shows the background.

In electron-doped NCCO, the results of recent μSR and magnetic susceptibility measurements suggest that the superconducting phase (see Fig. 3.8) enters into the antiferromagnetic region [32]. In this case, the phase diagram of NCCO looks similar to that of Fig. 9.1a, suggesting that superconductivity in NCCO is mediated by spin fluctuations.

The phase diagram of Bi2212 shown in Fig. 3.34 is depicted in Fig. 9.6 with additional data obtained in the heavy fermion $CePd_2Si_2$ [16]. In Fig. 9.6, the data obtained in $CePd_2Si_2$, which are also schematically shown in Fig. 9.1a, are scaled to fit the $T_c(p)$ dependence of Bi2212. The vertical scaling factor is 252, while it is impossible to define the horizontal scaling factor because the two origins on the horizontal axis in Fig. 9.6 do not coincide. In these two

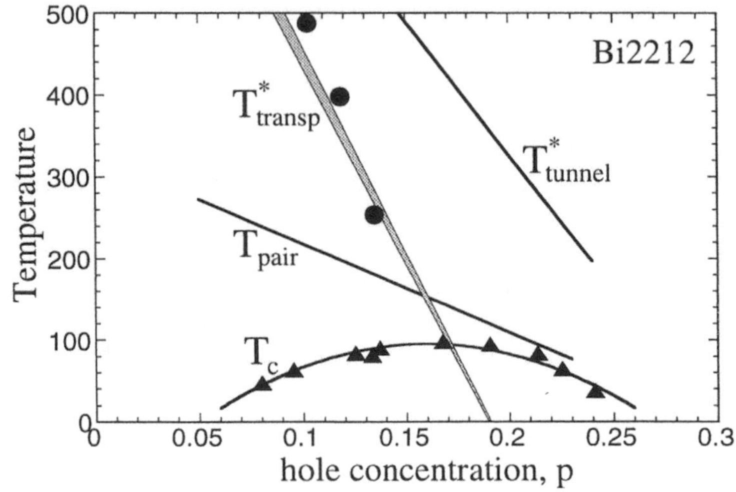

Figure 9.6. The phase diagram of Bi2212 from Fig. 3.34 and the scaled data obtained in the superconducting heavy fermion $CePd_2Si_2$ [16]. The values of the superconducting (triangles) and magnetic (dots) transition temperatures in $CePd_2Si_2$ are scaled vertically by a factor of 252.

cases, the variations of charge-carrier density are achieved differently: chemically in Bi2212 and by pressure in the heavy fermion $CePd_2Si_2$.

In Fig. 9.6, one can see that the $T_c(p)$ dependence of $CePd_2Si_2$ is very similar to the $T_c(p)$ dependence of Bi2212 and, thus, of any other superconducting cuprate. As discussed above, superconductivity in $CePd_2Si_2$ is mediated by spin fluctuations. Therefore, this striking similarity suggests that the phase coherence in cuprates, which sets in at T_c, is mediated by spin fluctuations. Another important conclusion which can be drawn from the data shown in Fig. 9.6 is that the pseudogap with the characteristic temperature T^*_{transp} has a magnetic origin. As a matter of fact, the magnetic origin of this pseudogap was recently established in YBCO by μSR measurements [33].

3.3 Phase-coherence properties

We now compare magnetic and phase-coherence characteristics of YBCO, Bi2212 and LSCO. The comparison will show that magnetic and coherence superconducting properties of the cuprates intimately relate to each other.

Figure 9.7a shows three temperature dependences of the superfluid density, $n_s(T)$, in near optimally doped single crystals of Bi2212 and YBCO, and in an overdoped LSCO (x = 0.2) single crystal as measured by microwave, μSR and ac-susceptibility techniques, respectively. The temperature dependences of the superfluid density are derived from the relation $n_s(T) \propto 1/\lambda^2(T)$, where $\lambda(T)$ is the magnetic penetration depth. Figure 9.7b shows three temperature

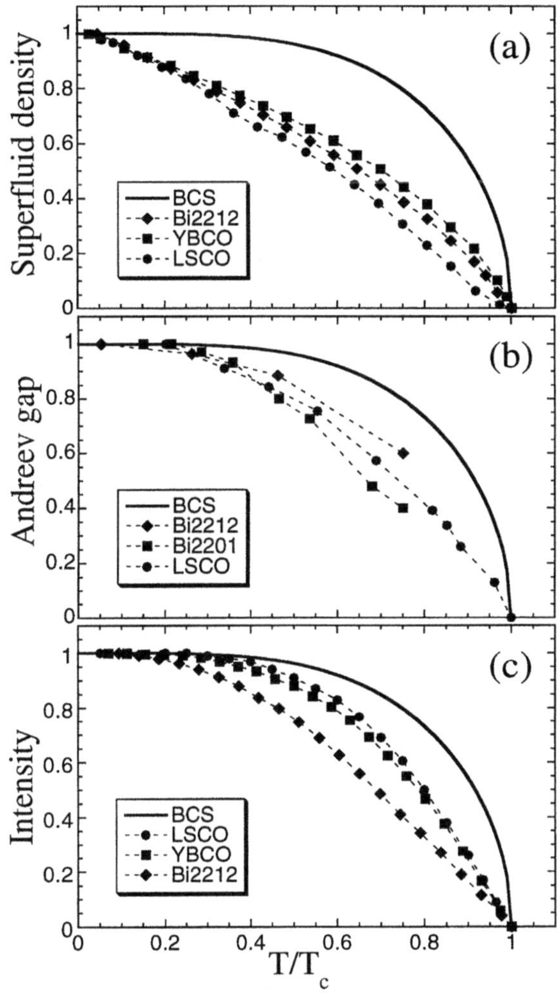

Figure 9.7 (a) Temperature dependences of the superfluid density in near optimally doped single crystals of Bi2212 (T_c = 93 K) and YBCO (T_c = 93 K), and in an overdoped LSCO (x = 0.2) single crystal (T_c = 36 K). (b) Temperature dependences of Andreev-reflection gap, $\Delta(T)/\Delta(T_{min})$, in an overdoped Bi2212 thin film (T_c = 80 K), and in overdoped single crystals of Bi2201 (T_c = 29 K) and LSCO (x = 0.2) (T_c = 28 K). (c) Temperature dependences of the peak intensity of the incommensurate elastic scattering in LSCO (x = 0) (T_c = 42 K) and the intensity of the magnetic resonance peak measured by INS in near optimally doped Bi2212 (T_c = 91 K) and YBCO (T_c = 92.5 K). The neutron-scattering data are averaged, the real data have a vertical error of the order of ±10–15%. In all plots, the BCS temperature dependence is shown by the thick solid line. The plot is taken from [24].

dependences of Andreev-reflection gap measured in an overdoped Bi2212 thin film, and in overdoped single crystals of Bi2201 and LSCO (x = 0.2). It is worth reminding that Andreev-reflection measurements reflect the coherence property of the condensate (see Chapter 2). Figures. 9.7a and 9.7b exhibit the good agreement found among temperature dependences of the coherence superconducting characteristics of these cuprates.

Figure 9.7c shows three temperature dependences of the peak intensity of the incommensurate elastic scattering in LSCO (x = 0) and the intensity of the commensurate resonance peak measured by INS in near optimally doped Bi2212 and YBCO. The data shown in Fig. 9.7c exclusively represent the *magnetic* properties of the cuprates.

In Fig. 9.7, all the temperature dependences exhibit below T_c a striking similarity. Moreover, they are all similar to the temperature dependence of

c-axis quasiparticle peaks in Bi2212, shown in Fig. 9.3, but different from the temperature dependence of in-plane quasiparticle peaks, also shown in Fig. 9.3. Thus, the data depicted in Figs. 9.7 and 9.2 clearly reveal that in cuprates the coherence superconducting and magnetic properties are intimately related to each other along the c axis.

Another important conclusion which can be drawn from the data shown in Figs. 9.7 and 9.2 is that the in-plane mechanism of superconductivity, at least, in Bi2212 has no or very little relation to magnetic interactions along the c axis.

3.4 Magnetic resonance peak

The magnetic resonance peak observed in YBCO, Bi2212 and Tl2201 was discussed in detail in Chapter 3. Figure 9.8 shows the phase diagram of the two energy scales in cuprates, depicted in Fig. 3.22. In Bi2212 and YBCO, the phase-coherence energy scale, Δ_c, is approximately proportional to T_c as $2\Delta_c \simeq 5.4k_BT_c$. Figure 9.8 also shows the energy position of the magnetic resonance peak, E_r, as a function of doping in Bi2212, YBCO (see [24] and references therein) and Tl2201 [34].

In Fig. 9.8, one can observe that, at different dopings, $E_r \simeq 2\Delta_c$. This relation suggests that spin excitations are responsible for establishing the phase coherence in these cuprates since such a relation is in good agreement with the theories of superconductivity mediated by spin fluctuations [35]. Thus, the relation $E_r \simeq 2\Delta_c$ indicates that the magnetic resonance peak observed in these cuprates is caused by a spin excitation which mediates superconductivity. Another recent experiment [36] also points in the same direction: in YBCO, modest magnetic fields applied below T_c significantly suppresses the intensity of the magnetic resonance peak. In other words, the effect of an external magnetic field is to disrupt the superconducting coupling mediated by the resonance mode.

As discussed in Chapter 3, a commensurate resonance mode was not observed in LSCO—spin fluctuations in LSCO are characterized by incommensurate peaks. We shall consider the case of LSCO later in Chapter 10.

3.5 Tunneling assisted by spin excitations in Bi2212

Electron tunneling assisted by phonons is well known. Tunneling can be also assisted by magnetic (spin) excitations [37]. In Fig. 9.6, one can see that, in Bi2212 having a hole concentration $p \geq 0.2$, pairing occurs just above T_c. Therefore, the double magnitude of the pairing gap in Bi2212 is approximately equal to the energy position of the magnetic resonance peak, i.e. in the heavily overdoped region, $2\Delta_p \approx 2\Delta_c = E_r$. Since, in a superconductor-insulator-superconductor (SIS) junction, electron tunneling occurs at double magnitude

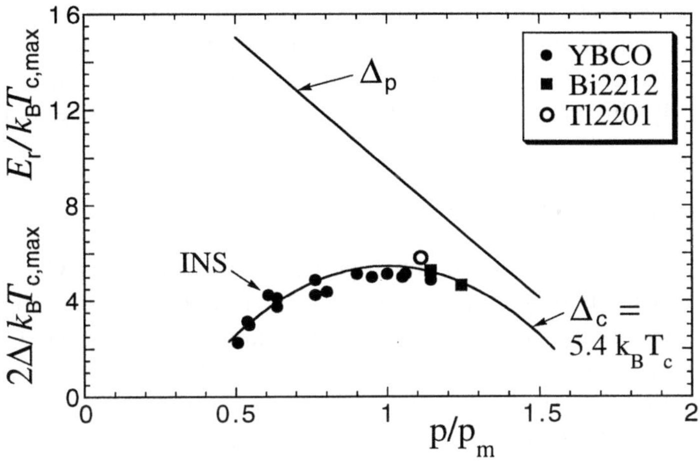

Figure 9.8. The phase diagram of cuprates from Fig. 3.22 and the energy position of the magnetic resonance peak, E_r, in Bi2212 (squares) YBCO (dots) and Tl2201 (circle) at different hole concentrations (p_m = 0.16). The plot is taken from [24], and the point for Tl2201 from [34].

of the energy gap, it is natural to expect that, in an SIS junction of Bi2212 having a doping level $p \geq 0.2$, spin excitations which cause the appearance of the resonance peak in neutron spectra can assist electron tunneling. This will lead to the appearance of additional structures in tunneling $dI(V)/dV$ and $I(V)$ characteristics.

In Fig. 6.5a, one can clearly see that the quasiparticle peaks in the conductances obtained in an SIS junction of Bi2212 with $p \simeq 0.2$ are composite: there is an additional contribution at the top of the peaks. The corresponding $I(V)$ characteristics also have a contribution of unknown origin at a bias value of 35 mV, as shown in Fig. 6.9a. In Figs. 6.5a and 6.9a, at temperatures $T \leq$ 42 K, this contribution appears at bias $V \simeq \pm 2\Delta_p/e$, where e is the electron charge. From the discussion presented in the previous paragraph, it is reasonable to associate this additional contribution to the tunneling spectra with electron tunneling assisted by spin excitations which cause the appearance of the magnetic resonance peak. In contrast, in tunneling junctions, phonons influence electron tunneling exclusively outside the gap structure, i.e. at bias $|V| > 2\Delta/e$.

If this interpretation is correct, then the tunneling data in Figs. 6.4b and 6.9a represent additional evidence that spin fluctuations mediate the long-range phase coherence in Bi2212. First, this shows that charge carriers are strongly coupled to spin excitations. Second, by comparing the values of Josephson I_cR_n product in the $I(V)$ characteristics shown in Figs. 6.3b and 6.4b (7.5 mV and 24.5 mV, respectively), the high value of Josephson product in Fig.

6.4b can be only caused by the contribution from electron tunneling assisted by this spin excitation: This contribution is indeed the *only* difference between the spectra depicted in Figs 6.3b and 6.4b (also compare Figs. 6.8c and 6.9a). Consequently, this means that spin fluctuations mediate the phase coherence in Bi2212.

3.6 Pr-doped YBCO

The anomalous properties of non-superconducting $PrBa_2Cu_3O_{6+x}$ (PrBCO) are well known. Its magnetic properties are strikingly different from those of other members of the RBCO family (R is a rare earth element). Long-range antiferromagnetic order in both the Cu and Pr sublattices persists along with semiconducting resistivity over the entire range of oxygen doping $0 < x < 1$. The magnetic transition in the Pr sublattice occurs at a temperature T_{Pr} an order of magnitude higher than that of other rare earths in RBCO.

Inelastic neutron scattering measurements performed on PrBCO single crystals show that the Pr-Pr interaction is unexpectedly strong which causes T_{Pr} to be substantially higher than the rare-earth ordering temperature in other RBCO [38]. In addition, the interlayer Cu-Cu exchange (i.e. along the c axis) in PrBCO is reduced by a factor larger than 2 relative to YBCO. Therefore, in PrBCO, the Cu-Cu spin fluctuations are not able to mediate interlayer phase coherence.

In the bilayer manganite $La_{1.2}Sr_{1.8}Mn_2O_7$, the Pr doping produces a similar effect [39].

3.7 Theory

In a model developed for cuprates [40, 41], which takes into account a competition between interlayer direct hopping and hopping assisted by spin fluctuations, the calculations show that, at least, in the underdoped region, the interlayer hopping assisted by spin fluctuations is dominant. Therefore, the interlayer direct hopping can be omitted [40, 41]. The model captures the main features of experimental data, for example, the anomalous behavior of the c-axis electronic conductivity in YBCO [41] and thermoelectric power in LSCO [40]. Therefore, this provides additional evidence that, in cuprates, spin fluctuations affect the c-axis transport properties and, are responsible for the anomalous c-axis transport properties in the underdoped cuprates [40, 41].

3.8 Concluding remarks

In superconducting cuprates, spin fluctuations mediate the long-range phase coherence and, in many aspects, are responsible for their anomalous properties. From Figs. 9.8 and 9.6, it is evident that the phase-coherence scale (gap) Δ_c has a magnetic origin. In addition, this energy scale is predominantly the c-

axis energy scale. At the same time, the data in Fig. 9.3 indicate that the in-plane pairing mechanism of superconductivity in Bi2212 has no or very little relation to the magnetic interactions, at least, along the c axis. In other words, the pairing gap Δ_p has a non-magnetic origin. Such a situation requires that, in Bi2212, the magnetic and the pairing order parameters are coupled to each other below T_c.

In cuprates, the doping level of $p = 0.19$ is the quantum critical point. Small islands with intact antiferromagnetic order completely vanish at this doping level $p = 0.19$, as shown in Fig. 3.16a.

In heavy-fermion and organic superconductors, the long-range phase coherence is also mediated by spin fluctuations but the pairing mechanism is apparently different.

Chapter 10

THE MECHANISM OF HIGH-T_C SUPERCONDUCTIVITY

A good theory of any complex system must be only a good "caricature" of the system, which exaggerates its typical properties while intentionally ignoring other, less important ones.

—Ya. I. Frenkel

The purpose of this Chapter is to present the mechanism of high-T_c superconductivity discovered in 1986 by J. G. Bednorz and K. A. Müller. This Chapter does not introduce the *exact* theory of superconductivity in cuprates, but concentrates on the *physics* of high-T_c superconductors. For a reader who has attentively read the previous Chapters, the mechanism of high-T_c superconductivity should be more or less obvious. This Chapter can be read independently of the preceding Chapters; however, it is better to start with Chapter 3.

As discussed in Chapter 8, the Cooper pairs in cuprates are not Davydov's bisolitons in their *initial* sense; however, we shall call them "bisolitons." Such a designation is the best choice because the Cooper pairs in cuprates consist of quasi-one-dimensional soliton-like excitations (see Chapter 6), and it exactly reflects the origin of the pairs.

The Chapter is organized as follows. A general description of the mechanism of high-T_c superconductivity is presented first, and then we shall consider separately all "ingredients" of the mechanism: element by element. At the end of this Chapter, we shall discuss superconductivity in organic and heavy-fermion superconductors.

1. A general description of the mechanism

Superconductivity in copper oxides (cuprates) is unconventional. In spite of this, the cuprates display a number of properties that are in many ways similar to those of conventional BCS superconductors. First, superconductivity in cuprates occurs due to electron pairing. Secondly, the electron pairing results in the appearance of an energy gap in the electronic excitation spectrum. Thirdly, the isotope effect is also present in cuprates; however, it is strongly dependent on hole concentration. Thus, the basic principles underlying the phenomenon of superconductivity in different materials are the same.

First, it is important to recall that, in conventional BCS superconductors, the pairing mechanism and the mechanism for the establishment of long-range phase coherence are different: the electron-phonon interaction is responsible for electron pairing, and the phase coherence among the pairs is established due to the overlap of their wave functions. In classical superconductors, the phase coherence can be mediated by the overlap of the Cooper-pair wave-functions since the average distance between Cooper pairs is much smaller than the coherence length (the size of a Cooper pair).

Cuprates become superconducting when they are slightly doped either by electrons or holes. Undoped cuprates are antiferromagnetic Mott insulators. The doping can be achieved chemically, by pressure, or in a field-effect transistor configuration. The crystal structure of copper oxides is highly anisotropic. Superconductivity occurs in two-dimensional copper-oxide planes. The CuO_2 layers are always separated by layers of other atoms, which are usually insulating or semiconducting. In chemical doping, these layers provide the charge carriers into CuO_2 planes, therefore they are often called charge reservoirs.

In cuprates, doped holes (electrons) are not distributed homogeneously into CuO_2 planes but forced to form quasi-one-dimensional charge stripes. The moderately strong, nonlinear electron-phonon interaction is *mainly* responsible for the formation of charge stripes. Charge stripes are the manifestation of self-trapped states. In superconducting cuprates, the stripes are half-filled and run along –O–Cu–O–Cu– bonds (see Fig. 3.2b). Charge stripes are separated by two-dimensional insulating antiferromagnetic stripes, as shown in Fig. 3.28a. Charge stripes are *dynamical*: they meander and can move in the transverse direction. Basically, charge stripes are insulating, i.e. there is a charge gap on the stripes. However, the presence of soliton-like excitations on the stripes makes them conducting. These soliton-like excitations are *one-dimensional polarons*, or *polaronic solitons*. They propagate at the Fermi level, thus inside the charge gap. On cooling, the polaronic solitons give rise to pairs coupled in a singlet state due to local deformation of the lattice. Thus, moderately strong, nonlinear electron-phonon interaction is responsible for electron pairing in cuprates. The pairs of polaronic solitons—bisolitons—are formed above T_c, and reside on charge stripes. The pairing occurs in the momentum space, not in a real space.

Figure 10.1 schematically shows the static picture of the striped phase of CuO_2 planes. The Cooper pairs (bisolitons) are formed on charge stripes at T_{pair}. The long-range phase coherence is established at a critical temperature T_c due to spin fluctuations in the antiferromagnetic stripes. The temperature dependences of the pairing, Δ_p, and phase-coherence, Δ_c, energy gaps (magnitudes of the order parameters) are schematically shown in Fig. 10.1. At any doping level, $\Delta_p > \Delta_c$ always. Below T_c, the superconducting pairing order parameter and the magnetic order parameter are coupled to each other.

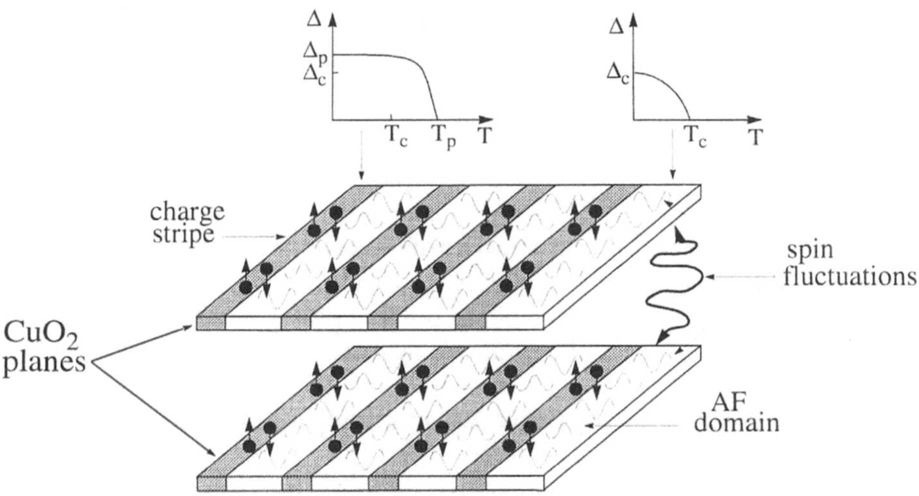

Figure 10.1. Naive (static) picture of superconductivity in cuprates near a doping level of $p \simeq 0.19$. In CuO_2 planes, the Cooper pairs are formed on charge stripes which are separated by insulating antiferromagnetic stripes. The phase coherence between the planes is mediated by spin fluctuations in the magnetic regions. The two plots show the temperature dependences of the pairing, Δ_p, and phase-coherence, Δ_c, energy gaps. See text for more details.

In all cuprates, the phase coherence is mediated by spin fluctuations. In different cuprates, however, the spin-fluctuation mechanism is slightly different. Let us classify all superconducting cuprates into two categories: *40 K cuprates* and *90 K cuprates*. All optimally doped cuprates having the T_c value less than 40 K belong to the first group, while with the T_c value more than 90 K to the second group. *De fait*, the 40 K cuprates have one CuO_2 layer per unit cell. The main difference between the two groups, which defines the T_c value, is the absence/presence of the so-called magnetic resonance peak in inelastic neutron scattering (INS) spectra. The phase coherence in 40 K cuprates is locked at T_c via a *slow*-oscillating spin-density-wave order. In 90 K cuprates, the phase coherence is mediated by a *fast* spin excitation which causes the appearance of the magnetic resonance peak in INS spectra.

The process of bisoliton condensation at T_c, as a matter of fact, is the Bose-Einstein condensation. Below T_c, the charge, spin and lattice degrees of freedom in the CuO_2 planes are coupled.

In first approximation, high-T_c superconductivity in cuprates can be described by a combination of two theoretical models: the bisoliton [1] and spin-fluctuation [2] theories. Some pairing characteristics of superconducting cuprates can be estimated by using the bisoliton model, while the spin-fluctuation model describes phase-coherence characteristics.

2. Important elements of high-T_c superconductivity

In this Section, we discuss in detail important elements of superconductivity in cuprates. Most of them were already considered in the previous Chapters. We start with the mechanism of electron pairing in cuprates.

2.1 Pairing mechanism

The electron pairing is a necessary condition for any type of superconductivity, and occurs in the momentum space. Electrons are fermions, while two coupled electrons represent a composite boson with a spin of $S = 0$ or 1. Since fermions and bosons obey different statistics, the electron pair behaves differently as a single electron. In conventional superconductors, only a 10^{-4} part of the electrons located near the Fermi surface participate in pairing. In cuprates, about 10% of all conduction electrons (holes) form the Cooper pairs. In addition to electron pairing, superconductivity also requires the phase coherence. The mechanism of phase coherence in cuprates will be considered in Section 2.3.

2.1.1 Phonons in cuprates

The absence of a substantial isotope effect on the transition temperature in optimally doped cuprates (see Fig. 3.17) has been used to rule out phonons as the bosons responsible for coupling the charge carriers. In fact, phonons are an essential part of the mechanism of high-T_c superconductivity. Simply, the electron-phonon interaction is able to provide, at least, two different mechanisms of electron pairing: *linear* and *nonlinear*. The electron-phonon interaction, that provides the electron pairing in conventional BCS superconductors, is linear and weak. In cuprates, the electron-phonon interaction is moderately strong and nonlinear.

It is a paradox: the isotope effect on the transition temperature is manifest itself when the electron-phonon interaction is weak, and can vanish as the electron-phonon interaction becomes stronger! The main result of the bisoliton model, considered in Chapter 7, is that the potential energy of a self-trapped electron state, which occurs due to local deformation of the lattice, does not depend on the mass of elementary lattice cell. This mass appears only in the kinetic energy of the self-trapped electron.

In solids, an electron or a hole can get self-trapped if the electron-phonon interaction is strong enough and nonlinear (see Chapter 5). A doped electron (hole) locally creates a lattice deformation which, in turn, attracts the electron (hole). In the same manner, two electrons (holes) can be paired in a singlet state due to the interaction with local lattice deformation created by them. The potential-well formed by a short-range deformation interaction of one electron (hole) attracts another electron (hole) which, in turn, deepens the well. Since the strength of the electron-phonon interaction in crystals is about 5 times weaker than the strength of the hole-phonon interaction (see Chapters 3 and 8), the electrons seldom relax into self-trapped states.

So, phonons interact with charge carriers in conventional superconductors and in cuprates in a different way. This can be shown by a few examples: Figure 10.2 depicts the logarithmic plot of pairing energy gap Δ_p versus the ratio ω_{ph}/E_F for different superconductors, where E_F is the Fermi energy and ω_{ph} is the energy of the highest phonon peak in the spectral function $\alpha^2F(\omega)$. The function $\alpha^2F(\omega)$ is the parameter of the electron-phonon interaction in the Eliashberg equations (see Chapter 2). The spectral functions for Nb and Bi2212 are shown in Figs. 2.6 and 8.6, respectively. In Fig. 10.2, one can clearly see that the interactions between phonons and charge carriers in conventional and unconventional superconductors lead to the formation of energy gaps with different magnitudes. In unconventional superconductors, the energy of phonons coupled to charge carriers, relative to E_F, is almost two orders of magnitude higher than that in conventional superconductors. Since the strength of the electron-phonon interaction in crystals is about 5 times weaker than the strength of the hole-phonon interaction, the magnitudes of pairing gaps in *hole*- and *electron*-doped unconventional superconductors also differ from each other approximately by the same value, as shown in Fig. 10.2. In unconventional superconductors, it is necessary to consider the pairing gap Δ_p rather than the T_c value because, in most unconventional superconductors, the T_c value and the magnitude of pairing gap do not correlate with each other. All superconductors shown in Fig. 3.20 (the Uemura plot) should fall within the two regions of unconventional superconductors in Fig. 10.2.

Consider other examples of different manifestations of the electron-phonon interaction in cuprates and in conventional superconductors: In metallic superconductors, the critical temperature T_c increases with lattice softening while in cuprates T_c increases with lattice stiffening, as shown in Fig. 3.24. As discussed in Chapter 3, the lattice and charge dynamics are inexorably mixed in superconducting cuprates, and structural phase transitions precede the pseudo-

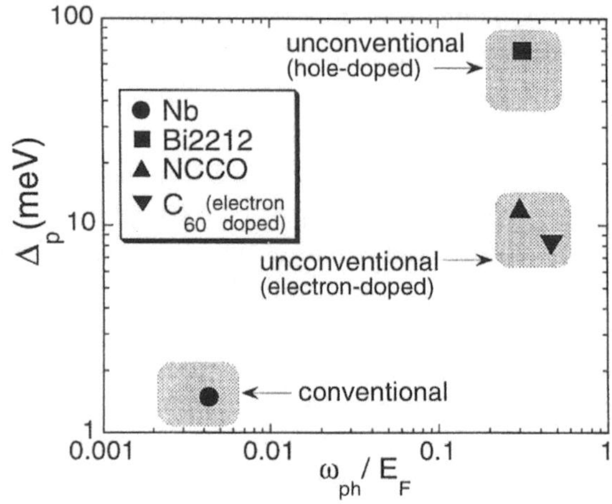

Figure 10.2. Pairing energy gap Δ_p versus the ratio ω_{ph}/E_F for various superconductors, where E_F is the Fermi energy, and ω_{ph} is the energy of the highest phonon peak in the Eliashberg function $\alpha^2F(\omega)$. The values of Δ_p for Bi2212 and NCCO are taken from the underdoped region.

gap formation and the transition into the superconducting state. Charge carriers in hole- and electron-doped cuprates are coupled not only to acoustic phonons, as in conventional superconductors, but also to optical phonons, in particular, to the 55 and 70 meV branches. In cuprates as well as in nickelates and manganites, phonons are responsible for the formation of charge stripes (see the next Section). All these facts provide evidence for extremely large electron-lattice coupling in cuprates (as well as in nickelates and manganites). Thus, the electron-phonon interaction in superconducting cuprates is different from the electron-phonon interaction in conventional superconductors.

The lattice involvement in the mechanism of high-T_c superconductivity in cuprates was shown in Chapter 3. The isotope effect in underdoped cuprates provides evidence for phonon participation in electron pairing: Figure 3.17 shows the oxygen (^{16}O vs ^{18}O) isotope-effect coefficient $\alpha_O \equiv d\ln(T_c)/d\ln(M)$ as a function of doping level, where M is the isotopic mass. In Fig. 3.17, one can see that, in underdoped LSCO and YBCO, there is a huge isotope effect. If phonons were not involved in electron pairing in these cuprates, the isotope effect should be absent or small.

In cuprates, three phonon branches are important for superconductivity: 20, 55 and 70 meV. The 20 meV phonons are acoustic. The other two phonon branches are associated with oxygen vibrations: the 55 meV phonons are attributed either to in-plane or out-of-plane Cu–O bond-bending (buckling) vibrations. The 70 meV branch is associated with half-breathing-like oxygen phonon modes that propagate in the CuO_2 plane. At this stage, the role of each phonon branch in the mechanism of superconductivity in cuprates is still undetermined. Usually, self-trapped states in crystals occur due to an interaction of quasiparticles with acoustic phonons.

2.1.2 Charge stripes

In nickelates, manganites and cuprates, the distribution of charge carriers doped into NiO_2, MnO_2 and CuO_2 planes, respectively, is inhomogeneous on a microscopic scale: they form quasi-one-dimensional charge stripes. Such a type of doping is called *topological.* Charge stripes in nickelates and manganites fluctuate in time and in space slowly, while in cuprates very fast. Therefore, it is much easier to observe the charge inhomogeneity in nickelates and manganites than in cuprates. As a consequence, there is ample evidence for charge stripes in nickelates and manganites in the literature (see Chapter 3). Nevertheless, there is also direct evidence for dynamical charge stripes in superconducting cuprates, such as LSCO [3], YBCO [4] and Bi2212 [5].

In cuprates, a structural phase transition precedes charge ordering, thus the lattice is responsible for the formation of charge stripes, not magnetic correlations, as shown in Fig. 3.29. The stripe phase is charge driven and the spin order between charge stripes is only subsequently enslaved. In nickelates and manganates, a structural phase transition, observed in acoustic measurements, also precedes charge ordering. Magnetic ordering in nickelates, manganites and cuprates occurs after the formation of charge stripes and, at low temperature, the charge, spin and lattice degrees of freedom are coupled. Due to a balanced interplay among the charge, spin and lattice degrees of freedom, the average distance between charge stripes in cuprates remains constant above $p \simeq 0.13$.

In nickelates, manganites and cuprates, charge stripes are insulating on a microscopic scale, i.e. there is a charge-density-wave order on the stripes. This charge-density-wave order is not conventional. On a macroscopic scale, however, charge stripes are conducting due to soliton-like excitations present on the stripes. Therefore, in-plane charge transport in cuprates is quasi-one-dimensional [6, 7], as well as in nickelates and manganites. As an example, Figure 10.3a shows the temperature dependences of longitudinal ρ_a and transverse ρ_c resistivities measured in *quasi-one-dimensional* $(TMTSF)_2PF_6$ organic conductor (the Bechgaard salt), where TMTSF and PF_6 stand for tetramethyltetraselenafulvalene and hexafluorophosphate, respectively. In Fig. 10.3a, the steep rises in ρ_a and ρ_c at low temperatures are due to the transition from a metal to an insulator, and occur at different temperatures: $T_{\rho_a} < T_{\rho_c}$. At high pressure, the Bechgaard salt becomes superconducting below 10 K.

Figure 10.3b shows the temperature dependences of in-plane ρ_{ab} and out-of-plane ρ_c resistivities obtained in undoped $TmB_2Cu_3O_{6.37}$ (TmBCO) single crystal. A visual inspection of Figs. 10.3a and 10.3b shows a striking simi-

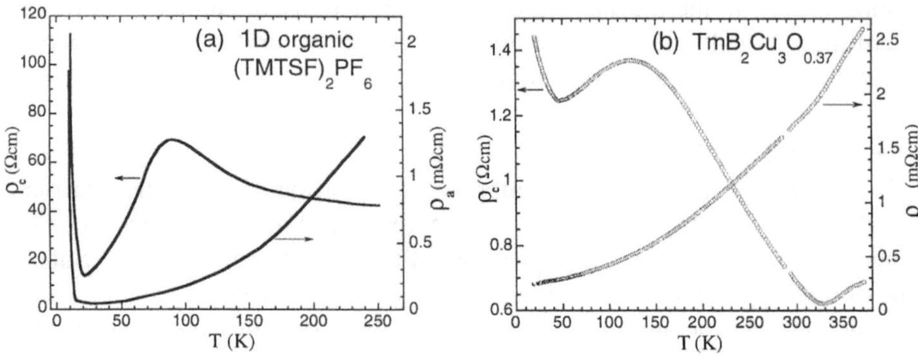

Figure 10.3. (a) Temperature dependence of longitudinal ρ_a and transverse ρ_c resistivities measured in one-dimensional (1D) $(TMTSF)_2PF_6$ organic conductor. (b) Temperature dependence of in-plane ρ_{ab} and out-of-plane ρ_c resistivities obtained in an undoped TmBCO ingle crystal. The plot is taken from [7].

larity between the two plots, meaning that, below 327 K, the in-plane electric transport in TmBCO is quasi-one-dimensional. In Fig. 10.3b, the in-plane resistivity ρ_{ab} does not show the rise at low temperature: the minimum temperature available in these measurements was not sufficiently low to observe the rise in ρ_{ab} [7].

Charge-stripe ordering in cuprates, nickelates and manganites is evidence for extremely large, nonlinear electron-lattice coupling: charge carriers are self-trapped. Bishop and co-workers [8] theoretically reproduced some experimental data obtained in cuprates by considering the phonon mechanism of charge-stripe ordering and assuming that the electron-phonon interaction is strong and nonlinear. In their model, one term in a Hamiltonian, which accounts for nonlinear effects, is taken from the Su-Schrieffer-Heeger model. The latter describes topological solitons in polyacetylene (see Chapter 5). In the calculations, Bishop and co-workers considered phonons with an energy from a frequency interval located on the border of acoustic and optical branches of lattice vibrations. This frequency range corresponds to in-plane Cu–O bond-bending vibrations. They showed that their model is self-consistent, and the strong nonlinear electron-phonon interaction alone is enough to account for experimental data.

Reading a book called "Spatial Solitons" which is entirely devoted to a description of self-trapped solitons in optical waveguides, I was surprised by one chapter of the book. The authors of this chapter discuss charge stripes in cuprates [9]. Being specialists in self-trapped states in optics, their expertise does not extend to the charge stripes in CuO_2 planes, but for them, it is more or less obvious that the charge stripes in cuprates are the manifestation of self-trapped electronic states. It is a gift of Nature that, in optics, self-trapped states can be studied directly: every little detail of the physics involved can be observed visually. In solid state physics, however, this is not the case.

It is assumed that the appearance of four incommensurate peaks (black circles and squares in Fig. 3.27a) in INS spectra is the consequence of stripe orientation in adjacent CuO_2 layers: the stripe orientation is alternately rotated by 90° layer by layer, as schematically shown in Fig. 3.28b. In fact, this assumption is not necessary. The charge-stripe phase is not distributed homogeneously in CuO_2 planes: At any doping level, the islands with the charge-stripe phase always coexist with islands containing either the intact antiferromagnetic phase, if $p < 0.19$, or the Fermi-sea phase, if $0.19 < p$. As a consequence, the charge-stripe orientation can be rotated by 90° domain by domain, as schematically depicted in Fig. 10.4.

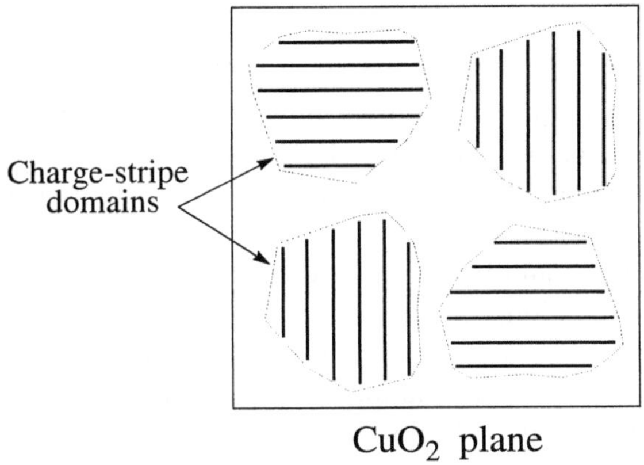

Figure 10.4. Sketch of charge-stripe domains in a CuO_2 plane, having two possible orientations in the same CuO_2 plane.

Charge stripes in cuprates are half-filled, that is, one positive electron charge per two copper sites along the stripes. Figure 10.5 shows two different types of charge stripes: full-filled and half-filled. In nickelates, charge stripes are full-filled. For half-filled charge stripes, there are two possible types of charge ordering: $4k_F$ and $2k_F$, where k_F is the momentum at the Fermi surface. Analysis of the data obtained in INS experiments suggests that the stripes in cuprates are of the $2k_F$ type [10]. The chains in YBCO have also the $2k_F$ charge modulation, as observed by tunneling spectroscopy [11]. The $2k_F$ ordering pattern on a stripe is schematically shown in Fig.10.5. In Bi2212, charge stripes are aligned with the Cu–O bonds, and the length of a separate stripe is about 100 Å [5]. Simulations of angle-resolved photoemission data obtained in Bi2212 show that the stripes in Bi2212 are most likely site-centered [12]. So, we consider only the $2k_F$ site-centered stripes.

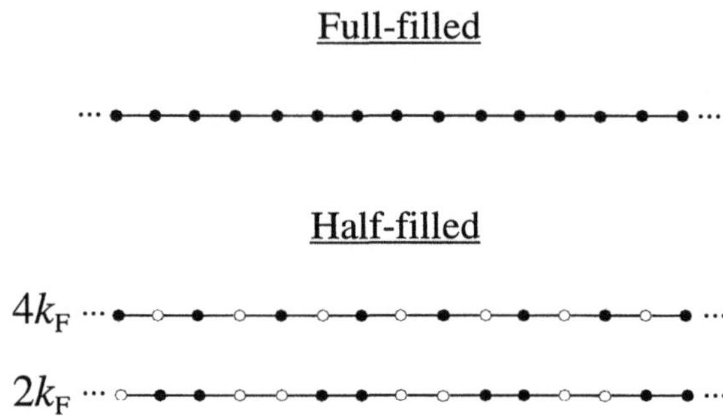

Figure 10.5. Schematic ordering patterns on full-filled and half-filled stripes. The half-filled stripes have two ordering patterns: $4k_F$ and $2k_F$. In plot, • (∘) denotes the presence (absence) of a hole.

2.1.3 Charge-stripe excitations and Cooper pairs

Charge stripes are quasi-one-dimensional. There are a few theories predicting the separation charge and spin excitations on the stripes. Experimental data obtained in underdoped YBCO show that there is no spin-charge separation in YBCO [13].

Polaronic solitons and bisolitons reside on charge stripes, thus they represent nonlinear excitations of the charge stripes. What type of charge-stripe excitations causes superconductivity in cuprates? To answer to this question, let us consider all possible excitations on the $2k_F$ charge stripes in hole-doped cuprates. In the case of hole-doped cuprates, we shall exclusively consider stripe excitations with a charge of $+e$, where e is the electron charge. Hence, a pair of such excitations, coupled in a singlet state, has a charge of $+2e$.

Figure 10.6 schematically shows three types of excitations on the $2k_F$ charge stripes: a soliton, a kink-up and a kink-down. In Fig. 10.6, the stripe excitations are shown at rest; however, in reality, they are dynamical and propagate

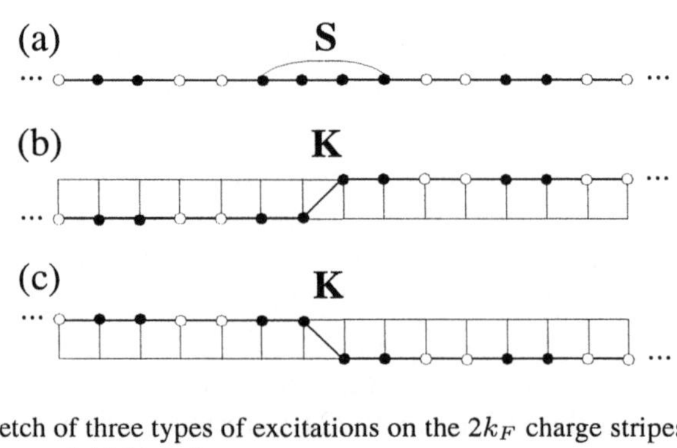

Figure 10.6. Sketch of three types of excitations on the $2k_F$ charge stripes, shown at rest: (a) soliton; (b) kink-up, and (c) kink-down.

along the stripes which themselves fluctuate into the CuO_2 planes. Basically, the excitations shown in Fig. 10.6 are solely charge excitations; however, as mentioned above, they also carry a spin of $\pm 1/2$. Zaanen and co-workers suggested that a kink-up and a kink-down may have different spin orientations [14]. For these three types of excitations, there are four *different* combinations of pairing: Figure 10.7 schematically shows these four combinations. Generally, charged solitons (polarons) repel each other. In conventional superconductors, however, two electrons forming a Cooper pair also repel each other. The occurrence of an attractive potential between electrons is central to the superconducting state. In cuprates, a local lattice deformation is responsible for electron pairing.

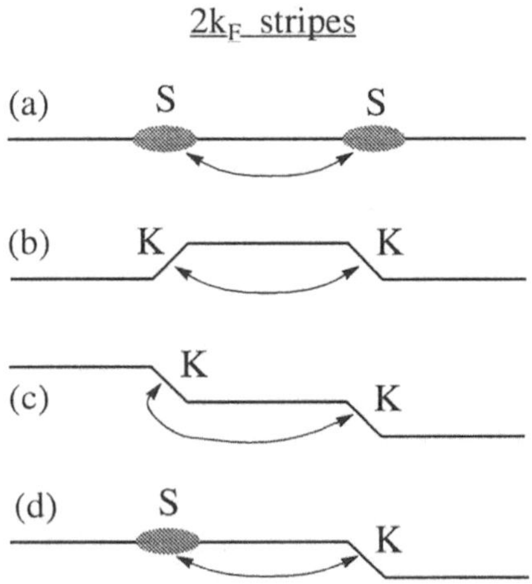

Figure 10.7. Naive sketch of possible coupling of charge-stripe excitations on $2k_F$ stripes: (a) two solitons; (b) a kink-up and a kink-down; (c) two kinks (up or down); and (d) a soliton and a kink (up or down). The arrows schematically show the coupling.

The combination (d) in Fig. 10.7 is asymmetrical and, generally speaking, less likely to be the case realized in cuprates. For example, the only combination which can be realized on the chains in YBCO, as well as in quasi-one-dimensional organic superconductors, is the combination (a) in Fig. 10.7. Since superconductivity on the chains in YBCO is induced, this indicates that, superconductivity in CuO_2 planes is most likely caused by another type of stripe excitations, thus either by (b) or (c) shown in Fig. 10.7. Indeed, the kink excitations have already been observed in LSCO by INS measurements [15]. Since two excitations in a bipolaron have opposite momenta, the stripe excitations on a single stripe, shown in Fig. 10.7, should move in opposite directions (alternatively, a stripe may be rotating around its center of gravity).

Taking into account that the length of charge stripes is about 100 Å, and assuming that the distance between bisolitons is at least $1.5\,\xi$, where $\xi \approx 15$–

20 Å is the coherence length, one obtains that a single charge stripe can carry maximum two bisolitons.

As shown in Chapter 6, the Cooper pairs (charge-stripe excitations) also exist at low temperature in manganate $La_{1.4}Sr_{1.6}Mn_2O_7$ (LSMO) which never exhibits superconductivity. Superconductivity requires not only the electron pairing but also the phase coherence. In LSMO, charge stripes are quasi-static, while in cuprates dynamical. This fact is crucial for the mechanism of high-T_c superconductivity (see Section 2.3).

2.1.4 Energy levels

The energy levels of polaronic solitons (electrosolitons) and bisolitons in Bi2212 are schematically shown in Fig. 8.2. In Bi2212, the energy level of polaronic solitons is located in the middle of charge gap, and coincides with the Fermi level. Thus, polaronic solitons propagate at the Fermi level and, on cooling, give rise to bisolitons. The "bonding" bisoliton energy level is situated below the Fermi level, while the corresponding "antibonding" energy level above the Fermi level. The energy level of bisolitons is defined by a magnitude of the pairing gap. At low temperature, the magnitude of the pairing gap is determined by a magnitude of charge gap (pseudogap) on fluctuating domain walls, and is about 1/3 of the magnitude of the charge gap: $\Delta_p \simeq \Delta_{cg}/3$, as shown in Fig. 8.2b. At low temperature, the antibonding energy level is empty and potentially available for occupation by quasiparticles.

As discussed in Chapter 5, the energy level of *topological* solitons is always situated in the middle of a gap, like that in Fig. 8.2a. This fact indicates that, being self-trapped, the charge-stripe excitations in cuprates are also topological in nature. In cuprates, the main factor which makes the charge-stripe excitations being topological is the strong electron-electron interaction [16].

2.1.5 A remark

Charge inhomogeneity in cuprates was first discussed by Gor'kov and Sokol in 1987 [17]. However, to the best of my knowledge, the existence of charge inhomogeneity in general was predicted by Krumhansl and Schrieffer in 1975 [18]. At the end of a paper in which they discuss the motion of domain walls in materials with Peierls transitions, they wrote: "Finally, we record a few speculative ideas, which may be worth further development. First, if these domain walls are present in the low-temperature phase of pseudo-one-dimensional crystals which have undergone Peierls transition, the Peierls energy gap in those walls could go to zero, the material becoming locally metallic. One could then have a distribution of conducting sheets (walls) in an insulating matrix. ..."

2.2 Phase diagram

Before we consider the phase-coherence mechanism of superconductivity in cuprates, it is necessary first to discuss the phase diagram. In this Section, a new magnetic energy/temperature scale will be introduced.

2.2.1 New temperature/energy scale

As discussed in Chapter 3, there are a few structural phase transitions in cuprates, which occur above a critical temperature T_c. In cuprates, as well as in manganites and nickelates, the charge ordering occurs just below a structural phase transition, as shown in Figs. 3.30 and 3.29. So, in first approximation, $T_{CO} \approx T_d$, where T_{CO} is the charge-ordering temperature and T_d is the structural-phase transition temperature. At the same time, as one can see in Figs. 3.29 and 3.30, the magnetic ordering occurs substantially below T_{CO} but above T_c, thus $T_c < T_{MO} < T_{CO}$, where T_{MO} is the magnetic-ordering temperature. This magnetic ordering occurs in insulating magnetic stripes which separate the charge stripes, as shown in Fig. 3.28a.

In slightly overdoped Bi2212 ($p \simeq 0.19$), tunneling measurements, which will be considered in Chapter 12, show that, at a temperature of about 210 K, the Bi2212 single crystals undergo a transition, while T_c = 88–89 K and $T_{CO} > 300$ K. Obviously, tunneling measurements alone cannot determine the type of this transition. From the discussion in the previous paragraph, it is reasonable to associate this transition with the magnetic ordering in Bi2212.

Figure 10.8 shows the phase diagram of Bi2212 which was already presented twice in the book: in Figs. 3.34 and 9.6. In the phase diagram of Bi2212 in Fig. 10.8, we use, however, two new designations: T_{CO} instead of T^*_{tunnel}, and T_{MT} instead of T^*_{transp}. In the designation T_{MT}, the letters "MT" stand for a magnetic transition. The fact that this temperature scale corresponds to a magnetic transition was shown in Chapter 9 (see Fig. 9.6). In Fig. 10.8, a new energy scale with a characteristic temperature T_{MO} is introduced and schematically shown by the dashed line. To construct the $T_{MO}(p)$ line, two points were used: $T_{MO}(p = 0.19) \simeq 210$ K and $T_{MO}(0.3) \simeq 0$. The doping level $p \simeq 0.3$ is a point where the extensions of two temperature scales marked by T_{CO} and T_{pair} approximately intersect the horizontal axis.

2.2.2 Different energy scales

Here we discuss all the temperature/energy scales shown in Fig. 10.8. The phase diagram of Bi2212 in Fig. 10.8 is not universal but reflects main features of the physics involved in other cuprates.

In Fig. 10.8, the energy scale with the characteristic temperature T_{CO} corresponds to charge ordering in shape of quasi-one-dimensional stripes. A structural phase transition precedes the charge ordering. The stripe order is short-

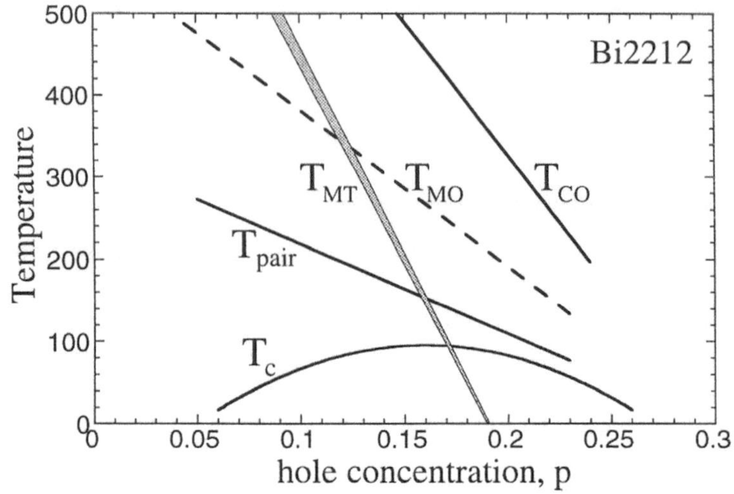

Figure 10.8. The phase diagram of Bi2212 from Fig. 3.34 with a new energy scale shown by the dashed line. In plot, T_{MO} is the onset temperature of a magnetic ordering occurring in insulating stripes which separate charge stripes. In this phase diagram: $T^*_{tunnel} \rightarrow T_{CO}$, and $T^*_{transp} \rightarrow T_{MT}$.

ranged and becomes long-ranged at T_{MO}. The charge stripes are dynamical and fluctuate in time and space. The characteristic temperature T_{CO} can be expressed through the doping level p as follows

$$T_{CO}(p) \simeq 980 \times \left[1 - \frac{p}{0.3}\right] \quad \text{(Kelvin)}. \tag{10.1}$$

The corresponding charge gap Δ_{cg}, observed in tunneling and angle-resolved photoemission (ARPES) spectra in form of humps, is also a function of doping level:

$$\Delta_{cg}(p) \simeq 251 \times \left[1 - \frac{p}{0.3}\right] \quad \text{(meV)}. \tag{10.2}$$

Nonlinear excitations on charge stripes (polaronic solitons) appear with the formation of the charge stripes, thus at T_{CO}.

In Fig. 10.8, the energy scale with the characteristic temperature T_{MO} was discussed in the previous subsection and corresponds to (antiferro)magnetic ordering which occurs in insulating stripes which separate the charge stripes. The magnetic ordering stabilizes the charge order which becomes long-ranged at T_{MO}. The characteristic temperature T_{MO} can be expressed through the doping level as follows

$$T_{MO}(p) \simeq \frac{T_{CO}}{\sqrt{3}} = \frac{980}{\sqrt{3}} \times \left[1 - \frac{p}{0.3}\right] \quad \text{(Kelvin)}. \tag{10.3}$$

Since charge stripes in CuO_2 planes fluctuate, this magnetic ordering rearranges itself fast and, in a sense, is dynamical. The dynamical two-dimensional magnetic stripes can be considered as the local memory effect of the antiferromagnetic phase.

Charge-stripe nonlinear excitations (polaronic solitons) give rise to bisolitons at T_{pair}, which are coupled in a singlet state due to local lattice deformation. In Fig. 10.8, the pairing energy scale can be expressed through the doping level as

$$T_{pair}(p) \simeq \frac{T_{CO}}{3} = \frac{980}{3} \times \left[1 - \frac{p}{0.3}\right] \quad \text{(Kelvin)}. \tag{10.4}$$

The bisoliton pairing gap Δ_p shown in Fig. 3.22 depends on doping level as follows

$$\Delta_{pair}(p) \simeq \frac{\Delta_{cg}}{3} = \frac{251}{3} \times \left[1 - \frac{p}{0.3}\right] \quad \text{(meV)}. \tag{10.5}$$

The pairing gap is manifest itself in tunneling and ARPES measurements. The extensions of three linear dependences $T_{CO}(p)$, $T_{MO}(p)$ and $T_{pair}(p)$ cut the horizontal axis approximately in one point, at $p = 0.3$.

In Fig. 10.8, the magnetic-transition temperature T_{MT} is analogous with the magnetic transition temperature of a long-range antiferromagnetic phase in heavy fermions, as shown in Figs. 9.6 and 9.1a. The doping dependence of this temperature scale, $T_{MT}(p)$, can be expressed as

$$T_{MT}(p) \simeq T_0 \times \left[1 - \frac{p}{0.19}\right], \tag{10.6}$$

where $T_0 \simeq$ 970–990 K [19]. This energy scale is obtained in resistivity (see Fig. 3.33b), nuclear magnetic resonance (NMR) and specific heat measurements [19, 20]. By analogy with heavy fermions, the "magnetic phase" situated below T_{MT} is, in a sense, a phase with static, long-range antiferromagnetic order. In cuprates, the doping level $p = 0.19$ is the quantum critical point at which magnetic fluctuations are strongest. Magnetic fluctuations are also strong along the transition temperature $T_{MT}(p)$.

What is interesting and unexpected is that, in first approximation, the two temperature/energy scales with the characteristic temperatures T_{CO} and T_{MT} intersect the vertical axis in one point, at $T \approx 980$ K. This fact indicates that, in addition to a magnetic ordering below T_{MO}, the charge ordering also induces another magnetic "phase" with the transition temperature T_{MT}. These two magnetic phases will be discussed in the next subsection.

The superconducting phase appears at a critical temperature T_c, as shown in Fig. 10.8. In Bi2212, the doping dependence $T_c(p)$ can be expressed [19, 20] as

$$T_c(p) \simeq T_{c,max}[1 - 82.6(p - 0.16)^2]. \tag{10.7}$$

For Bi2212, $T_{c,max}$ = 95 K. At T_c, the phase coherence among bisolitons in cuprates is mediated by spin fluctuations in insulating antiferromagnetic stripes. The phase-coherence energy gap Δ_c shown in Fig. 3.22 scales with T_c as follows

$$2\Delta_c \simeq 5.4 k_B T_c, \tag{10.8}$$

where k_B is the Boltzmann constant. The phase-coherence gap is manifest itself in Andreev-reflection and penetration-depth measurements (also in tunneling measurements, see Chapter 12).

As discussed in Chapter 3, in some cuprates, independently of their doping level, a structural phase transition is observed at $T_{c,max}$, where $T_{c,max}$ is the maximum critical temperature in each compound. This fact indicates that, in each compound, the maximum value of T_c is determined by the lattice. The mechanism of phase coherence will be discussed in the following subsection.

2.2.3 Two magnetic "phases"

In the phase diagram of Bi2212 shown in Fig. 10.8, one can see that there are two types of magnetic ordering. As mentioned above, the magnetic ordering at the characteristic temperature T_{MO} is, in a sense, dynamical, while the magnetic "phase" below T_{MT} is static.

As discussed in Chapter 3, the charge distribution in CuO_2 planes is inhomogeneous on a microscopic and a macroscopic scales. Even in the superconducting CuO_2 planes, there always exist small islands with intact antiferromagnetic order or small Fermi-sea islands, as shown in Fig. 3.14. Therefore, the two coexisting magnetic "phases," shown in Fig. 10.8, are spatially separated in CuO_2 planes. Figure 10.9 schematically shows three types of small islands existing at certain doping levels in CuO_2 planes. As the doping level increases, islands with intact antiferromagnetic order, marked by the letter A in Fig. 10.9, completely vanish at p = 0.19, and Fermi-sea islands (C) start appearing. Domains with charge-stripe phase, marked by the letter B in Fig. 10.9, exist throughout the superconducting phase.

In Fig. 10.9, the magnetic ordering in charge-stripe domains occurs at T_{MO}, while in the virgin antiferromagnetic islands at T_{MT}. The commensurate antiferromagnetic ordering observed in YBCO, shown in Fig. 9.5, occurs most likely in the intact antiferromagnetic islands.

2.2.4 Special points in doping

The phase diagram of each superconducting copper-oxide compound has two special doping points, p = 0.16 and 0.19. The phase diagram of LSCO has an additional special point at $p \simeq \frac{1}{8}$. The doping point p = 0.16 manifests

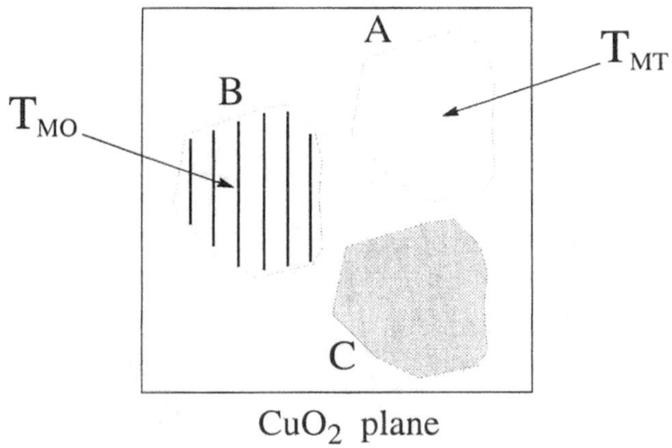

Figure 10.9. Sketch of three types of domains existing at certain dopings in CuO_2 planes. The domain A contains the intact antiferromagnetic order; B—charge stripes, and C—the Fermi sea. The magnetic ordering in the domain A occurs at T_{MT}, and in domain B at T_{MO}.

itself at high temperatures; the point $p = 0.19$ at low temperature, and $p = \frac{1}{8}$ at intermediate temperatures. Consider the cause of each point in detail.

High-temperature point: $p = 0.16$. In most cuprates, the maximum transition temperature is observed near this doping level. Why at $p = 0.16$, and not at another doping? In all cuprates, the maximum critical temperature T_c scales with the magnetic transition temperature T_{MT}. This fact can be easily understood since, for a doping level $p < 0.19$, magnetic fluctuations are strongest near the $T_{MT}(p)$ scale. As an example, Figure 10.10 shows the $T_c(p)$ dependences measured in Co-doped Bi2212 [20]. In Fig. 10.10, one can see that, with increasing concentration of Co, the bell-like curves of the $T_c(p)$ dependences collapse while their maximums scale with $T_{MT}(p)$. The collapse of $T_c(p)$ curves in YBCO and LSCO is very similar to that in Fig. 10.10 [19, 20].

Low-temperature point: $p = 0.19$. In cuprates, the doping level $p = 0.19$ is the quantum critical point where magnetic fluctuations are strongest. Since magnetic fluctuations mediate the phase coherence in cuprates, superconductivity at low temperature is most robust at this doping level $p = 0.19$, and not at $p = 0.16$. The superconducting condensation energy as a function of doping has a maximum at $p = 0.19$, as shown in Fig. 3.21b. The superconducting-phase fraction is maximal too, as schematically shown in Fig. 3.16c.

Intermediate-temperature point: $p = \frac{1}{8}$. In Fig. 10.8, the $T_c(p)$ dependence has a nearly bell-like shape. In LSCO, however, the curve has a dip at doping level $p = \frac{1}{8}$. This dip is the $\frac{1}{8}$ anomaly and inherent exclusively to LSCO. The dip at $\frac{1}{8}$ is not present in the $T_c(p)$ dependence of other high-T_c super-

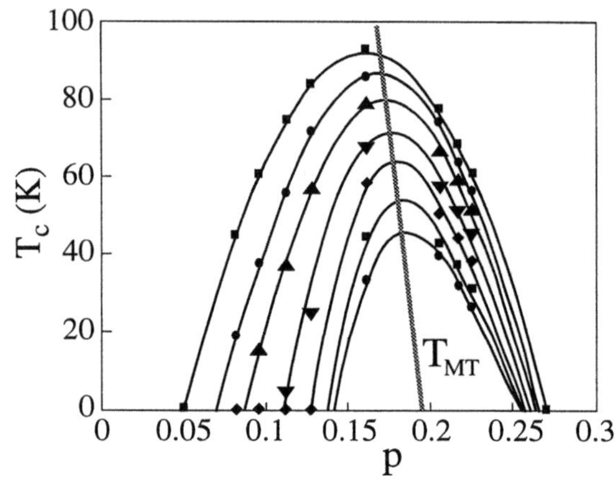

Figure 10.10. T_c plotted as a function of hole concentration p for Bi2212 with 0, 1, 2, 4, 6, 8, and 10 % Co substitution for Cu [20].

conductors, however, was observed in Zn-doped Bi2212 [21]. The local *static* magnetic order is responsible for the $\frac{1}{8}$ anomaly [22], which is caused by the collective pinning of the LTT lattice distortion [23], *thus* by the lattice.

In the phase diagram of Bi2212, obviously, it is not possible to explain the presence of the $\frac{1}{8}$ anomaly since structural phase transitions are not indicated in Fig. 10.8. What is, however, interesting is that, from the $T_{MT}(p)$ and $T_{MO}(p)$ dependences given explicitly above, one can obtain that the two magnetic T_{MO} and T_{MT} energy/temperature scales meet at $p \simeq 0.126$. This is most likely by accident since the crossing temperature $T \simeq 330$ K is too high to be associated with temperatures near T_c.

2.3 Phase-coherence mechanism

In cuprates, the magnetic origin of the phase-coherence mechanism was determined in the previous Chapter. However, the magnetic mechanism was not considered in detail—this the purpose of this Section.

The moderately strong, nonlinear electron-phonon interaction in cuprates is mainly responsible for self-trapped electronic states in shape of quasi-one-dimensional charge stripes. An interplay among the charge, spin and lattice degrees of freedom stabilizes charge stripes. The formation of Cooper pairs in form of bisolitons occurs on charge stripes due to local lattice deformation. The phase coherence among bisolitons is mediated by spin fluctuations. But, what is the magnetic mechanism on a microscopic scale? Why do manganites and nickelates never become superconducting? What is the main difference between, for example, LSCO and YBCO? These questions will be discussed in this Section.

2.3.1 Magnetism and magnetic excitations

In general, there are two sources of magnetism in metals—localized magnetic moments and the Fermi-sea of conduction electrons. Local magnetism occurs due to the incomplete filling of electrons in the inner atomic shells. This leads to a well-defined magnetic moment at every fixed atomic site, which in turn produces long-range magnetic coupling due to the exchange of conduction electrons. The second type of magnetism—known as band magnetism—arises from the magnetic moments of the conduction electrons. In insulators, including undoped cuprates, local magnetism can only occur due to localized magnetic moments. In cuprates, Cu^{2+} ions with a spin of 1/2 produce localized magnetic moments which are ordered antiferromagnetically. So, we shall consider below only magnetism caused by localized magnetic moments. Localized magnetic moments also exist in many superconducting heavy-fermion metals.

In materials with localized magnetic moments, spin fluctuations represent magnetic excitations which propagate into the magnetic environment. Three types of elementary magnetic excitations are particularly interesting for us: *magnons*, *excitons* and solitons. Magnons and excitons are non-interacting plane waves. To propagate, a magnon uses only the ground spin states—antiferromagnetic or ferromagnetic—while an exciton excited spin states. Magnetic solitons (Bloch and Néel walls) were discussed in Chapter 5. Basically, solitons are one-dimensional objects. Nevertheless, solitons can propagate as a plane wave: As an example, the first encountered soliton was observed by Russell in a water channel having a *finite* width (see Chapter 5). So, the width of this soliton in a direction perpendicular to the propagation vector was finite, as if it were a plane wave.

2.3.2 Stripe phase and antiferromagnetic order

As discussed in Chapter 3, apart from the magnetic resonance peak which is commensurate, INS measurements have found four incommensurate peaks at some energy transfers, as schematically shown in Fig. 3.27a. The similarity of spin dynamics in two different cuprates, LSCO and YBCO, demonstrate that the spin dynamics do not depend on the details of the Fermi surface, but have an analogous form with that for the stripe phase. Thus, the incommensurate peaks are caused by a spin wave that is locally commensurate, but whose phase jumps by π at a periodic array of charge stripes, as shown in Fig. 3.28a.

In Fig. 10.8, below T_{MO}, insulating stripes between charge stripes are antiferromagnetically ordered. The spin direction in antiferromagnetic stripes rotates by 180° on crossing a domain wall, as shown in Fig. 3.28a. Therefore, below T_{MO}, charge stripes *always* carry spin excitations at their ends, as schematically shown in Fig. 10.11. At the end of each charge stripe, two antiferromagnetic stripes separated by the charge stripe come in contact with each other. Because of the spin direction in the two antiferromagnetic stripes rotates by 180° on crossing the charge stripe, at the charge-stripe end spin orientations in the two domains are opposite. Therefore, any spin orientation—up or down—at the charge-stipe end induces a local spin excitation, as shown in Fig. 10.11. So, charge stripes always carry spin excitations at their ends.

It is important noting that the occurrence of antiferromagnetic ordering between charge stripes does not imply the presence of antiferromagnetic order around charge stripes. However, the antiferromagnetic order always exists in regions where charge-stripe domains (domain B in Fig. 10.9) are in contact with domains containing intact antiferromagnetic order (domain A in Fig. 10.9).

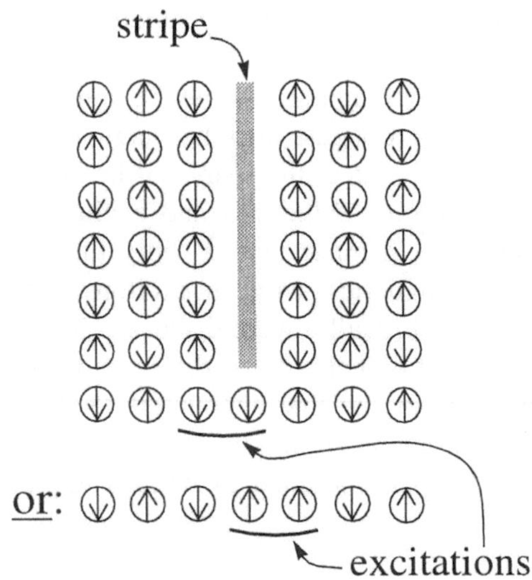

Figure 10.11. Sketch of a charge stripe in the antiferromagnetic environment. Independently of spin orientation, each end of a charge stripe always carries a spin excitation.

2.3.3 Spin-fluctuation mechanism

Spin-fluctuation mechanism of superconductivity was initially proposed for explaining superconductivity in heavy fermions. In the framework of this mechanism, magnons mediate superconducting correlations, realizing an energy transfer. For a long time the existence of magnetically-mediated superconductivity remained unconfirmed. Only recent studies of heavy-fermion superconductors, especially ferromagnetic ones, have shown that spin fluctuations are indeed responsible for the appearance of superconductivity in these materials (see Chapter 9). However, the necessary conditions for occurrence of magnetically-mediated superconductivity remained unknown. In this sense, cuprates are an excellent testing ground for understanding the magnetic mechanism of superconductivity.

In cuprates, spin-fluctuation mechanism of superconductivity requires not only the presence of magnetic excitations but also they have to propagate fast. Second, these magnetic excitations are not magnons. Third, frequency (energy) of these excitations has to be above a certain value. Apparently, these are the basic conditions for spin-fluctuation mechanism of superconductivity in any compound.

As discussed in Chapter 9, the phase coherence along the c axis in cuprates is mediated by spin fluctuations. From c-axis far-infrared measurements, Tajima and co-workers concluded [23]: "The superconductivity coexisting with the static stripe order is a state with a suppressed c-axis phase coherence." This means that charge-stripe *fluctuations* in cuprates are a crucial element of the magnetic mechanism of phase coherence. In other words, fluctuations of charge stripes into an antiferromagnetic background of CuO_2 planes induce magnetic excitations which mediate the phase coherence. From Chapter 9, the magnetic resonance peak observed in INS spectra of some cuprates is caused by these magnetic excitations. Evidence for the activation of domain-wall motion at 38 K, obtained in LSCO by acoustic measurements, is presented in Chapter 6 (see Fig. 6.11).

Figure 10.12 shows the frequency scale of charge-stripe fluctuations in cuprates, which is divided into three regions. Charge-stripe fluctuations oscillating with a frequency $\omega_{crit} \leq \omega$ induce fast magnetic excitations which mediate the phase coherence, where ω_{crit} is the critical frequency. These magnetic excitations are commensurate and cause the appearance of the resonance peak in INS spectra. This case is realized in the 90 K cuprates, such as YBCO, Tl2201 and Bi2212. Charge-stripe fluctuations oscillating with a frequency $\omega_{min} \leq \omega < \omega_{crit}$ excite dynamical spin fluctuations still capable of mediating the phase coherence, where ω_{min} is the minimum frequency. These dynamical spin excitations are *locally* commensurate with the lattice. This case is realized in the 40 K cuprates: LSCO and Bi2201. Finally, charge-stripe fluctuations oscillating with a frequency $\omega < \omega_{min}$ induce spin fluctuations which are too slow to mediate the phase coherence. This is the reason of the absence of superconductivity in manganites and nickelates.

What are the values of frequencies ω_{min} and ω_{crit} in cuprates? I can only give estimates. As discussed in Chapter 9, superconductivity in heavy fermion UPt_3 is mediated by spin fluctuations and reminiscent of superconductivity in the 40 K cuprates. In UPt_3, the spin-density wave order oscillates with finite frequencies larger than those of the NMR frequency ($\sim 10^{6-7}$ Hz) and smaller than those of the neutron diffraction ($\sim 10^{11-12}$ Hz) [24]. These values are good estimates for cuprates: $\omega_{min} \sim 10^7$ Hz and $\omega_{crit} \sim 10^{11-12}$ Hz.

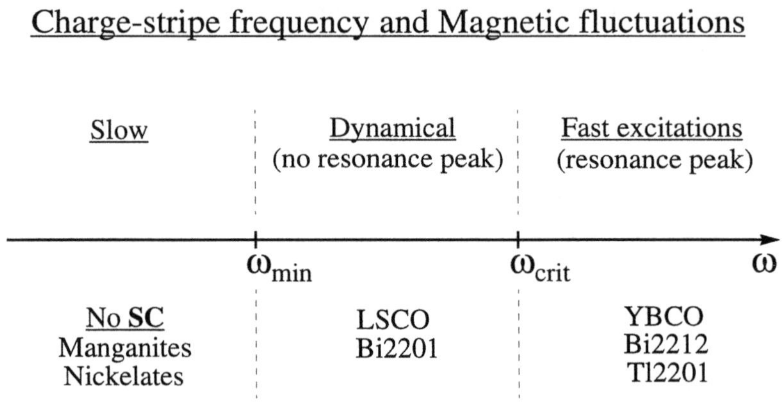

Figure 10.12. Frequency scale of charge-stripe fluctuations in cuprates: ω_{crit} is the critical frequency and ω_{min} is the minimum frequency. Charge-stripe fluctuations with different frequency induce different types of magnetic excitations (for more details, see text).

Why is the frequency of charge-stripe fluctuations different in different cuprates? Because the lattice deformation is not the same in different cuprates. As discussed in Chapter 3, the highest critical temperatures are observed in cuprates having flat and square CuO_2 planes. Contrary to this, the CuO_2 planes in LSCO are buckled. Therefore, in LSCO, charge stripes cannot fluctuate fast—the plane buckling slows them down. Moreover, the maximum critical temperature $T_{c,max}$ in each compound is also determined by the lattice. As mentioned above, in some cuprates independently of their doping level, a structural phase transition is observed at $T_{c,max}$. This means that the lattice triggers fluctuations of charge stripes through a structural phase transition.

To summarize, above T_c in cuprates, charge stripes carrying bisolitons fluctuate but not fast enough for exciting spin fluctuations capable of mediating phase coherence. At T_c or slightly above T_c, there is a structural phase transition which flattens the CuO_2 planes and/or makes them more tetragonal. As a consequence, charge stripes start to fluctuate faster, and induce spin excitations capable of mediating the phase coherence.

2.3.4 Magnetic resonance peak

In Fig. 3.26, one can see that the doping dependence of the *intensity* of the magnetic resonance peak is very similar to the doping dependence of the temperature scale $T_{MT}(p)$ in Fig. 10.8. This fact suggests first that, in INS measurements, the resonance peak is, first of all, noticeable in domains with intact antiferromagnetic order (domain A in Fig. 10.9). Second, the fast commensurate spin excitations are capable of being disengaged from charge-stripe domains, travelling across domains with intact antiferromagnetic order. Apparently, the dynamical spin excitations cannot be disengaged from charge-stripe domains: they exclusively reside into charge-stripe domains because they are damped at the borders of these domains—their energy is not enough to be self-sustained.

What type of spin excitations is preferable for mediating superconducting correlations? Theoretically, magnon-like excitations. What kind of spin excitations causes the magnetic resonance peak in INS spectra? Magnons cannot be responsible for the occurrence of the magnetic resonance peak because the resonance peak is too narrow in energy. Magnons have a large degree of dispersion, if they were the cause of the resonance peak, the peak should be quite wide in energy. The width of the resonance peak is also narrow in real space: in YBCO, the width of the resonance peak in real space is $\sim$ 15 Å [25], thus about 3 lattice units in diagonal direction. Sato and co-workers proposed that magnetic *excitons* cause the appearance of the resonance peak in INS spectra [26]. Generally speaking, excitons are also plane waves, consequently, they should have also a large degree of dispersion.

On the other hand, the spin excitation causing the resonance peak has features peculiar to solitons. Experimentally, this magnetic excitation is very robust. Second, from Fig. 10.12, it appears exclusively above frequency ω_{crit}, thus it has a threshold. Third, below T_c, its energy position does not depend on temperature (only its intensity). So, it is possible that the spin excitation which causes the appearance of the resonance peak in INS spectra is a magnetic soliton. What is interesting is that magnetic solitons were recently observed by electron-paramagnetic resonance measurements in the manganite $Nd_{0.5}Sr_{0.5}MnO_3$ [27]. In the copper metaborate CuB_2O_4 which has a three dimensional magnetic lattice, magnetic solitons appear spontaneously below 10 K, i.e. without external disturbance, as observed by neutron scattering measurements [28].

2.3.5 Static stripe order

As discussed above, the $\frac{1}{8}$ anomaly in LSCO is caused by the local static magnetic order [22]. This is a good sign that the spin-density wave order oscillates in LSCO with low frequency. In LSCO, the lattice distortion pins charge stripes [23]. Nevertheless, superconductivity can coexist with static stripe order if nearby there are domains with fluctuating charge stripes. Fluctuations of charge stripes in these domains will generate a spin excitation necessary for mediating the phase coherence. Thus, static or quasi-static stripe order in a few charge-stripe domains will not affect superconductivity on a macroscopic scale as long as charge stripes continue to fluctuate in the neighboring domains.

Manganites and, probably, nickelates will superconduct if one can increase the frequency of charge-stripe fluctuations in these perovskites. For potential superconductivity in manganites, it is not important if spin order is ferromagnetic or antiferromagnetic (at different dopings, manganites can be ordered ferromagnetically or antiferromagnetically). It is worth noting that two effects—colossal magnetoresistance in manganites and superconductivity in cuprates—are two different manifestations of the same phenomenon.

2.3.6 "Granular" superconductivity in cuprates

Superconductivity in cuprates has a *domain structure*. Bisolitons reside on charge stripes which are located in charge-stripes domains (domain B in Fig. 10.9). In the underdoped region, charge-stripes domains coexist with domains containing virgin antiferromagnetic order (domain A in Fig. 10.9). Below T_c, the Cooper pairs tunnel through the insulating domains (the dc Josephson effect). Since spin-fluctuations mediate the phase coherence, the Cooper-pair phase in domains with intact antiferromagnetic order remains unchanged.

In the highly overdoped region, charge-stripes domains coexist with conducting (Fermi-sea) domains (domain C in Fig. 10.9). Due to the proximity effect, superconductivity can be induced into the conducting domains. At low temperature, (bulk) superconductivity in cuprates is the most homogeneous near $p = 0.19$ (see Fig. 3.16a).

It is worth noting that isolated superconducting domains may exist between $0.27 \leq p < 0.3$. In the phase diagram shown in Fig. 10.8, the superconducting phase with long-range phase coherence disappears at $p = 0.27$. However, the pairing scale collapses at $p \simeq 0.3$. This fact suggests that the pairs may exist locally in the interval $0.27 \leq p < 0.3$.

2.3.7 Spin-glass phase

In the phase diagram of LSCO shown in Fig. 3.3, the spin-glass phase appears at low doping level and low temperature. As discussed in Chapter 3, this spin-glass phase is not conventional. Spin glass does not occur at zero doping $p = 0$. On increasing doping level, thus still in the antiferromagnetic state, this phase appears at a transition temperature which is linearly proportional to p, as shown in Fig. 3.3. This fact reveals that the spin-glass phase appears in charge-stripe domains, not in domains with intact antiferromagnetic order. In charge-stripe domains, spins freeze most likely due to a static stripe order.

Apparently, the presence of this spin-glass phase into the superconducting phase at very low temperature, as shown in Fig. 3.3, is due to the presence of charge-stripe domains with *diagonal* charge-stripe order, which represent a small fraction relative to domains containing *vertical* charge-stripe order. As the doping level increases, the domains containing diagonal charge-stripe order vanish, as shown in Fig. 3.3.

2.3.8 Coupling between spin fluctuations and phonons

It is known that, in one-dimensional insulating systems, antiferromagnetic fluctuations can couple to longitudinal acoustic phonons [29]. In cuprates, spin fluctuations are coupled not only to longitudinal phonons but also to transverse phonons. In Fig. 3.31, the transverse elastic coefficient $(C_{11} - C_{12})/2$ obtained in acoustic measurements in LSCO ($x = 0.14$) depends below T_c on the amplitude and the direction of applied magnetic field, revealing that the in-plane transverse phonons are coupled to spin fluctuations below T_c. Acoustic measurements performed in YBCO showed that the longitudinal sound velocity along the c axis has a jump at T_c [30]. Since spin fluctuations mediate the phase coherence in cuprates along the c axis, this means that longitudinal c-axis phonons (C_{33}) are coupled to spin fluctuations.

In fact, spins and phonons in cuprates start to interact, at least, at a temperature $T_{MO}(p)$ shown in Fig. 10.8. The magnetic ordering occurring at T_{MO} stabilizes long-range charge-stripe order, as shown in Fig. 3.30. Therefore spins and phonons interact with each other below T_{MO}.

2.4 Symmetry of the order parameters

The wave function $\Psi(\vec{r})$ describing a superconducting state is also called the order parameter, which is characterized by an amplitude and a phase. As discussed above, cuprates and most unconventional superconductors have two superconducting order parameters—the pairing and phase-coherence order parameters shown for Bi2212 in Fig. 3.22. In fact, in all superconductors, including conventional, the pairing and phase-coherence mechanisms are different. In spite of this, conventional superconductors have only one order parameter—the pairing one. This is because the phase coherence in conventional superconductors is established due to the overlap of Cooper-pair wave functions—the process which is not governed by a order parameter. In cuprates, the phase coherence is not mediated by the overlap of Cooper-pair wave functions. Therefore, superconducting cuprates have two distinctive order parameters: Phonons are responsible for pairing, and spin fluctuations mediate the phase coherence.

In a sense, the pairing order parameter Δ_p in cuprates is "internal", while the phase-coherence order parameter Δ_c is "external", shown in Fig. 3.22. It is worth recalling that the pairing wave function in cuprates is very localized in space (see Chapters 8 and 5). The process of bisoliton condensation at T_c is the Bose-Einstein condensation.

In a superconductor, as shown by Gor'kov, the amplitude of pairing order parameter is proportional to the energy gap in electronic excitation spectrum, arising at the Fermi surface. In conventional superconductors, the symmetry of order parameter, thus the energy gap, is isotropic or slightly anisotropic s-wave. What are the symmetries of the two order parameters in cuprates?

In cuprates, all phase-sensitive and Andreev-reflection measurements show that hole- and electron-doped cuprates have the $d_{x^2-y^2}$ symmetry of the order parameter (see Fig. 3.23). Since spin fluctuations mediate superconductivity exclusively with a $d_{x^2-y^2}$ ground state, this fact reveals that the phase-sensitive techniques probe the symmetry of the "external" order parameter in cuprates, thus the symmetry of Δ_c.

What is the symmetry of the pairing order parameter? It must have an s-wave symmetry, and not a d-wave. Otherwise, it is impossible to explain tunneling measurements performed between s-wave superconductors and cuprates along the c axis (see Chapter 3). Moreover, tunneling along the c axis was observed between two parts of a Bi2212 single crystal, turned relatively to each other by 45° [31]. In the latter case, if the symmetry of the pairing order parameter of Bi2212 is a d-wave, the tunneling current has to be zero. The measurements, however, show that this is not the case [31]. Tunneling measurements in electron-doped cuprates also show the presence of an s-wave condensate [32]. Fortunately, we know that phonons are responsible for pairing in cuprates, and they do not favor pairing with a d-wave symmetry.

In Bi2212, angle-resolved photoemission and tunneling measurements show that the pairing order parameter, thus Δ_p, is anisotropic and has the four-fold symmetry. In addition to a usual anisotropic s-wave symmetry, Δ_p can also have an extended s-wave symmetry schematically shown in Fig. 10.13. In the case of an extended s-wave symmetry, the order parameter is everywhere positive (negative) except four small lobes where the order parameter is negative (positive). In Fig. 10.13, the case of s-wave order parameter with nodes is the intermediate case between the two cases—anisotropic and extended.

In heavy fermions, the situation is very similar. For example, in UBe_{13}, Andreev-reflection measurements show the presence of a d-wave order parameter [33], while tunneling measurements show that UBe_{13} is an s-wave superconductor [34]. Tunneling measurements performed in UPt_3 indicate that the s-wave order parameter is anisotropic [35].

A magnetic ordering and phonons usually compete with each other. In cuprates, an estimate of the energy value of hole-hole coupling via phonons gives $E_p \simeq 0.6$ eV (the Fröhlich interaction) [36]. This energy is four times larger than the magnetic exchange energy in cuprates, $J_m \simeq 0.15$ eV. Therefore, in the doping interval of interest, antiferromagnetic interactions in cuprates cannot break pairs coupled via phonons, or influence the symmetry of the pairing order parameter.

Let us return to ruthenocuprates $RuSr_2RCu_2O_8$ (R = Eu, Gd and Y) considered in Chapter 9. The crystal structure of $RuSr_2RCu_2O_8$ is similar to that of YBCO except for the replacement of one-dimensional CuO chains by two-dimensional RuO_2 layers. It is reasonable to assume that, in ruthenocuprates, the symmetry of the pairing order parameter is the same as in YBCO, since

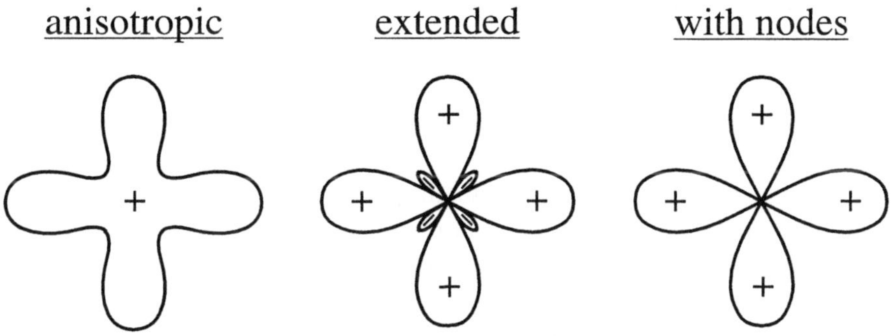

Figure 10.13. Possible types of s-wave symmetry of the pairing order parameter in cuprates: (a) anisotropic, (b) extended and (c) with nodes. The case (c) is intermediate between (a) and (b).

the pairing occurs in CuO_2 planes. In ruthenocuprates, a weak ferromagnetic ordering arises in the RuO_2 layers due to Ru magnetic moments. It was assumed that superconductivity and ferromagnetism in ruthenocuprates coexist in different planes of the lattice. Recent measurements, however, show that superconductivity and ferromagnetism do occur in the same location. The superconducting gap coexists with the ferromagnetic component in the RuO_2 planes on a *microscopic* scale [37, 38]. Such a coexistence is possible if the two following conditions are satisfied: First, the pairing force that holds two electrons has a non-magnetic origin. Second, the pairing energy is larger than the energy of the ferromagnetic interaction in the RuO_2 layers. From the first condition, the pairing energy must also be larger than the energy of the antiferromagnetic interaction in the CuO_2 layers.

In borocarbides, the situation is similar to that of ruthenocuprates. As also discussed in Chapter 9, in $ErNi_2B_2C$ borocarbide, an antiferromagnetic ordering and superconductivity *microscopically* coexist at low temperature with spontaneous weak ferromagnetism [39]. At the same time, there is evidence that, in borocarbides, superconductivity is s-wave and mediated by phonons. However, recent thermal-conductivity measurements performed in $LuNi_2B_2C$ borocarbide show that the energy gap is highly anisotropic, with a gap minimum at least 10 times smaller than the gap maximum, $\Delta_{min} \leq \Delta_{max}/10$, and possibly going to zero at nodes [40]. In $LuNi_2B_2C$, the magnetic-field dependence is very similar to that of UPt_3, and very different from the exponential dependence characteristic of s-wave superconductors. To reconcile all these data, one has to assume that the mechanism of superconductivity in borocarbides is similar to that in cuprates and heavy fermions: The pairing and phase-coherence order parameters have different symmetries and different origins; and different experimental techniques probe different order parameters.

2.5 In-plane coherence lengths

The coherence length in conventional superconductors is the average distance between two electrons in a Cooper pair. In cuprates, besides the c-axis coherence length, there are two in-plane characteristic lengths: the average distance between quasiparticles in a bisoliton, d, and the phase-coherence length ξ_{sf} which is determined by spin fluctuations into CuO_2 planes. Let us find out what experimental techniques measure d and ξ_{sf}.

In cuprates, the "external" order parameter Δ_c can be tested by any phase-sensitive technique. Penetration-depth measurements also probe the Δ_c energy scale, and this can be shown by a simple example. The energy scale Δ_c has an inverse parabolic dependence on doping, $\Delta_c \propto \frac{1}{p^2}$, shown in Fig. 3.22. By using the standard BCS expression cited in Eq. (2.8), $\xi(0) \propto \frac{1}{\Delta}$, one can obtain that $\xi(0) \propto p^2$, as schematically shown in Fig. 10.14. In a d-wave BCS superconductor, the magnitude of penetration depth can be estimated from the following expression

$$\frac{\lambda^2(0)}{\lambda^2(T)} \approx 1 - \frac{(2\ln 2)T}{\Delta(0)}, \tag{10.9}$$

where $\lambda(0)$ is the magnitude of the penetration depth at $T = 0$. Then, by using this equation and $\Delta_c \propto \frac{1}{p^2}$, one can easily obtain that $\lambda(0) \propto p^2$, as depicted in Fig. 10.14. Indeed, the doping dependence of $\lambda(0)$, obtained in LSCO and shown in Fig. 3.21a, is similar to the dependence in Fig. 10.14. This means that penetration-depth measurements, indeed, probe Δ_c energy scale, therefore the value of ξ_{sf} in cuprates can be inferred from in-plane penetration-depth measurements.

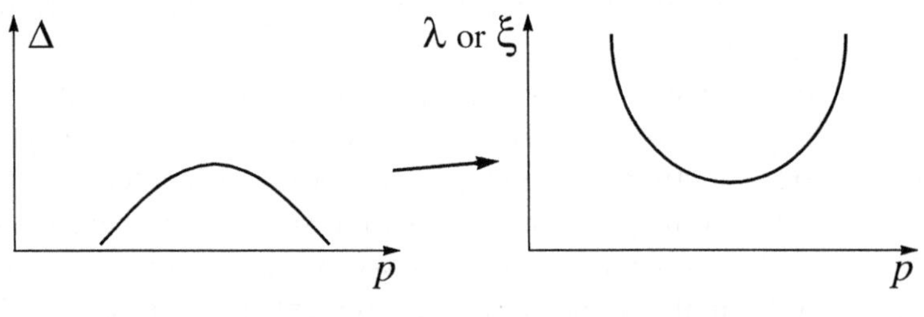

Figure 10.14. Sketch of the doping dependence of phase-coherence gap in cuprates, and the corresponding doping dependences of penetration depth and coherence length.

Figure 10.15 shows the doping dependences of in-plane and out-of-plane coherence lengths obtained in LSCO by applying an external magnetic field. In Fig. 10.15, the values of coherence lengths, inferred from Eq. (2.11), are the Ginzburg-Landau values. By comparing Figs. 8.5 and 10.15a, one can conclude that, in cuprates, magnetic-field measurements measure d, i.e. the size of bisolitons. The latter result is in fact surprising since, by applying a

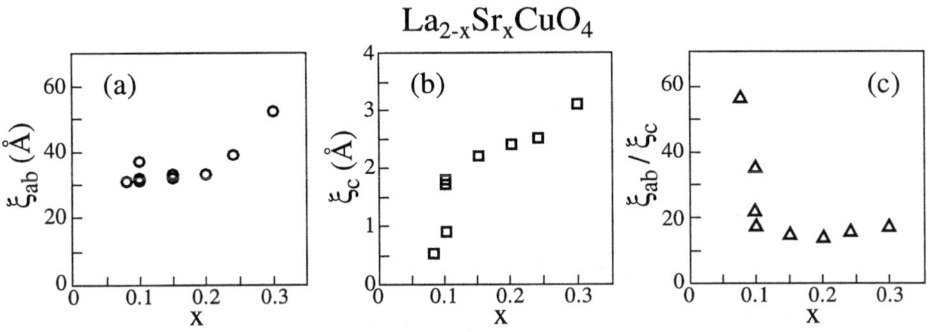

Figure 10.15. (a) In-plane coherence length $\xi_{ab}(0)$ for $La_{2-x}Sr_xCuO_4$ as a function of x. (b) Out-of-plane coherence length $\xi_c(0)$, and (c) anisotropy factor $\xi_{ab}(0)/\xi_c(0)$ as functions of x [41].

magnetic field to a system with two order parameters, it is anticipated that the weaker order parameter will be suppressed first. As discussed above, in cuprates the strength of electron-phonon interactions can be four times stronger than the strength of magnetic interactions. Nevertheless, in cuprates, an applied magnetic field first suppresses the phonon pairing, not the phase coherence mediated by spin fluctuations.

2.6 Effect of impurities

The charge distribution in superconducting cuprates is inhomogeneous either on a microscopic scale or a macroscopic scale: charge-stripe domains always coexist either with insulating antiferromagnetic domains or conducting Fermi-sea domains, shown in Fig. 10.9. Therefore, the same foreign atom substituting Cu into CuO_2 planes will eventually produce different effects on superconductivity, if it is located in different domains—insulating; charge stripes, or conducting. However, it is reasonable to assume that any impurity or lattice defect will attract charge-containing domains—depending on doping level—either charge-stripe or conducting domains. Since most studies of Cu substitution have been carried out in underdoped, optimally doped or slightly overdoped regions of the phase diagram, the conclusions made in these studies, first of all, reflect the effect on superconductivity by impurities located in charge-stripe domains (see Chapter 3).

2.7 Key experiments

Key experiments for identifying bisoliton superconductivity in superconducting materials were discussed in Chapter 8, and they are:

- measurements of coherence length (d);

- tunneling measurements, and
- acoustic measurements.

The short coherence length (the size of a Cooper pair, d) is a good sign of bisoliton superconductivity.

Characteristic features of magnetically-mediated superconductivity were discussed in Chapter 9, and they are:

- the presence of a quantum critical point;
- a bell-like shape of the superconducting phase;
- the $d_{x^2-y^2}$ symmetry of the order parameter;
- a specific temperature dependence of superconducting characteristics;
- the enhancement of spin fluctuations at T_c, and
- the presence of the resonance mode in INS spectra (not always).

In compounds with magnetically-mediated superconductivity, superconducting characteristics have the temperature dependence reminiscent of the squared BCS temperature dependence (see Fig. 9.2). These characteristic features of magnetically-mediated superconductivity can be detected/measured by various experimental techniques.

2.8 Future theory

The BCS theory of superconductivity is "one-step" theory: when the wave function of a Cooper pair is constructed, the phase coherence occurs "automatically." The future theory of high-T_c superconductivity is multi-step theory. It must deal with the following processes:

- the formation of fluctuating charge stripes;
- the formation of antiferromagnetic stripes between charge stripes;
- charge-stripe excitations;
- pairing of these excitations;
- spin excitations induced by fluctuating charge stripes, and
- the phase coherence mediated by the spin excitations.

The bisoliton theory and the spin-fluctuation theory can be used as a starting point, but both have to be modified: The Coulomb repulsion must be taken into account in the framework of the bisoliton model. The spin-fluctuation model has to deal with spin excitations different from magnons: excitons and/or solitons.

2.9 Interpretation of some experiments

Knowledge of the mechanism of high-T_c superconductivity offers ample scope for interpretation of experimental data obtained in cuprates. Interpretation of great deal of data obtained by tunneling and angle-resolved photoemission measurements, mainly, in Bi2212 is presented in Chapter 12. Interpretation of acoustic measurements can be found in Chapter 8. In this Section, we consider a few examples. In fact, at this stage, the reader should be capable to interpret experimental data independently.

Above T_c, measurements of the Nernst effect in LSCO showed the presence of mysterious vortex-like excitations [42]. The Nernst effect is similar to the Hall effect, but a Nernst voltage is generated by a temperature gradient. In the underdoped region, the onset temperature of these vortex-like excitations is around 150 K. The doping dependence of the onset temperature is reminiscent of that of Δ_p in Bi2212. So, it is obvious that these vortex-like excitations are charge-stripe excitations. The authors are correct in indicating that the observed excitations are vortex-like because vortices are solitons (see Chapter 5).

In near optimally doped YBCO, NMR measurements show that, by applying a magnetic field of 14.8 T, the onset temperature of spin gap (pseudogap) remains unchanged, while the critical temperature is shifted down by 8 K [43]. In cuprates, NMR probes the *local* environment of Cu atoms in domains with intact antiferromagnetic order (domain A in Fig. 10.9), ordered below T_{MT} (see Section 2.2). The spin gap in these domains has no relations to the T_c onset.

In-plane torque anisotropy measurements performed on Tl2201 thin films show the existence of two order parameters having different symmetries: s-wave and d-wave [44]. What is interesting is that the amplitude of the s-wave order parameter is about two times larger than the amplitude of the d-wave order parameter. It is worth mentioning that, in contrast to phase-sensitive techniques, the torque measurement is a *bulk* experiment, and independent of grain boundaries and interfaces. From the phase diagram of Bi2212, it is clear that the s-wave order parameter in Tl2201 is the pairing order parameter, while the order parameter with the d-wave symmetry is the phase-coherence order parameter.

In transport measurements, the pseudogap temperature T^* is the temperature at which the temperature dependence of resistivity, on cooling, deviates from linear, as shown in Fig. 3.33b. This pseudogap temperature corresponds to the temperature T_{MT} in the phase diagram of Bi2212 (YBCO and LSCO). The temperature T_{MT} designates the onset temperature of antiferromagnetic order in insulating domains (domain A in Fig. 10.9). In charge-stripe domains or conducting domains (domains B and C in Fig. 10.9), electrons can move easily. To sustain electrical current on a macroscopic scale, quasiparticles have to tunnel through insulating domains. It is mistakenly believed that an antiferromagnetic ordering blocks electron transport. The opposite is true: electrons tunnel through a thin layer of NiO_2 ordered antiferromagnetically about five times more easily than through a thin layer of non-magnetic metal oxides [45]. This is the reason why the temperature dependence of resistivity in cuprates has a kink at magnetic transition temperature T_{MT}: the resistivity falls at T_{MT}.

There are a few reports suggesting that, in the overdoped region, superconductivity in cuprates has a trend towards becoming "less d-wave" and "more s-wave", compared with that in the underdoped region. This may indeed be true. In the underdoped region, superconducting domains in cuprates coexist with antiferromagnetic domains, while in the overdoped region with metallic domains. Due to the proximity effect, superconductivity can be induced into the metallic domains where it will tend to be s-wave rather than d-wave.

2.10 Interesting facts

Magical number "3." From the discussion above, the temperature scales $T_{CO}(p)$, $T_{MO}(p)$ and $T_{pair}(p)$ in Fig. 10.8 relate to each other as follows $T_{CO}(p) \simeq \frac{T_{MO}(p)}{\sqrt{3}} \simeq \frac{T_{pair}(p)}{3}$. Thus, these temperature scales are related to each other through the number "3", which can also be a π. I cannot explain why the number is "3."

In Chapter 6, the functions $q_s(V)$ and $q_n(V)$ were used to fit quasiparticles peaks in tunneling conductances obtained in Bi2212, and the functions $c_s(V)$ and $c_n(V)$ to fit $I(V)$ curves. Independently of junction type—SIN or SIS—the gap magnitude Δ in these functions determines the voltage position of quasiparticle peaks, while the voltage V_0 the width of the peaks. What is interesting is that, by applying these fits to tunneling data obtained in underdoped, overdoped and Ni-doped Bi2212, thus in different doping regions, I noticed one pattern. The ratios $\frac{\Delta}{eV_0}$ and $\frac{2\Delta}{eV_0}$ (depending on junction type) were always equal 3±0.5. So, in first approximation, knowledge of gap magnitude to fit the data is not necessary, and the functions can be expressed as follows

$$q_n(V) = q_s(V) \approx A_q \times \left[sech^2\left(\frac{V}{V_0} + 3\right) + sech^2\left(\frac{V}{V_0} - 3\right)\right], \quad (10.10)$$

$$c_n(V) = c_s(V) \approx A_c \times \left[\tanh\left(\frac{V}{V_0} + 3\right) + \tanh\left(\frac{V}{V_0} - 3\right)\right] \quad (10.11)$$

Mathematically, this means, first, that the width of quasiparticle peaks in conductances obtained in an SIS junction is about twice wider than the width of quasiparticle peaks in conductances obtained in an SIN junction. This fact can be easily understood, and we shall discuss it in Chapter 12. Second, the width of quasiparticle peaks linearly scales with the magnitude of pairing gap. *Physically*, this means that the gap magnitude determines the life-time of Cooper pairs. The life-time of Cooper pairs is inversely proportional to the gap magnitude: the more gap is wider, the more life-time is shorter. This statement sounds also logic, but why the factor "3"? The meaning of the number "3" remains a mystery (at least, to me).

High-T_c superconductivity in cuprates is a *relativistic-like phenomenon*. Charge-stripe excitations in cuprates are topological in nature. As discussed in Chapter 5, the kinetic energy of topological solitons, E, depends on its velocity v as a relativistic particle does, namely,

$$E = \frac{m_s v^2}{\sqrt{1 - \frac{v^2}{c_o^2}}}, \tag{10.12}$$

where m_s is the soliton mass, and c_o is the longitudinal sound velocity. Thus, topological solitons behave like relativistic-like particles, and they cannot propagate faster than c_o. Since charge-stripe excitations in cuprates are topological, the phenomenon of high-T_c superconductivity is a relativistic-like phenomenon!

One interesting fact more. In Chapter 3, discussing magnetic properties of YBCO, it was mentioned that, at any doping level of YBCO, in the even (optical) channel, there is a "gap" in the spin-wave excitation spectrum, as shown in Fig. 3.25. This is not a true spin gap but the absence of magnetic excitations in the optical channel below a certain frequency. What is interesting is that this "gap" has the same magnitude and the doping dependence as the pairing energy gap Δ_p [46]. What is even more surprising is that, in the *undoped* region, this gap remains and continues increasing as p decreases. What can be in common between Δ_p and this "gap"? The pairing gap Δ_p is a product of lattice vibrations in a CuO_2 plane. At the same time, the optical channel of magnetic excitations reflects the presence/absence of magnetic excitations in adjacent planes (spins in adjacent planes rotate in opposite directions, thus sensing the restoring force from the interplane coupling $J_\perp$). So, these two physical quantities have nothing in common, nevertheless, at different dopings, they have same magnitudes. This fact may be a pure coincidence. On the other hand, this may be due to the fact that, in cuprates, spin fluctuations are coupled to longitudinal and transverse phonons (see Section 2.1).

3. Organic and heavy-fermion superconductors

Many experimental data obtained in quasi-one-dimensional organic superconductors and heavy fermions reveal that the mechanism of superconductivity in these compounds is similar to that of cuprates. Probably, Uemura and coworkers were first to show this: the Uemura plot in Fig. 3.20 reveals that "all of these systems belong to a special group of superconductors with some fundamental features in common and, possibly, share a common condensation mechanism."

In quasi-one-dimensional organic superconductors and heavy fermions, the phase coherence is mediated by spin fluctuations (see Chapter 9). The superconducting phase in these compounds is always located in the vicinity of, or into, an antiferromagnetic phase (in a few heavy fermions, into a ferromagnetic phase). In one organic compound, superconductivity is induced by an applied magnetic field (see Chapter 9).

In quasi-one-dimensional organic superconductors, quasiparticles are by definition quasi-one-dimensional excitations, and so are the Cooper pairs. Some quasi-one-dimensional organic superconductors show the presence of charge-density-wave order [47], some structural and electronic instabilities [48]. Chage carriers participating in pairing are strongly coupled to acoustic phonons [49, 50], while the isotope effect is small [49]. Depending on experimental technique, many studies carried out in organic superconductors report the d-wave symmetry of order parameter, while some the s-wave symmetry [51]. So, there are *many* analogies between superconductivity in cuprates and in organic superconductors.

Spin-ladder cuprates are, in a sense, "the bridge" between cuprates and quasi-one-dimensional organic superconductors. Recently, gate-induced superconductivity was observed in the spin-ladder cuprate $[CaCu_2O_3]_4$ [52], indicating that the mechanism of superconductivity in cuprates, spin-ladder cuprates and quasi-one-dimensional organic superconductors is basically the same.

Phonon effects in heavy fermions, as well as in organic superconductors, are often masked by spin-fluctuation effects. Nevertheless, acoustic measurements performed in heavy fermions show the presence of structural phase transitions [53], revealing that the lattice effects are important for superconductivity in heavy fermions. As discussed above (see Section 2.4), an s-wave superconducting order parameter is found in some heavy fermions [34, 35]. Spin fluctuations cannot be responsible for occurrence of superconductivity with an s-wave order parameter, but phonons do. So, the mechanism of superconductivity in heavy fermions seems to be similar to that in cuprates. This implies that, below T_c, heavy fermions are not three-dimensional.

Chapter 11

HIGH-T_C SUPERCONDUCTIVITY IN CUPRATES COULD HAVE BEEN PREDICTED

In the field of observation, chance only favors those minds which have been prepared.
—Louis Pasteur

After reading the title of this Chapter, it is possible to *predict* the first reaction of most specialists working in the field of superconductivity: "No way." Einstein once said: "Never say never." Read the first part of this Chapter and then to reread the title. I am sure that your second reaction will be different from the first one.

In the first part of this Chapter we shall that superconductivity in cuprates could have been predicted in the early-mid eighties. For that, we only need one assumption, namely, that in the eighties the transport properties of cuprates were known to be quasi-one-dimensional, and not two-dimensional. That is all. The assumption is reasonable because transport properties of cuprates could be known, for example, from geology: perovskites were intensively studied by geophysicists in the seventies.

In the second part of this Chapter, we shall discuss the main principles of superconductivity as a phenomenon, independently of characteristic properties of a material where it occurs. Then, we shall consider the different mechanisms of electron pairing and phase coherence, and some superconducting materials where these mechanisms are realized.

1. Back in 1985

Imagine we travel back in time, say, to 1985, and we are looking for new superconducting materials. Persistent search for new superconducting materials is mainly concerned with two factors: a high value of T_c and a high value of critical current. The latter means the high value of critical magnetic field B_{c2}.

Generally speaking, search for new materials is usually, but not always, based upon analysis of specific properties of existing materials. In the following two Sections, we shall analyze superconducting properties of materials well known in 1985: A-15 superconductors and Chevrel phases.

1.1 A-15 superconductors

Intermetallic compounds of transition metals of niobium (Nb) and vanadium (V) such as Nb_3B and V_3B, where B is one of the nontransitional metals, have the structure of beta-tungsten (β-W) designated in crystallography by the symbol A-15. As a consequence, superconductors having the structure A_3B (A = Nb, V, Ta, Zr, and B = Sn, Ge, Al, Ga, Si) are called *the A-15 superconductors.*

Figure 11.1 shows the structure of the binary A_3B compounds. In Fig. 11.1, the atoms B form a body-centered cubic sublattice, while the atoms A are situated on the faces of the cube forming three sets of non-interacting orthogonal *one-dimensional* chains. The distance between atoms A on the chains are shorter than the distance between chains.

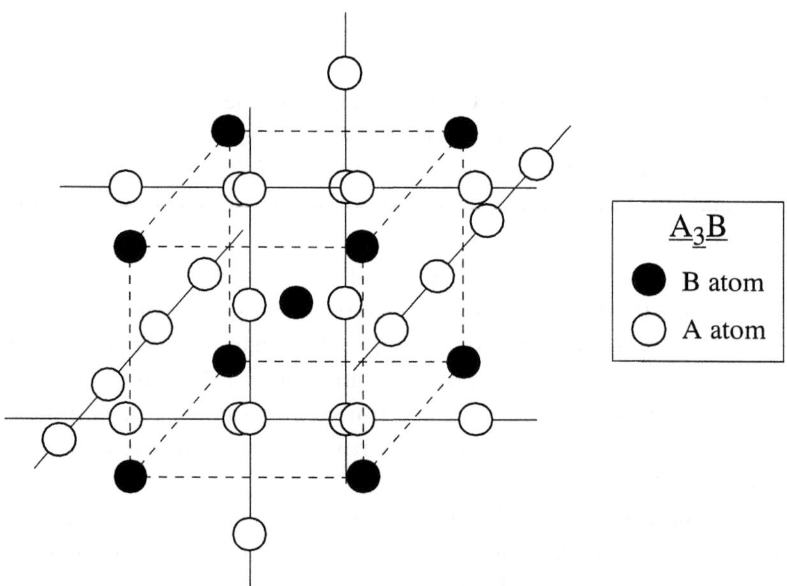

Figure 11.1. Crystal structure of A_3B compounds (A-15 superconductors). The atoms A form one-dimensional chains on each face of the cube. Chains on the opposite faces are parallel, while on the neighboring faces orthogonal to each other.

The first A-15 superconductor V_3Si was discovered by Hardy and Hulm in 1954. Nearly 70 different A-15 superconductors were already known in 1985. Before the discovery of superconductivity in cuprates, A-15 superconductors had the highest value of T_c. Table 11.1 lists some characteristics of six A-15 superconductors with the highest T_c values. The T_c value of A-15 superconductors is very sensitive to changes in the 3:1 stoichiometry.

Table 11.1. Characteristics of A-15 superconductors: the critical temperature T_c; the upper critical magnetic field B_{c2}; and the gap ratio $\frac{2\Delta}{k_B T_c}$ inferred from infrared measurements.

Compound	T_c *(K)*	B_{c2} *(T)*	$\frac{2\Delta}{k_B T_c}$
Nb_3Ge	23.2	38	4.2
Nb_3Ga	20.3	34	-
Nb_3Al	18.9	33	4.4
Nb_3Sn	18.3	24	4.2-4.4
V_3Si	17.1	23	3.8
V_3Ga	15.4	23	-

In addition to unusually high T_c values, A-15 superconductors display a few superconducting properties which cannot be explained in the framework of the BCS theory. The coherence length of A-15 superconductors is very short, $\xi_0 \simeq$ 35–200 Å. They have extraordinary soft acoustic and optical phonon modes. A lattice instability preceding a structural phase transition and then followed by superconductivity is typical for A-15 compounds [1]. This structural phase transition is called *martensite*. The presence of densely packed one-dimensional chains is believed to be responsible for this crystalline instability. The transition temperature T_m for V_3Si and Nb_3Sn is 20.5 K and 43 K, respectively. The symmetry transition, from cubic to tetragonal, is accompanied only by a rearrangement of the crystal lattice, with the volume of the crystal remaining unchanged. Immediately after the transition, there is a softening of the elastic coefficients of the lattice, as observed in acoustic measurements. The Debye temperature of A-15 superconductors is moderate, 300–500 K. It has been established that the more metastable the lattice, the higher the T_c value. One of the main factors in increasing the T_c value is the softening of the phonon spectrum, i.e. the phonon spectrum shifts to lower phonon frequencies [2].

It is important to note that before 1986 A-15 superconductors were the only superconductors having one-dimensional chains. Back in 1985, in the search for new superconductors with a high value of T_c, it was obvious to look for materials which are quasi-one-dimensional and have an unstable lattice.

As a matter of fact, the search of high-temperature superconductors in materials exhibiting structural instabilities was already proposed in 1971 [3]. Even at that time, it was known that metals with strong electron-lattice coupling tend towards structural instabilities.

1.2 Chevrel phases

In 1985, A-15 superconductors had the highest values of T_c, but another class of "conventional" superconductors—Chevrel phases—were the record holders in exhibiting the highest values of upper critical magnetic field B_{c2}.

In 1971 Chevrel and co-workers discovered a new class of ternary molybdenum sulfides, having the general chemical formula $M_xMo_6S_8$, where M stands for a large number of metals and rare earths. They were called the *Chevrel phases*. These superconductors have unusually high values of B_{c2} given in Table 11.2. Chevrel phases with S substituted for Se or Te also display superconductivity.

Table 11.2. The critical temperature and the upper critical magnetic field of Chevrel phases.

Compound	T_c *(K)*	B_{c2} *(T)*
$PbMo_6S_8$	15	60
$LaMo_6S_8$	7	44.5
$SnMo_6S_8$	12	36

As discussed in Chapter 9, superconductivity in Chevrel phases coexists with antiferromagnetism of the rare earths [4] (in fact, superconductivity in Chevrel phases is mediated by magnetic fluctuations). Since in 1985 Chevrel phases had the highest values of B_{c2}, it was evident that the search for new superconductors with a high value of B_{c2} should concentrate on materials which are antiferromagnetic.

It is worth mentioning that in the normal state Chevrel phases exhibit a strong softening in the elastic constants as a function of temperature for the longitudinal and transverse modes [5].

Now we are in a position to "predict" superconductivity in cuprates.

1.3 Cuprates

In 1985, if looking for new superconductors having high values of T_c and B_{c2}, it is necessary, as shown by the two previous Sections, to search materials which are:

- quasi-one-dimensional;
- structurally-unstable, and
- antiferromagnetic.

In 1985, it was already known that cuprates are structurally-unstable and antiferromagnetic. The fact that transport properties of cuprates are quasi-one-dimensional was not yet known to physicists, but could be known in other sciences. For example, in geology—perovskites were intensively studied by geophysicists. If we assume that physicists working in the field of superconductivity in 1985 knew these three characteristic properties of cuprates, then superconductivity in cuprates could have been *consciously* predicted.

Obviously, this does not mean that following this prediction superconductivity would be found in cuprates. There are too many reasons for this. First, nickelates and also manganites at a certain doping level are quasi-one-dimensional, structurally-unstable and antiferromagnetic but they never exhibit superconductivity. So, cuprates could be associated with these materials and not investigated. Secondly, cuprates could be tested but what would be the result of these tests if the doping level of these cuprates was above 0.27 or below 0.05? Third, during the measurements, someone could simply forget to plug, for example, a voltmeter or any other device etc. So, there are many reasons why cuprates would not have been discovered as superconductors. Nevertheless, the possibility for the existence of superconductivity in cuprates could have been *consciously* predicted.

In fact, in the early 1970s, a compound of lanthanum and copper oxides was synthesized in Moscow (Russia). This was not driven by a search for superconductivity—the researchers were looking for a good, cheap conductor. Obviously, at low temperatures the conductivity of this new material showed an abnormal behavior. The researchers realized the significance of this abnormality but they could not continue the tests because in the 1970s, the only liquid helium available in Moscow was under the control of Kapitza at another institute. They did not dare approach him over so controversial a piece of research. The mysterious compound was put away in a cupboard and forgotten. In 1986, Berdnorz and Müller discovered high-T_c superconductivity in a compound virtually identical to that of the Russians [6]. So, cuprates *could* have been discovered a few years earlier than heavy fermions and organic superconductors.

From the discussion in this Section, it is possible to make one useful conclusion, namely, that A-15 compounds and Chevrel phases are not conventional superconductors. Indeed, the Uemura plot in Fig. 3.20 clearly indicates that Chevrel phases are not conventional superconductors: the mechanism of superconductivity in Chevrel phases is similar to that in cuprates, quasi-one-dimensional organic superconductors and heavy fermions. The mechanism of superconductivity in A-15 compounds is in a sense "more" conventional than that in Chevrel phases. There are two superconducting subsystems in A-15 superconductors: one is conventional and the other is unconventional. The mechanism of superconductivity in A-15 compounds is similar to that in MgB_2 and we shall discuss it in Section 3.

2. Principles of Superconductivity

In this Section, we summarize the main principles of superconductivity as a phenomenon, independently of the characteristic properties of a material where it occurs. The underlying mechanism of superconductivity in different materials can be different but these three principles must by satisfied. The first principle of superconductivity:

Principle 1: **Superconductivity requires electron pairing**

The electron (quasiparticle) pairing occurs in the momentum space, not in a real space. The first principle of superconductivity is obvious and can be called *the Cooper principle*. The second principle of superconductivity is also obvious but was never emphasized:

Principle 2: **The transition into the superconducting state is the Bose-Einstein condensation**

In general, the Bose-Einstein condensation is a macroscopic quantum phenomenon that was first predicted by Einstein. The Bose-Einstein condensate is formed when the quantum wave packets of particles overlap at low temperature and the particles condense, almost motionless, into the lowest quantum state. In other words, the wavelengths of the matter waves associated with the particles—the Broglie waves—become similar in size to the mean particle distances in a cold and dense sample. The first Bose-Einstein condensate different from superconductivity was found in 1995 for rubidium atoms. The superconducting state, as a matter of fact, is the Bose-Einstein condensate; however, conceptually different from that predicted by Einstein.

In conventional superconductors, the electron pairing and the Cooper-pair condensation occur almost simultaneously, and hence it was unnecessary to discuss separately the Cooper-pair condensation. In high-T_c superconductors, however, this is not the case: quasiparticles can be paired above T_c, while the Cooper pairs start forming a Bose-Einstein condensate at T_c. Hence, it is necessary to distinguish these two different phenomena—the electron pairing and the Cooper-pair condensation. Superconductivity requires both. The Cooper-pair condensation is usually called the onset of long-range phase coherence.

The third principle of superconductivity, in a sense, is new:

Principle 3: **The mechanism of electron pairing and the mechanism of Cooper-pair condensation must be different**

In all known cases, this principle is satisfied. In conventional superconductors, phonons mediate the electron pairing, while the Josephson coupling provides the Cooper-pair condensation. In cuprates, organic superconductors and heavy fermions, phonons also mediate the electron pairing, while spin fluctuations provoke the Cooper-pair condensation. Generally speaking, in a superconductor in which the same "field" mediates the electron pairing and the long-range phase coherence (the Cooper-pair condensation), this will simply lead to the collapse of superconductivity. The same "field" cannot mediate the electron pairing and the Cooper-pair condensation: the mechanism of electron pairing and the mechanism of phase coherence (Cooper-pair condensation) in a superconductor must be different.

Of course, one can suggest, at least, two cases in which phonons *or* spin fluctuations mediate the electron pairing **and** the Cooper-pair condensation: For example, the acoustic channel is responsible for electron pairing, while the optical for Cooper-pair condensation, or vice versa. Theoretically, such a situation is possible; however, the main problem is that, usually, these two channels—acoustic and optical—compete rather than cooperate with each other. In fact, even if we assume that these two cases can be realized in some superconducting systems, the mechanisms of electron pairing and phase coherence (Cooper-pair condensation) are, in a sense, again different.

3. Different Types of Superconductivity

In this Section, we discuss different types of superconductivity. There are three different types of superconductivity known today. Each type is characterized by its specific mechanism of electron pairing and by its specific mechanism of Cooper-pair condensation. By convention, these three types of superconductivity are:

- the BCS type (conventional),
- the A-15 type (half-conventional), and
- the cuprate type (unconventional).

The first and third types are known to the reader since they were extensively discussed throughout the book. The second type—the A-15 type—is the case realized in A-15 compounds, MgB_2 and $Ba_{1-x}K_xBiO_3$. We shall discuss these superconductors later on, but let us first review different mechanisms of electron pairing and different mechanisms of Cooper-pair condensation. Such a sequence will simplify the understanding of the main differences among the three types of superconductivity.

3.1 Pairing mechanisms

There are two types of electron pairing known today—linear and nonlinear. In both cases, the lattice (phonons) is responsible for electron pairing.

3.1.1 Phonons: Linear coupling

The linear type of electron pairing due to phonons is the classical BCS type. In the BCS theory, the electron-phonon interaction, by definition, is weak and linear. Historically, this type of electron pairing was discovered first, experimentally and theoretically. The characteristic features can be found in Chapter 2. The two main features of the linear type of electron pairing are (i) the isotope effect on the critical temperature and (ii) an s-wave symmetry of the pairing order parameter.

3.1.2 Phonons: Nonlinear coupling

The nonlinear type of electron pairing due to phonons, in a sense, is new. However, this type is experimentally known since 1954 when superconductivity was found in A-15 compounds, almost a half of a century ago! In all known cases, the nonlinear type of electron pairing occurs in materials having quasi-one-dimensional structures. Chapters 6–8 and 10 present a description of this type of electron pairing: The electron-phonon interaction is moderately strong and nonlinear. In such an "uncomfortable" environment, quasiparticles become self-trapped. The characteristic features of the nonlinear type of electron pairing can be found in Chapters 8 and 10. The main two features are (i) the short coherence length (the size of a Cooper pair) and (ii) compounds in which this type occurs tend towards structural instabilities. Apparently, the pairing order parameter always has an s-wave symmetry (but highly anisotropic).

3.1.3 Other pairing mechanisms

The problem is that there are no other pairing mechanisms known today. Theoretically however, spin fluctuations may lead to an electron pairing. Personally, I am sceptical that this can be realized. The reason is simple: if we assume that, in some materials, spin fluctuations can indeed mediate the electron pairing, what mechanism will provide the Cooper-pair condensation (the long-range phase coherence)? This mechanism must have a non-magnetic origin. The magnetic interactions are long-ranged and this is the problem. Independently of the phase-coherence mechanism—the Josephson coupling or the electron-phonon interaction—the magnetic interactions will interfere with the "field" providing the Cooper-pair condensation and, virtually, will destroy the superconducting state.

I am also sceptical of other exotic pairing mechanisms proposed theoretically, for example a spinon pairing (see Introduction) or charge fluctuations.

Nevertheless, let us "leave the door open"; we shall see in the future what surprises Nature has prepared for us.

3.2 Phase-coherence mechanisms

There are two mechanisms of Cooper-pair condensation (phase coherence) known today: the Josephson coupling (the overlap of Cooper-pair wave functions) and spin fluctuations (magnetic). The first mechanism is characteristic of the BCS and A-15 types of superconductivity, while the second of unconventional superconductors such as cuprates, organic quasi-one-dimensional superconductors, heavy fermions and Chevrel phases.

3.2.1 Josephson Coupling

The Josephson-coupling mechanism of Cooper-pair condensation is realized when the distance between Cooper pairs is smaller than the average size of Cooper pairs. In this case, the Cooper-pair wave functions become overlapped resulting in a Bose-Einstein condensation of the Cooper pairs. This mechanism of Cooper-pair condensation is the simplest and does not lead to the appearance of an additional order parameter. However, the Josephson coupling can easily be destroyed either by a strong magnetic field or an electrical current. A characteristic feature of this mechanism of Cooper-pair condensation is that the Cooper pairs condense immediately after their formation, thus the two processes occur almost simultaneously.

3.2.2 Spin fluctuations

This mechanism of Cooper-pair condensation was discussed in detail in Chapters 9 and 10 as well as its characteristic features. It is worth noting that, in materials where this phase-coherence mechanism can be realized, the type of magnetic ordering—antiferromagnetic or ferromagnetic—is not very important. In both cases, the electron pairing will occur due to phonons, while spin fluctuations exclusively provide the Cooper-pair condensation, without differentiating the symmetry of the pairing order parameter. The superconducting state arising due to the magnetic mechanism of Cooper-pair condensation has always two different order parameters: the pairing and phase-coherence order parameters. The phase-coherence order parameter has either a d-wave symmetry (in antiferromagnetic materials) or a p-wave symmetry (in ferromagnetic materials).

3.2.3 Others

At the present, only the Josephson coupling and the magnetic mechanism are known to provide the Cooper-pair condensation. Theoretically, phonons can mediate the phase coherence but then the problem is that what mechanism can mediate the electron pairing? Magnetic? Personally, I am sceptical (see the discussion above).

3.3 Different combinations

Theoretically, the number of different combinations among two mechanisms of electron pairing and two mechanisms of Cooper-pair condensation is four. They are the following:

- Linear phonons + the Josephson coupling.
- Linear phonons + Spin fluctuations.
- Nonlinear phonons + the Josephson coupling.
- Nonlinear phonons + Spin fluctuations.

In these four combinations, the pairing mechanism is indicated first, and the phase-coherence mechanism second. There is however one combination more:

- Nonlinear phonons + (Linear phonons + the Josephson Coupling),

in which there are two kinds of Cooper pairs, and they condense due to the Josephson coupling of the BCS-type superconducting condensate.

Thus, we have five different combinations. However, not all of them can be realized in practice—the second and third combinations are unrealistic. In the second case, the linear pairing mechanism is too weak for Cooper pairs to be coupled to spin fluctuations—the magnetic channel will simply overwhelm the "fragile" linear pairing mechanism. The third combination was discussed in detail in Chapter 8: the Josephson coupling cannot mediate the phase coherence among the Cooper pairs coupled by the nonlinear electron-phonon interaction because the distance between Cooper pairs is larger than their size. It is worth recalling that the wave function of any self-trapped state is very localized. The other three combinations represent the real cases.

We are now in a position to discuss real superconducting materials.

3.3.1 Linear + Josephson Coupling

This combination is the classical BCS type of superconductivity and was discussed in Chapter 2. Historically, this type of superconductivity was discovered first. All superconducting metals exhibit the BCS type of superconductivity. This is more or less obvious because, by definition, the electron-phonon interaction in the BCS-type superconducting materials must be weak.

3.3.2 Nonlinear + (Linear + Josephson Coupling)

The second type of superconductivity—the A-15 type—is realized in non-magnetic materials such as A-15 compounds, MgB_2 and $Ba_{1-x}K_xBiO_3$ (BKBO). It is not the purpose of this book to discuss superconductivity in these compounds. Nevertheless, it is useful to briefly review superconducting and structural properties of A-15 compounds, MgB_2 and BKBO.

Acoustic measurements performed in BKBO [7] show that many physical properties of BKBO are quite similar to those of A-15 superconductors such as a structural instability just above T_c, soft phonon modes (acoustic in A-15 compounds and optical in BKBO), an anharmonicity of some phonons and large softening of the elastic constants. The results of acoustic measurements performed in MgB_2 are not yet available in the literature.

In A-15 compounds and BKBO, slightly above T_c there is a structural phase transition [7] resulting in the appearance of Cooper pairs coupled due to the nonlinear electron-phonon interaction. This process is similar to that in cuprates. The density of this type of Cooper pairs is low, so they cannot condense like those in cuprates. Magnetic interactions in these materials are absent. The "nonlinear" Cooper pairs induce electron pairing in the sample due to weak electron-phonon interaction. The "linear" Cooper pairs condense due to the Josephson coupling, leading also to the condensation of the "nonlinear" Cooper pairs. As a consequence, some characteristics of these superconductors are reminiscent of BCS-type characteristics, while some of unconventional superconductors. Thus, the type of superconductivity in these compounds is literally between the conventional type and the unconventional type.

The plot in Fig. 10.2 is shown again in Fig. 11.2 with the data for MgB_2, BKBO and Nb_3Ge (A-15 compound). In Fig. 11.2, one can notice that this group of superconductors is indeed situated between conventional and unconventional *electron*-doped superconductors. In MgB_2, the relaxation rate is $\sigma \simeq$ 8–10 μs^{-1} [8]. So, in the Uemura plot shown in Fig. 3.20, MgB_2 is literally situated between the large group of unconventional superconductors and Nb. At the same time, BKBO is located in the Uemura plot closer to unconventional superconductors. In Table 11.3, one can see that the characteristics of these superconductors have similar values.

Table 11.3. Characteristics of Nb_3Ge (A-15 compound), BKBO ($x \simeq 0.4$) and MgB_2: T_c; ω_{ph} is the energy of the highest phonon peak in the Eliashberg function $\alpha^2F(\omega)$; the Fermi velocity v_F; the coherence length; the gap ratio $\frac{2\Delta}{k_BT_c}$ and the upper critical magnetic field B_{c2}.

Compound	T_c *(K)*	ω_{ph} *(meV)*	v_F *(10⁷ cm/s)*	ξ_0 *(Å)*	$\frac{2\Delta}{k_BT_c}$	B_{c2} *(T)*
Nb_3Ge	23	25	2.2	35–50	4.2	38
BKBO	31	60	3	35–50	4.5	32
MgB_2	39	90	4.8	35–50	4.5	39

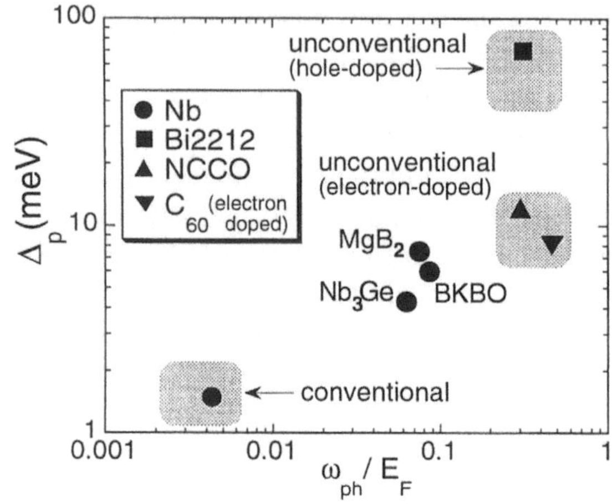

Figure 11.2. Same as Fig. 10.2 with the data for MgB_2, $Ba_{1-x}K_xBiO_3$ ($x \simeq 0.4$) (BKBO) and Nb_3Ge (A-15 compound).

The "nonlinear" Cooper pairs (bisolitons) need the presence of one dimensionality in the structure or a charge-density-wave order. In A-15 superconductors, the presence of one dimensionality is obvious (see Fig. 11.1). Undoped and superconducting BKBO has a charge-density-wave order. MgB_2 is structurally two-dimensional. However, compounds having a hexagonal lattice tend to the formation of solitons (see Section 3.5). It is worth noting that in MgB_2 bisolitons may reside in the Mg layers, not in the B layers, in spite of the fact that the Mg isotope effect is *very* small.

A-15 compounds, BKBO and MgB_2 have two pairing order parameters. The gap ratios in Table 11.2 correspond to the "nonlinear" order parameter which has a *very* anisotropic s-wave symmetry (probably with nodes, as that in cuprates). The gap ratio $\frac{2\Delta}{k_B T_c}$ for the induced BCS-type pairing order parameter is smaller than the BCS value of 3.52. For example, this ratio in MgB_2 is around 1.7. The "linear" order parameter has the s-wave symmetry being either isotropic or slightly anisotropic.

The case of superconductivity in MgB_2 may indicate that the BCS-type superconductivity has a maximum critical temperature of around 40 K. It is important to note that this is possible only in the presence of "nonlinear" Cooper pairs. Apparently, the BCS-type superconductivity alone cannot occur above 10 K.

3.3.3 Nonlinear + Spin Fluctuations

This is the cuprate type of superconductivity or the unconventional type which occurs in cuprates, organic quasi-one-dimensional superconductors, heavy

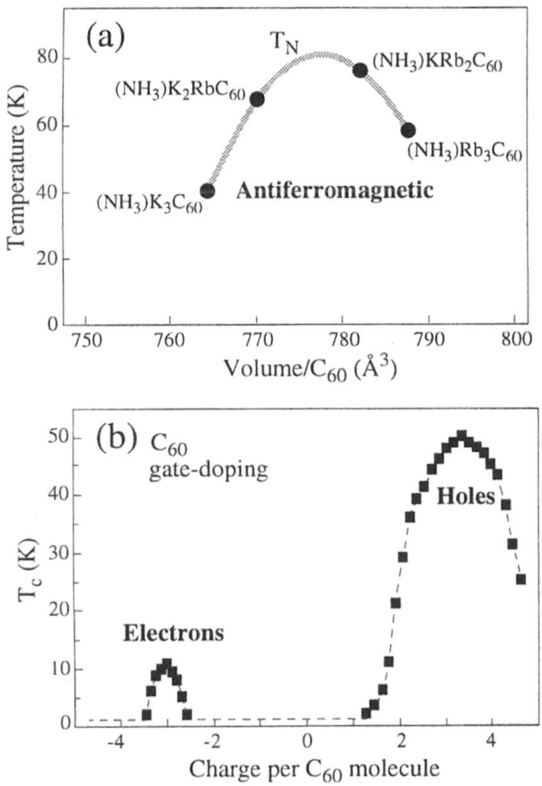

Figure 11.3 T-V phase diagram of the orthorhombic non-superconducting $(NH_3)A_3C_{60}$ [10]. The horizontal axis V represents the volume/C_{60}. T_N is the Néel temperature. The grey line is a quadratic fit. (b) The transition temperature as a function of charge per C_{60} molecule. The doping is achieved in the field-effect transistor geometry (gate-doping) [11].

fermions, Chevrel phases, borocarbides, carbon nanotubes and doped DNA. The mechanism of superconductivity in ferromagnetic heavy fermions UGe_2 and $ZrZn_2$ is most likely the same as that in cuprates with the exception of the symmetry of the "external" (magnetic) order parameter. In the ferromagnetic heavy fermions, it has a p-wave symmetry, not a d-wave. Chapter 10 is devoted to a description of the mechanism of this type of superconductivity.

In C_{60}, the mechanism of superconductivity is also of the unconventional type: Let us briefly discuss the mechanism of superconductivity in C_{60}. In Figs. 3.20 (the Uemura plot) and 11.2 (also 10.2), C_{60} is situated among other unconventional superconductors. In C_{60}, there is evidence that phonons take part in the mechanism of superconductivity [9]. The coherence length is short, $\sim$ 30 Å. The C isotope effect in some C_{60} compounds is larger than 0.5 (the BCS value), similar to that in cuprates (see Fig. 3.17). In Rb_3C_{60}, the C isotope effect can be as large as 2 [9]. *At the same time*, in antiferromagnetic non-superconducting C_{60}, the Néel temperature as a function of crystal volume has a bell-like shape, as shown in Fig. 11.3a. In superconducting C_{60}, the doping dependence of critical temperature has also a bell-like shape either in electron- or hole-doped C_{60}, as shown in Fig. 11.3b. Such dependences are the fingerprints of spin fluctuations taking part in the mechanism of superconductivity in C_{60}.

From the Uemura plot in Fig. 3.20, one may suggest that superconductivity in BKBO is of the unconventional type. This is impossible because, first, BKBO is non-magnetic. Second, if BKBO would have the cuprate type of superconductivity, then the value of its B_{c2} should be at least two or three times larger than 32 T (see Table 11.3).

3.3.4 Type-II Superconductors

It is possible that, in all type-II conventional superconductors, there exists a small amount of "nonlinear" Cooper pairs. In Nb, for example, acoustic measurements revealed the presence of two energy gaps [12].

3.4 Superconductivity in Two Dimensions

Bisoliton superconductivity in two-dimensional compounds such as cuprates, two-dimensional organic superconductors and MgB_2 is responsible for high values of their critical temperature. Bisoliton superconductivity occurs in one dimension. So, superconductivity in two-dimensional compounds is in fact one-dimensional. It seems that superconductivity does not like two dimensions and rather prefers either one or three dimensions. Obviously, this statement is very general but nevertheless contains some truth.

3.5 Room-Temperature Superconductivity

There are at least two papers reporting superconductivity above room temperature. In the first report, superconductivity is observed in a thin surface layer of the complex material $Ag_{\beta}Pb_6CO_9$ ($0.7 < \beta < 1$) at 240–340 K [13]. The second paper claims that superconductivity exists in carbon-based multiwall nanotubes above 400 K [14]. It is worth mentioning that both these materials have the hexagonal lattice (honeycomb). Superconductivity was already observed in single-wall nanotubes at low temperature. So, the hexagonal lattice "encourages" the occurrence of superconductivity. In MgB_2, both the layers—the Mg layer and the B layer—have the two-dimensional hexagonal structure. In a C_{60} molecule, the majority of cells are hexagonal. How does Nature find one dimensionality in a two-dimensional hexagonal lattice?

As a matter of fact, bisolitons (Cooper pairs) exist at 300 K in living tissues (see Chapter 7). The question is, is it possible to make them communicate with each other?

In 2011, superconductivity will be a hundred years old. How will the value of T_c evolve in the future?

Chapter 12

ANALYSIS OF TUNNELING MEASUREMENTS IN CUPRATES

There is nothing more flexible and, at the same time, more stubborn than the human mind.
—From the text

This Chapter is not a review of tunneling measurements in cuprates published in the literature between 1987 and 2002. The main purpose of this Chapter is to discuss the physics of high-T_c superconductors, which can be inferred from tunneling data. This Chapter is mainly addressed to specialists working in the field of high-temperature superconductivity; however, non-specialists can also benefit from its reading.

In tunneling measurements performed in cuprates, one can observe two superconducting gaps and one normal-state gap (charge gap). All these gaps are anisotropic, and they can interfere with each other. Thus, in cuprates, the situation is much more complicated than that in conventional superconductors. Therefore, for a better understanding, we shall discuss these gaps either one or two at a time. We shall also consider angle-resolved photoemission (ARPES) measurements. In general, this technique provides information which is very similar to that obtained in tunneling measurements, *but* not the same.

In this Chapter we shall mainly discuss data obtained in Bi2212. However, tunneling data measured in other cuprates such as YBCO, NCCO and Tl2201 will be considered too. Experimental details of the growth and characterization of Bi2212 single crystals used here can be found in Appendix A. Tunneling measurements were performed by break junctions and in-plane point contacts. In the first technique, the type of tunneling junctions is a superconductor-insulator-superconductor (SIS), while in the second the superconductor-insulator-normal metal (SIN) type. The description of break-junction and point-contact setups can be found in Appendix B.

1. Introduction

As mentioned in the Preface, I began running tunneling measurements in cuprates in 1997. Fortunately, I was alone, without an instructor. I began doing tunneling measurements in Bi2212. In early 1998, I regularly noticed the presence of two energy gaps in tunneling spectra of Bi2212. At that moment, however, I could not find a single paper reporting the presence of two superconducting gaps in Bi2212 single crystals. Nevertheless, I believe in information inferred from measurements. Thus, in November 1998, I put a paper on the web-site of Los Alamos National Laboratory, in which I showed the presence of *two* superconducting gaps, *one* normal-state gap and *one* subgap in Bi2212 [1].

Figure 12.1 shows two tunneling conductances obtained in SIS junctions of Bi2212 [1]. The two spectra are measured in the same Bi2212 single crystal. In Fig. 12.1, one can clearly see that these two spectra are completely different: they have different shapes and different magnitudes. The upper conductance has a subgap and a zero-bias peak due to the Josephson current, while the other does not. Instead, the lower curve is very smooth between the quasiparticle peaks. The upper curve is measured in a junction with low resistance, while the lower curve in a junction with high resistance.

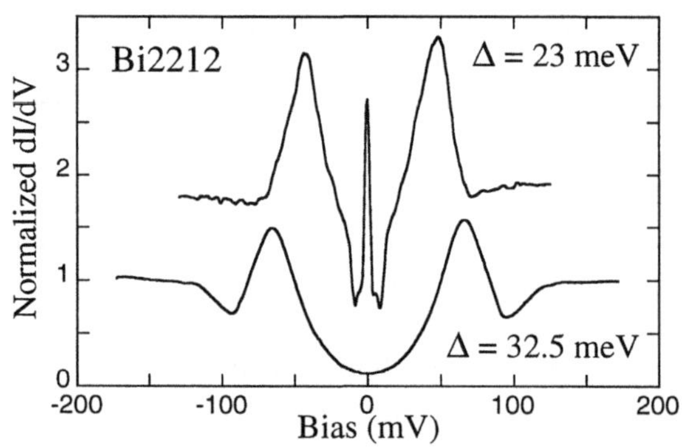

Figure 12.1. Two tunneling conductances obtained in a Bi2212 single crystal with T_c = 89.5 K ($p \simeq 0.19$) by SIS junctions [1]. The upper curve is offset vertically for clarity.

In February 1999, Nature published a paper written by Deutscher [2]. Surprisingly, in his paper, the results concerning the two superconducting gaps were almost identical to the data in my own paper. Obviously, I was very pleased to read Deutscher's paper. In February 2000, I met Deutscher at the M^{2}S-HTSC-VI conference in Houston. I told him that I clearly see two superconducting gaps in tunneling measurements, as those in his paper. He replied that "you cannot see these two gaps in *tunneling* measurements: you must see one gap in tunneling and the other in Andreev-reflection measurements." Generally speaking, his answer was correct; nevertheless, I do belief in what I saw in my measurements—his reply did not change my point of view.

My results also contradicted numerous ARPES measurements showing the presence of only one energy gap—the pairing gap, as depicted in Fig. 12.2. Therefore, from the beginning I was very sceptical of ARPES measurements. In spite of the general belief that tunneling and ARPES measurements provide similar information, I felt that something was faulty wrong with the ARPES technique. At the same conference in Houston, by accident, I talked with someone from Shen's group (Stanford University). I said directly that I do not rely on ARPES data and I explained why: ARPES measurements probe only one energy gap and fail to detect the second.

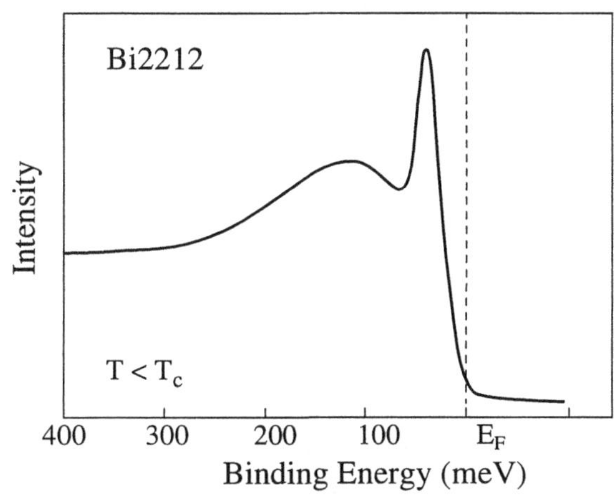

Figure 12.2. ARPES spectrum obtained in a slightly overdoped Bi2212 single crystal having $T_c = 91$ K [3].

The "official" specialists in tunneling also rejected these two gaps. My regular referees in *Physical Review B* (E. L. Wolf) and *Physical Review Letters* (J. F. Zasadzinski) have exclusively written negative reports in replies to my papers submitted to these journals. Once, in correspondence with Wolf, I wrote that my data are in good agreement with Deutscher's paper published in Nature. He replied that "publication in Nature is not proof of correctness." I have no further comments regarding this. The other referee once responded to my paper: "This manuscript is not acceptable for publication in any journal." A few months later, this paper was published in *Europhysics Letters*. This book is devoted to a young generation of scientists: learn from the above examples, they are instructive. Read the prolog to this Chapter again.

In this Chapter we shall see that tunneling measurements performed in cuprates **can** detect two energy gaps. Secondly, tunneling and ARPES measurements **do not** provide identical information. Finally, nothing is wrong with

ARPES measurements—ARPES data are correct and so are the tunneling data. In fact, these two excellent techniques supplement each other. In the following Section we shall see that the spectra shown in Figs. 12.1 and 12.2 reflect the physics of a Bose-Einstein condensate with two energy gaps $\Delta_p > \Delta_c$, where Δ_p is the pairing gap and Δ_c is the phase-coherence gap (the Bose-Einstein condensate gap).

2. Excitation spectrum of a Bose-Einstein condensate

Before we discuss tunneling measurements in cuprates, let us first consider the excitation spectrum of a Bose-Einstein condensate made of composite bosons—electron pairs. Assume that the pairing energy of two electrons, Δ_p, is larger than the condensation energy of the pairs, Δ_c, i.e. $\Delta_p > \Delta_c$. For simplicity, assume that both energy gaps are isotropic. What then is the excitation spectrum of this condensate? Deep below the transition temperature, there are no excitations if the supplied energy is lower than Δ_c (per electron). What will happen if $E = \Delta_c$? The electron pairs become excited and can escape the condensate, being still paired. Thus, the energy $E = \Delta_c$ is not enough to break the composite bosons; however, being already enough to excite them, they can leave the coherent quantum state. The peak in the excitation spectrum at $E = \Delta_c$ is schematically shown in Fig. 12.3a. The next peak is at energy $\sqrt{\Delta_c^2 + \Delta_p^2}$, as shown in Fig. 12.3a, which corresponds to single-electron excitations. Thus, the energy $\sqrt{\Delta_c^2 + \Delta_p^2}$ is enough to break up the condensed pairs. In Fig.12.3a, the first peak corresponds to Cooper-pair excitations, while the second to single-electron excitations. In Fig. 12.3a we assume that below

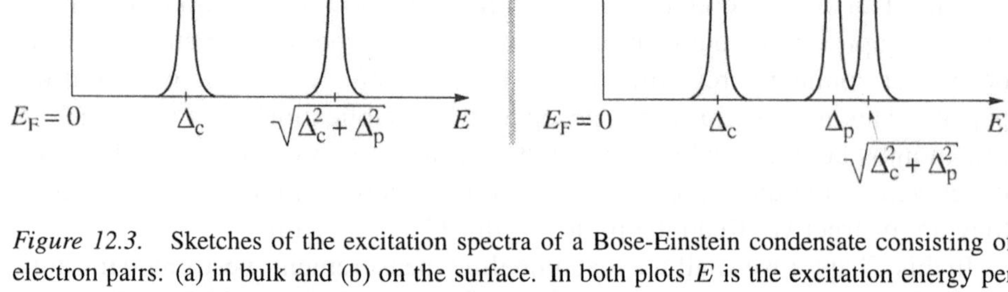

Figure 12.3. Sketches of the excitation spectra of a Bose-Einstein condensate consisting of electron pairs: (a) in bulk and (b) on the surface. In both plots E is the excitation energy per electron, E_F is the Fermi level, Δ_c is the Bose-Einstein condensation energy, and Δ_p is the pairing energy of electrons. For simplicity, $2\Delta_c \simeq \Delta_p$.

the transition temperature, all the composite bosons are condensed.

In Fig. 12.3a, the spectrum, in fact, represents bulk excitations. We are interested in to know the excitation spectrum on the surface: is it different from that in the bulk? This depends on the "field" which mediates the Bose-Einstein condensation. If, for example, the Josephson coupling mediates the long-range phase coherence, then the excitation spectrum on the surface will be identical to that in the bulk. In the case of magnetic fluctuations, the excitation spectrum on the surface will differ from the bulk-excitation spectrum. This is due to the fact that in some magnetic materials, the surface layer is magnetically "dead" [4]. In other magnetic materials the surface layer is still "alive", but the spectrum of magnetic excitations on the surface is different from that in the bulk [4]. Therefore, in the case of a magnetic condensation "field", the electron pairs may exist on the surface being uncondensed. As a consequence, the *averaged* excitation spectrum on the surface will have an additional peak at $E = \Delta_p$ (per electron), between Δ_c and $\sqrt{\Delta_c^2 + \Delta_p^2}$, as shown in Fig. 12.3b. The peak at $E = \Delta_p$ corresponds to breaking up the uncondensed electron pairs present on the surface.

In Fig. 12.3, the peaks are narrow because we assumed that the energy gaps are isotropic. If one or both energy gaps are anisotropic, then each peak in Fig. 12.3 will have a "tail" on its left-hand side. Thus, in this case, the excitation spectrum will be continuous or nearly continuous, having some distribution in energy.

3. Two energy gaps in cuprates

The excitation spectra shown in Fig. 12.3 are directly related to cuprates. In a sense, such a situation is unusual because, in conventional superconductors, the excitation spectrum has only one peak at $E = \Delta_p$ (per electron) corresponding to single electron excitations. *In cuprates, it is easier to excite a Cooper pair than to break it.* In a classical view of a superconducting condensate, this fact is new.

Since both energy gaps Δ_c and Δ_p in cuprates are highly anisotropic the tunneling spectra obtained in cuprates may not have two or three narrow peaks, as those in Fig.12.3, but one wide hump below $E = \sqrt{\Delta_c^2 + \Delta_p^2}$.

It is worth recalling that tunneling measurements probe the *local* density of states of quasiparticle excitations, while ARPES measurements test the density of states averaged over the surface.

3.1 Bi2212

From the discussion in the previous Section, it is clear that the two tunneling spectra in Fig. 12.1 represent two different tunneling regimes: the upper conductance corresponds to Cooper-pair tunneling, while the lower curve to single-electron tunneling. This is the reason these two curves have different shapes.

In fact, in Bi2212 there is a distribution of the energy gap [1, 5]. Figure 12.4 shows three sets of tunneling conductances obtained in Bi2212 single crystals having $T_c \simeq$ 87–90 K. In Fig. 12.4, the spectra in each plot are measured in one single crystal. One can see clearly that there is a gap distribution in Bi2212. In plot (c) the lower curve has two peaks. The doping level of these Bi2212 single crystals is around $p \simeq 0.19$. Consequently, from Fig. 3.22, one can observe that at $p = 0.19$, $2\Delta_c \simeq 42$ meV; $2\Delta_p \simeq 61$ meV, and $2\sqrt{\Delta_c^2 + \Delta_p^2} \simeq 74$ meV. Thus, the measured values of gap magnitude, given in Fig. 12.4, are in good agreement with the calculated values. In Fig. 12.4, the largest gap magnitudes seem to correspond to the value $2\sqrt{\Delta_c^2 + \Delta_p^2}$, while the smallest magnitudes to the value $2\Delta_c$.

In Figs. 12.1 and 12.4, the conductances are obtained in SIS junctions. The Copper-pair tunneling in SIS junctions is the usual phenomenon. However, as mentioned in Section 1, the Cooper-pair tunneling near $2\Delta_c/e$ can only be observed if the normal resistance of the junction is relatively low ($<$ 5 kΩ), where e is the electron charge. At a very highly resistive SIS junction, only single-electron peaks are observable; thus, the peaks at/near $2\Delta_p/e$ and $2\sqrt{\Delta_c^2 + \Delta_p^2}/e$ in Fig. 12.3b. This is one of the reasons why the conductances with large gap magnitude in Figs. 12.1 and 12.4 do not have a zero-bias peak due to the Josephson current—the normal resistance of these junctions was very large ($>$ 0.1 MΩ).

In Bi2212, the gap distribution is observable not only in SIS junctions but in SIN junctions as well [5]. What is interesting is that even in SIN junctions, the lower-energy peak corresponds to Cooper-pair tunneling! Such a situation is unusual for SIN junctions. However, this may occur if the resistance of an SIN junction is low. It is possible to verify this statement—one must measure the charge magnitude of tunneling quasiparticles at a bias of Δ_c/e and Δ_p/e: The charge magnitude of quasiparticles tunneling at a bias of Δ_c/e must be $2e$, whilst that of a bias Δ_p/e is just e.

The discussion in the previous paragraph can explain why ARPES measurements can observe only one energy gap, as shown in Fig. 12.2. In a sense, the ARPES technique is analogous with an SIN tunneling junction having a *very* large normal resistance. Therefore, depending on surface conditions of a measured sample, the magnitude of the energy gap, inferred from ARPES measurements, is either Δ_p or $\sqrt{\Delta_c^2 + \Delta_p^2}$. The ARPES technique is not able to observe the Cooper-pair excitation peak shown in Fig. 12.3. However, ARPES measurements have one unquestionable advantage with respect to tunneling measurements, their high angular resolution. In addition, ARPES is at present the only available way to probe the momentum of electrons in cuprates. Hence,

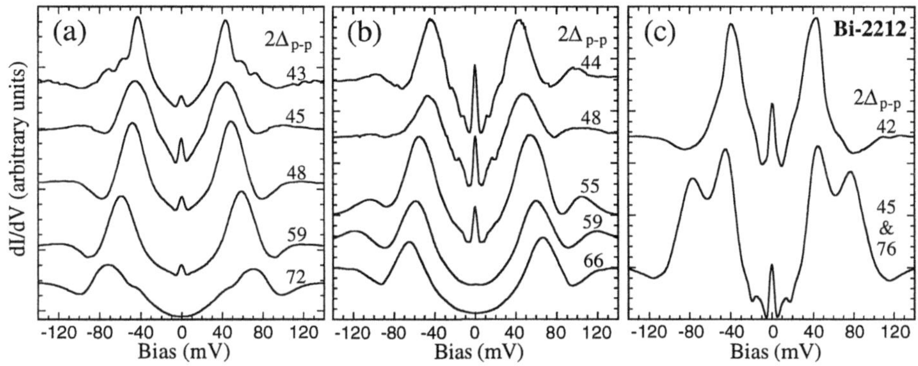

Figure 12.4. Three sets of tunneling conductances measured in Bi2212 single crystals with $T_c \simeq$ 87–90 K by SIS junctions [4]. The spectra in each plot are obtained in one single crystal. In each plot the curves are offset vertically for clarity. The peak-to-peak values $2\Delta_{p-p}$ are given in millielectronvolts. In plot (c) the lower conductance has the double-peak structure.

one has to remember that ARPES data mainly provide information about the magnitude and the symmetry of the *pairing* energy gap Δ_p.

3.2 YBCO and Tl2201

The values of T_c in optimally doped Tl2201, YBCO and Bi2212 are very similar, 93–95 K. So it is possible to compare directly the magnitudes of the two energy gaps in these three cuprates. Figure 12.5 shows three sets of tunneling conductances obtained in slightly overdoped single crystals of Tl2201, YBCO and Bi2212. These data are obtained in SIN junctions. In Fig. 12.5c, the two conductances obtained in Bi2212 represent the smallest and the largest magnitudes of energy gap (see Fig. 12.4).

In Fig. 12.5b, one can see that the two conductances obtained in YBCO clearly display the presence of two energy gaps with magnitudes similar to those in Bi2212. In Fig. 12.5b, the subgap in both conductances at ± 5 mV is attributed to an induced superconducting gap on Cu–O chains [6] (see also Fig. 6.10).

Tunneling measurements performed in Tl2201 exhibit only the presence of one energy gap, as shown in Fig. 12.5a. The magnitude of this gap is in good agreement with Δ_c of Bi2212 and YBCO. Does it mean that the pairing gap Δ_p is absent in Tl2201? Not at all. Simply, Tl2201 is *magnetically* more homogeneous than Bi2212 and YBCO. Second, the interlayer distance in Tl2201 is the largest in all cuprates, therefore, the nearest CuO_2 plane to the surface is situated much deeper than those in Bi2212 and YBCO (the measurements in Tl2201 were performed along the *c* axis). These two factors taken together may explain the *apparent* absence of pairing gap in Tl2201. At the same time,

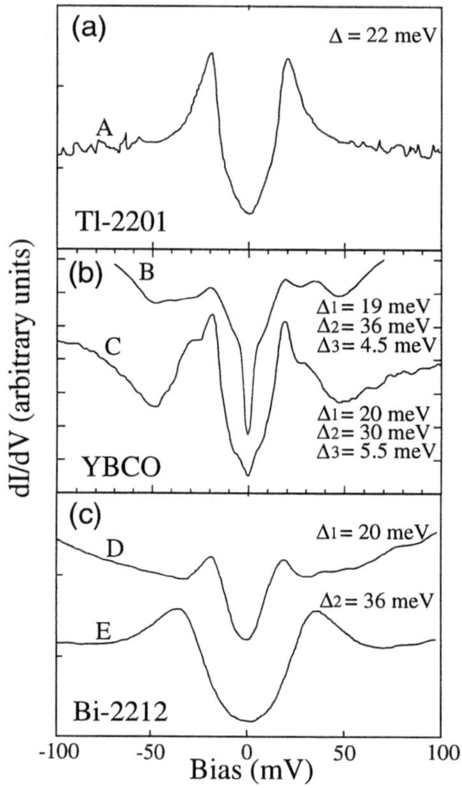

Figure 12.5 Tunneling conductances obtained in slightly overdoped single crystals of (a) Tl2201 (T_c = 91 K); (b) YBCO (T_c = 89 K, curve B; and T_c = 91 K, curve C), and (c) Bi2212 (T_c = 91 K). The data are obtained in SIN junctions. The curves D and E are measured within the same sample. The curves B and D are shifted up for clarity. The graph is taken from [6].

in-plane torque anisotropy measurements performed in Tl2201 show the existence of two energy gaps: the amplitude of one gap is about two times larger than the amplitude of the other [7]. It is worth recalling that, in contrast to the tunneling technique, the torque measurement is a *bulk* experiment. In addition, these torque measurements showed that the two gaps in Tl2201 have different symmetries (see Chapter 10).

It is worth noting that, in all these three cuprates, the long-range phase coherence is mediated by a magnetic resonance peak. The energy position of the resonance peak, E_r, in these three cuprates is slightly different. As a consequence, the values of Δ_c in YBCO, Bi2212 and Tl2201 are slightly different too. The ratio E_r/k_BT_c is about 5.2–5.3 in YBCO; 5.5 in Bi2212, and 5.9 in Tl2201. As discussed in Chapter 9, the ratio $2\Delta_c/k_BT_c$ in each of the cuprates must coincide with the corresponding number above. One then obtains that, in *optimally* doped YBCO, Bi2212 and Tl2201, the magnitude of Δ_c is around 21 meV, 22.5 meV and 24 meV, respectively.

3.3 Phase diagram

Figure 12.3b is adjusted for cuprates, and the result is shown in Fig. 12.6. In Fig. 12.6, three energy scales Δ_c, Δ_p and $\sqrt{\Delta_c^2 + \Delta_p^2}$ are shown as a function of doping. In cuprates, depending on the type of experiment (bulk or surface-layer sensitive), measurements *may* exhibit one, two or three energy scales shown in Fig. 12.6.

In Fig. 12.6, one can observe that the two energy scales Δ_p and $\sqrt{\Delta_c^2 + \Delta_p^2}$ have similar magnitudes.

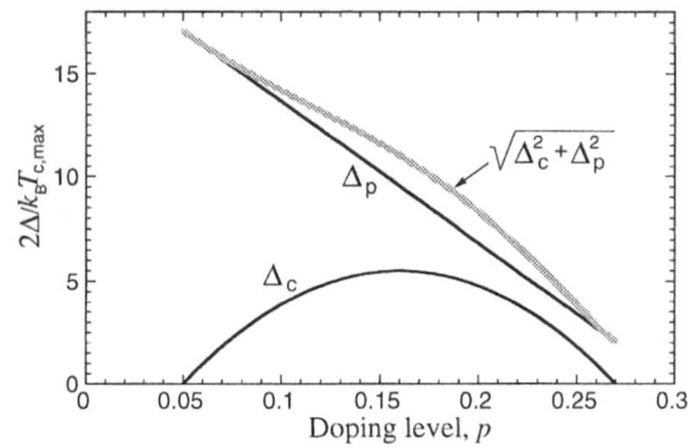

Figure 12.6. Low-temperature phase diagram of superconducting cuprates: the pairing energy scale Δ_p and the Cooper-pair condensation scales Δ_c (see also Fig. 3.22).

3.4 Two energy gaps in magnetic field

How does an applied magnetic field affect Cooper-pair and single-electron tunneling in cuprates?

Figure 12.7 shows the conductance from Fig. 12.5b (curve B), obtained in an YBCO/Pb junction. All the features of this conductance were already discussed in Section 3.2. In Fig. 12.7, the dotted line shows the conductance taken in a magnetic field of 8 T. In Fig. 12.7, one can observe that an applied magnetic field of 8 T slightly suppresses the Cooper-pair tunneling at ± 19 mV. At the same time, the single-electron tunneling at ± 36 mV is not affected by this field. Thus, in YBCO, a magnetic field of 8 T is not strong enough to suppress single-electron tunneling, enough to affect the Cooper-pair tunneling.

In Bi2212, even a magnetic field of 20 T does not affect the single-electron tunneling, as shown in Fig. 12.8. At the same time, this field is strong enough to suppress Cooper-pair tunneling at zero bias (the *dc* Josephson effect). In Fig. 12.8 an apparent increase of the gap magnitude in the magnetic field can be an artifact. On the other hand, this effect can be due to a "switching" from a surface-situated Δ_p energy scale to a bulk-situated $\sqrt{\Delta_c^2 + \Delta_p^2}$ energy scale.

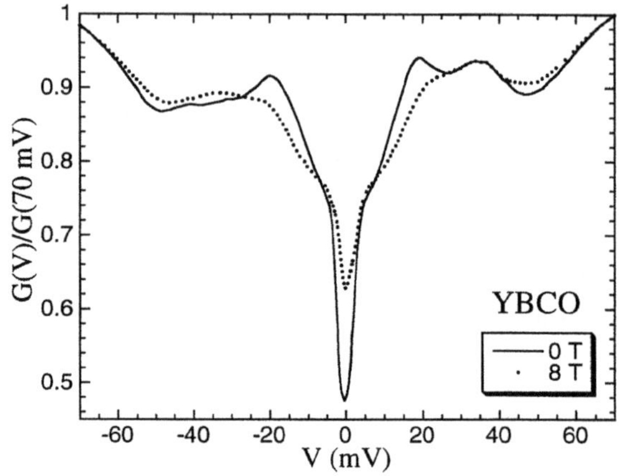

Figure 12.7. Tunneling conductances for YBCO/Pb junction in $B = 0$ T (solid line) and in 8 T (dotted line), obtained at 10 K. The slightly underdoped YBCO has $T_c = 89$ K. The applied field is parallel to the c axis [8].

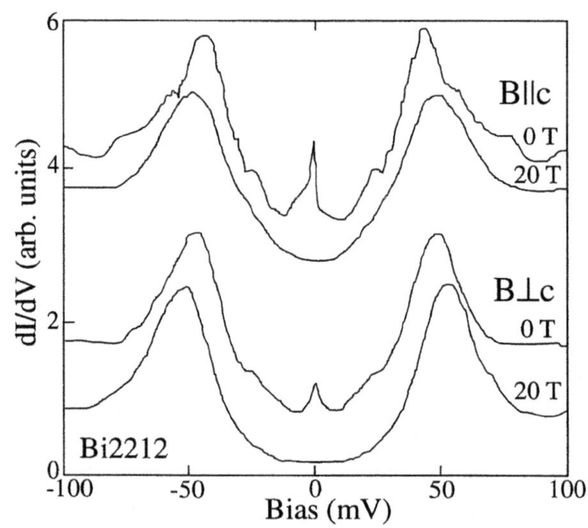

Figure 12.8. Tunneling conductances obtained at 4.2 K in two Bi2212 break junctions (SIS junctions) in $B = 0$ T and 20 T. The slightly overdoped Bi2212 has $T_c \simeq 80$ K [9]. The orientation of the magnetic field is indicated. Three curves are offset vertically for clarity.

4. Pseudogap

The pseudogap which can be observed in tunneling and ARPES measurements is a charge gap on charge stripes, being some kind of charge-density-wave gap. In this Chapter, the word "pseudogap" refers to this charge gap. The temperature dependence of a pseudogap in a combination with a pairing electron gap will be considered in Section 7 below. In this Section, we discuss two questions—the pseudogap renormalization at T_c and how the pseudogap affects tunneling $I(V)$ characteristics.

Consider first the pseudogap renormalization at T_c. Figure 12.9a schematically shows the pseudogap above and deep below T_c. Above the critical temperature, there are quasiparticle excitations at the Fermi level (at zero bias in Fig. 12.9a). As the temperature is lowered through T_c, quasiparticle excitations at the Fermi level, thus inside the pseudogap, are strongly renormalized: the pseudogap deepens at low bias, as shown in Fig. 12.9a. Simultaneously, as one can see in Fig. 12.9a, the maximum magnitude of pseudogap (humps) above and below T_c is in first approximation unchanged. Analysis of ARPES data in [3] suggests that a full gap opens up at the Fermi level, as shown in Fig. 12.9b. At $T \ll T_c$, the magnitude of this full gap at the Fermi level is slightly smaller than the magnitude of the pairing gap Δ_p, and its temperature dependence is reminiscent of the BCS temperature dependence [3]. At this moment, there is a lack of evidence showing what case is realized in cuprates—the case in Fig. 12.9a or the case in Fig. 12.9b. The main point, however, is that, as the temperature is lowered through T_c, the pseudogap deepens at low bias, while the magnitude of the pseudogap is practically unchanged.

In the case shown in Fig. 12.9b, inferred from ARPES data [3], one can conclude that, deep below T_c the pseudogap has an anisotropic s-wave symmetry and an anisotropy ratio of about 3: If at $T \ll T_c$ the minimum magnitude of the pseudogap is $\Delta_{pg,min} \approx \Delta_p$, then $\Delta_{pg,max}/\Delta_{pg,min} \simeq 3\Delta_p/\Delta_p = 3$.

We now discuss the second issue raised above—how the pseudogap affects tunneling $I(V)$ characteristics in cuprates. In Chapter 4, we already discussed $I(V)$ characteristics obtained above T_c in overdoped Bi2212 single crystals. In Fig. 4.3 the $I(V)$ curves have a small "negative" offset from the straight line at high bias. Is the pseudogap responsible for this "negative" offset? The answer is "yes."

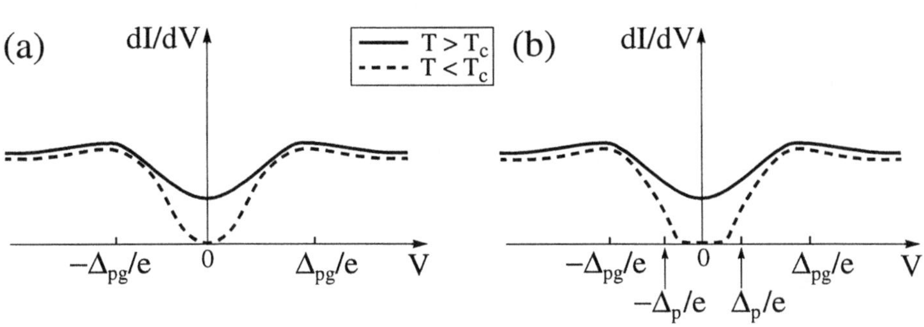

Figure 12.9. Pseudogap (charge gap) above and deep below T_c, inferred from (a) tunneling and (b) ARPES [3] measurements. In both plots, the solid lines are slightly shifted up for clarity.

Figure 12.10a shows an $I(V)$ curve obtained at 98 K in an underdoped Bi2212 single crystal having $T_c = 51$ K. It is worth emphasizing that the tunneling characteristics in Fig. 12.10a are measured at a temperature which is near the double T_c value. For comparison, Figure 12.10b presents the plot (a) from Fig. 4.3. The data in Fig. 12.10b are taken at 119 K in an overdoped Bi2212 single crystal having $T_c = 88$ K. In Fig. 12.10, one can clearly see that, in underdoped Bi2212, this "negative" offset is larger than that in overdoped Bi2212. From Fig. 10.8, the magnitude of the tunneling pseudogap (the T_{CO} scale) decreases as the doping increases. Consequently, the amplitude of this "negative" offset scales with the magnitude of the pseudogap. Therefore, it is reasonable to associate this "negative" offset with the pseudogap (the charge gap).

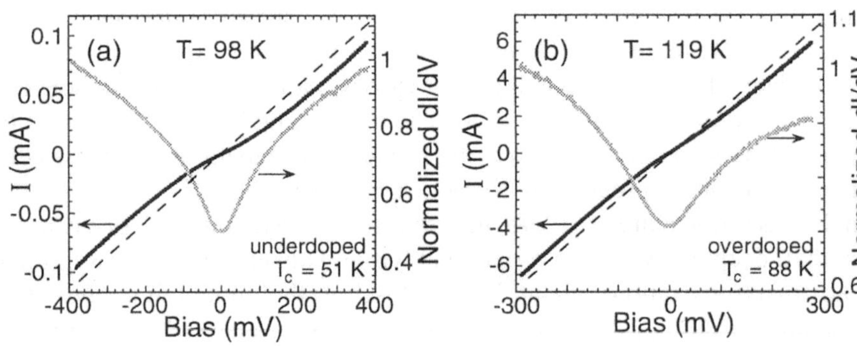

Figure 12.10. Tunneling pseudogap in Bi2212. (a) $I(V)$ (black curve) and $dI(V)/dV$ (grey curve) characteristics obtained at $T = 98$ K in an SIS junction of an underdoped Bi2212 single crystal with $T_c = 51$ K. (b) $I(V)$ (black curve) and $dI(V)/dV$ (grey curve) characteristics obtained at $T = 119$ K in an SIS junction of an overdoped Bi2212 single crystal with $T_c = 88$ K, shown also in Fig. 4.3a. In both plots the dashed lines are parallel to the $I(V)$ curves at high bias. Note the difference in horizontal scales in the plots.

It is important noting that the "pseudogap" in Fig. 6.1b is in fact not a pseudogap below T_c—it is a product of tunneling between a pseudogap and bisolitons. This is because the data in Fig. 6.1b are obtained in an SIS junction. In the following Section, we shall discuss the process of SIS-junction tunneling in cuprates in detail. *Nevertheless*, the $I(V)$ characteristic in Fig. 6.1b still has this "negative" offset at high bias with an amplitude similar to that in Fig. 12.10a.

5. Pairing gap and pseudogap

In this Section, we consider exclusively single-electron tunneling in cuprates, thus tunneling in SIN and SIS junctions having a very large value of normal resistance. In such a tunneling regime, only two components of the quasiparticle-excitation spectrum of cuprates can be tested—the pseudogap and bisoliton excitations. At the same time, the Cooper-pair excitations cannot be tested.

5.1 Two contributions to tunneling spectra

As discussed in Chapters 4 and 6, tunneling $dI(V)/dV$ and $I(V)$ characteristics consist of two contributions caused by the superconducting condensate (bisolitons) *and* by the pseudogap. In SIN-junction conductances, these two contributions are superimposed *linearly.* The position of quasiparticle peaks in the conductance is defined by a magnitude of the pairing gap, $V = \pm\Delta_p/e$. Independently, the position of pseudogap humps in the conductance is defined by a magnitude of pseudogap, $V = \pm\Delta_{pg}/e$. As discussed in Chapters 3, 8 and 10, the magnitude of the pseudogap is about three times larger than the magnitude of the pairing gap, $\Delta_{pg} \simeq 3\Delta_p$. Thus, in an SIN junction, the quasiparticle peaks appear at bias $V = \pm\Delta_p/e$ while the pseudogap humps at bias $\pm 3\Delta_p/e$.

At the same time, in SIS-junction conductances, these two contributions are superimposed *nonlinearly* because, in SIS junctions, the pseudogap and bisolitons interact with each other. We are going to discuss this issue in the following subsection.

It is worth mentioning that in SIN-junction conductances, the dips which appear between quasiparticle peaks and pseudogap humps, thus approximately at $V \simeq \pm 2\Delta_p/e$, **have no physical meaning** [1]. In SIS junctions, these dips appear at $\pm 3\Delta_p/e$.

5.2 SIN and SIS junctions of cuprates

In conventional superconductors, the distance between quasiparticle peaks in a conductance obtained in an SIN junction is defined by a magnitude of energy gap, and equals $2\Delta/e$. In an SIS junction, this distance becomes $4\Delta/e$. In cuprates, this "rule" is satisfied too. However, the pseudogap humps in conductances obtained in an SIN junction *and* in an SIS junction do not follow this "rule." In an SIN-junction conductance, the distance between pseudogap humps equals $6\Delta_p$. Then the hump-hump distance in SIS-junction conductances must be $12\Delta_p/e$. In reality, however, the distance between "pseudogap humps" in an SIS-junction conductance equals $8\Delta_p$. This clear disagreement between experiment and "theory" puzzled specialists working in the field of tunneling in cuprates during a long period of time (including myself). The purpose of this subsection is to unravel this puzzle.

As discussed above, in an SIN junction, the quasiparticle peaks appear at bias $V = \pm\Delta_p/e$, while the pseudogap humps at bias $\pm 3\Delta_p/e$. As an example, the energy position of quasiparticle peak in Fig. 12.2 is about 40 meV, while the hump has a maximum at 120 meV. Figure 12.2 depicts, in fact, an

ARPES spectrum but, as shown above, ARPES spectra are analogous with tunneling data obtained in an SIN junction having a large value of normal resistance. So, the ARPES spectrum in Fig. 12.2 represents all features of SIN-junction tunneling data.

Figure 12.11a *schematically* shows the density of states of quasiparticle excitations in a cuprate at zero temperature. At $T = 0$ electron tunneling between a cuprate and a normal metal simply reproduces the density of states of the cuprate, as shown in Fig. 12.11b. Thus, the principle of tunneling in an SIN

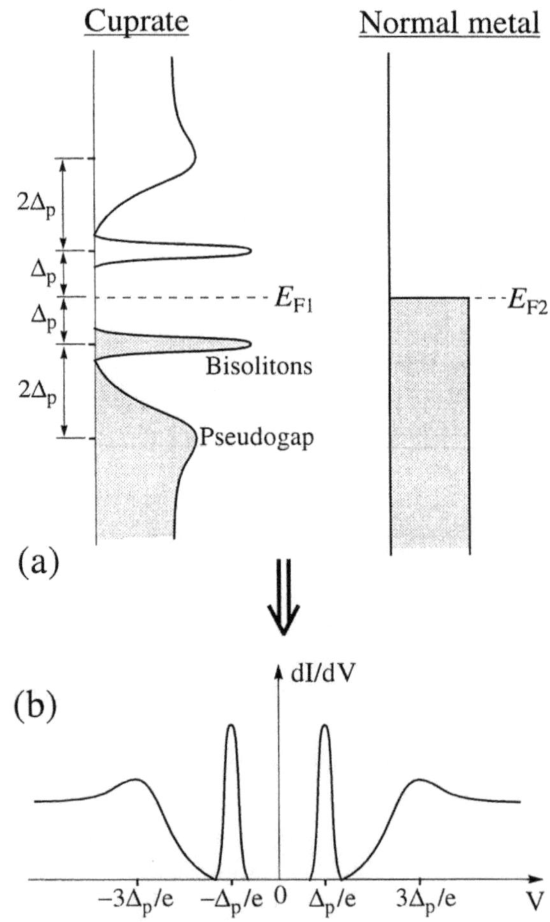

Figure 12.11. Sketch of tunneling between a cuprate and a normal metal (SIN junction) at $T = 0$. (a) The density of states of quasiparticle excitations in a cuprate and in a normal metal. In the cuprate the pseudogap and the bisoliton peaks are shown schematically. The Fermi levels E_{F1} and E_{F2} are shown by the dashed lines. (b) The tunneling conductance for this junction is shown schematically. The magnitude of the pseudogap is $3\Delta_p$, where Δ_p is the pairing energy gap.

junction is the same for cuprates and for conventional superconductors (see Figs. 2.7 and 2.13). This, however, is not the case for SIS junctions.

Figure 12.12 shows the conductance obtained in a slightly overdoped Bi2212 single crystal with T_c = 88 K by an SIS junction. In Fig. 12.12, one can see that, in addition to the quasiparticle peaks at bias $V = \pm 60$ mV (= $2\Delta/e$), the curve has humps at ± 120 mV and ± 180 mV. The humps at bias $V = \pm 120$ mV (= $4\Delta/e$) are sufficiently large, while at $V = \pm 180$ mV (= $6\Delta/e$) small. In this subsection, the humps at $V = \pm 4\Delta/e$ in SIS-junction conductances we shall call *the main humps*, while the humps at $V = \pm 6\Delta/e$ *the small humps*. In most cases, the small humps in SIS-junction conductances are absent. The conductance in Fig. 12.12 is presented intentionally. Now we are in a position to discuss the main question: If the pseudogap magnitude equals $3\Delta_p$, then the humps in an SIS-junction conductance, corresponding to the pseudogap, should appear at bias $V = \pm 6\Delta_p/e$, and not at $\pm 4\Delta_p/e$. Why is this?

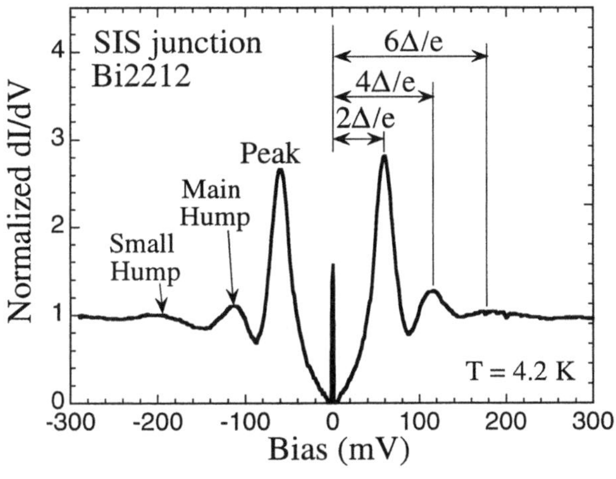

Figure 12.12. Tunneling conductance obtained in a slightly overdoped Bi2212 single crystal with T_c = 88 K by an SIS junction. The curve has the quasiparticle peaks at $V = \pm 2\Delta/e$, the main humps at $\pm 4\Delta/e$, and the small humps at $\pm 6\Delta/e$, where Δ is the pairing energy gap, and e is the electron charge. The origins of the main and small humps as well as the peaks are explained in Figs. 12.13–12.15.

Figure 12.13 explains the origin of small humps in SIS-junction conductances: The small humps at bias $V = \pm 6\Delta_p/e$ are due to pseudogap-to-pseudogap tunneling, as schematically shown in Fig. 12.13. As the bias decreases, an "additional" tunneling occurs at bias $V = \pm 4\Delta_p/e$, as shown in Fig. 12.14, resulting in the appearance of the main humps in SIS-junction conductances. Thus, the main humps in SIS-junction conductances originate from peak-to-pseudogap tunneling. In a sense, the peak-to-pseudogap tunneling at bias $\pm 4\Delta_p/e$ is "double", as schematically depicted in Fig. 12.14. This is one

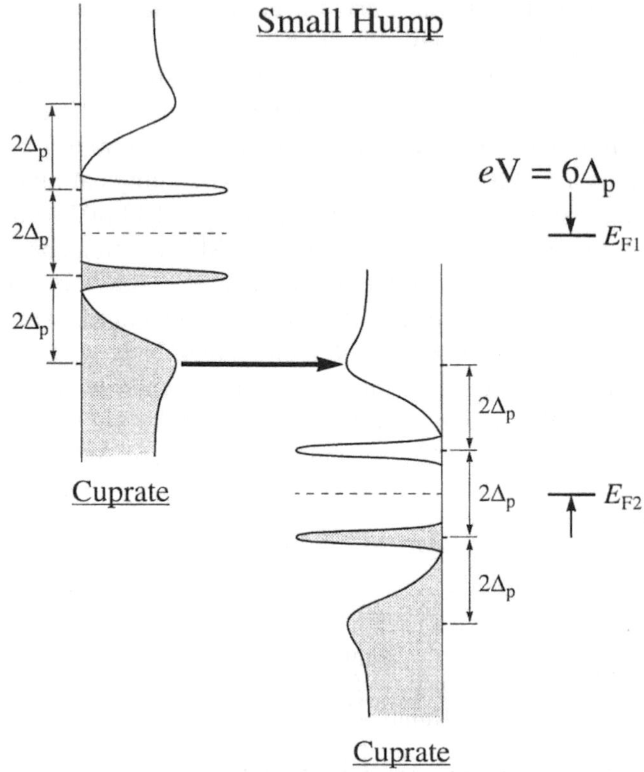

Figure 12.13. Cuprates: tunneling in an SIS junction at $T = 0$. The sketch explains the origin of small humps in conductances, appearing at bias $V = \pm 6\Delta/e$, as shown in Fig. 12.12. The thick arrow indicates the tunneling process.

of two reasons why the main humps in SIS-junction conductances are more pronounced than the small humps, as illustrated in Fig. 12.12. The second reason is that the amplitude of the bisoliton peaks is larger than the maximum amplitude of the pseudogap.

Finally, Figure 12.15 shows the process of peak-to-peak tunneling in an SIS junction. It is obvious that peak-to-peak tunneling occurs at bias $V = \pm 2\Delta_p/e$. As discussed in Chapter 5, near a peak bias, the bisoliton density of states has a $sech^2$-function shape. Then, the bisoliton peaks in an SIN-junction conductance must also have the same shape (see Fig. 12.11b). The shape of bisoliton peaks in an SIS-junction conductance will be considered in Section 14.3 below.

Thus, the puzzle is solved. One can now understand why the SIS-junction conductance in Fig. 6.1b, obtained below T_c, does not correspond directly to the pseudogap: it is a product of peak-to-pseudogap tunneling. In Fig. 6.1b,

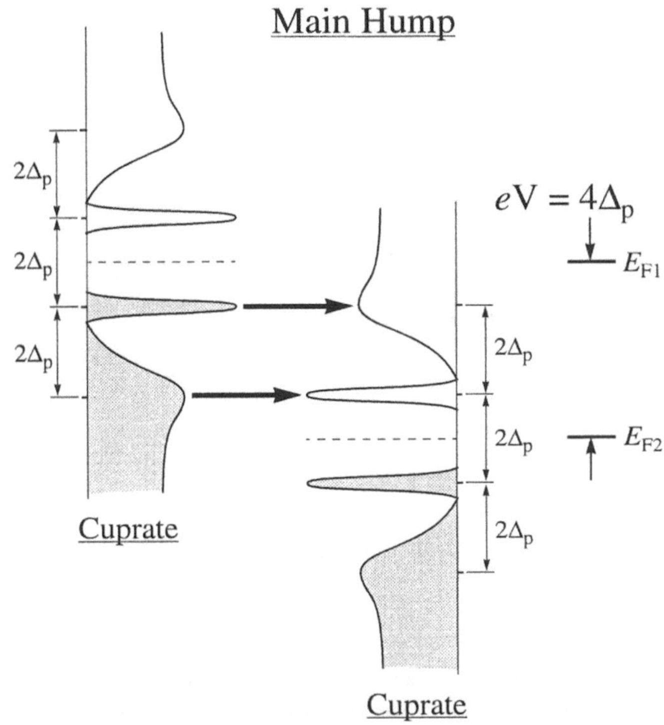

Figure 12.14. Cuprates: tunneling in an SIS junction at $T = 0$. The sketch explains the origin of the main humps in conductances, appearing at bias $V = \pm 4\Delta/e$ as shown in Fig. 12.12. The thick arrows indicate the tunneling process.

the shape of the pseudogap can be deduced from the conductance by deconvoluting it with the bisoliton density of states shown in Fig. 6.2. In Figs. 6.6b and 6.6c, the small humps in SIS-junction conductances at a bias equal to the double value of peak bias have the same origin—they are the product of peak-to-pseudogap tunneling. At the same time, in Fig. 6.6b (Fig. 6.6c), the humps at ± 130 mV (± 110 mV), thus at a bias near the triple value of peak bias, originate from a pseudogap-to-pseudogap tunneling. The fact that the conductance in Fig. 6.1b is not the pseudogap does not affect the conclusions of Chapter 6. The opposite is true: this fact fully supports all data transformations made in Chapter 6.

6. Subgap

The SIS-junction tunneling conductances often have subgaps. As an example, Figure 12.16 shows a conductance obtained in a slightly overdoped Bi2212 single crystal, which has a subgap. We are interested in to know the origin of this subgap. Many specialists were in fact puzzled by this subgap

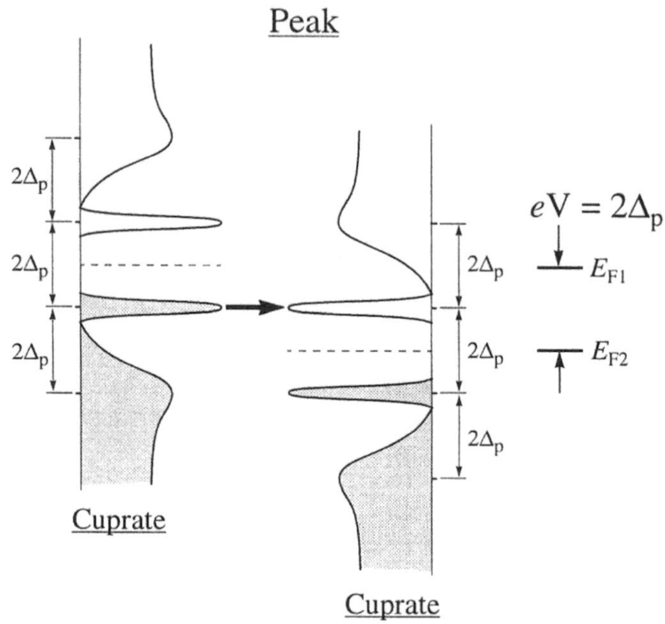

Figure 12.15. Cuprates: tunneling in an SIS junction at $T = 0$. The sketch explains the origin of conductance peaks at $V = \pm 2\Delta/e$. The thick arrow indicates the tunneling process.

in SIS-junction conductances, but nobody dared to give an explanation for its origin.

Analysis of tunneling data indicates that the subgap exclusively appears in conductances obtained in SIS junctions having a low resistance, thus when Cooper-pair excitations are tested (see Section 2), and never in junctions with a high resistance. The subgap appears in SIS-junction conductances inside the gap structure. The subgap temperature dependence is similar to that of Δ_c. So, from the beginning, it was clear to me that this subgap is directly related to the "field" which mediates the phase coherence in cuprates, thus to spin fluctuations. I even proposed that the subgap can be caused by a g-wave magnetic order parameter [1], since theoretically, the magnitude of a g-wave order parameter (if such exists) must be about half of the magnitude of d-wave order parameter, thus a half of Δ_c.

One may suggest that the subgap in SIS-junction conductances having a low resistance is caused by Andreev reflections of normal electrons. In cuprates, for example, the pseudogap excitations are normal-state excitations. In Fig. 12.16, however, one can see that near the tops of quasiparticle peaks there is the "second" harmonic of this subgap. Andreev reflections cannot cause this "second" harmonic.

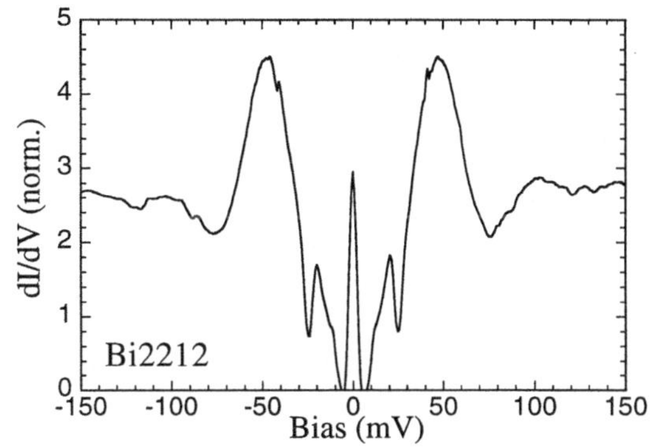

Figure 12.16. Conductance obtained in the same Bi2212 single crystal as those in Fig. 12.1.

In fact, the subgap peaks are not important, the dips outside the subgap are responsible for their existence. What can cause the dips in conductances? In Fig. 12.16, the dips have a shape similar to the shape of dips appearing in conductance when electrons tunnel through an antiferromagnetic medium [10]. So, it is most likely that these dips originate from an interaction of tunneling *normal* electrons with an antiferromagnetic medium which they tunnel through. It is worth noting that the dips sometimes appear in conductances outside the gap structure.

7. Temperature dependence

In cuprates, the temperature dependence of quasiparticle peaks in tunneling conductances is a complex question. This is because there are three different gaps in cuprates, observable in tunneling measurements—two superconducting gaps, Δ_p and Δ_c, and one normal-state pseudogap. All the three gaps have different temperature dependences. In addition, all these gaps are anisotropic. So, performing tunneling measurements in cuprates, it is not easy to know exactly what gap (or combination of two gaps) is tested. Moreover, since tunneling measurements probe the *local* density of states of quasiparticle excitations, in addition to such intrinsic properties of cuprates, there is a technical problem. Since optimally-doped cuprates have a high critical temperature, therefore, performing the measurements at low and then at high temperatures, one cannot be sure that, at different temperatures, the tunneling junction remains the same. In an SIS junction, a small thermal expansion of a test setup can change the distance between two parts of a single crystal, leading to a drift of tunneling current from one spot to another.

In spite of the aforementioned difficulties, here we shall discuss real data obtained in Bi2212, trying to recover the physics of the superconducting state and of the normal state from the data. First we concentrate on a temperature region from deep below the critical temperature to a temperature slightly above T_c. Then, we focus our attention on tunneling data obtained in the normal state.

7.1 Superconducting state

Figure 12.17 shows the temperature dependence of the conductance from Fig. 12.1, obtained in a slightly overdoped Bi2212 single crystal by an SIS junction. As discussed in Section 3, the quasiparticle peaks in this conductance are caused by Cooper-pair tunneling. Then, it is logical to anticipate that the quasiparticle peaks in Fig. 12.17 must vanish at T_c and have a temperature dependence similar to that of phase-coherence gap Δ_c, shown in Fig. 9.7.

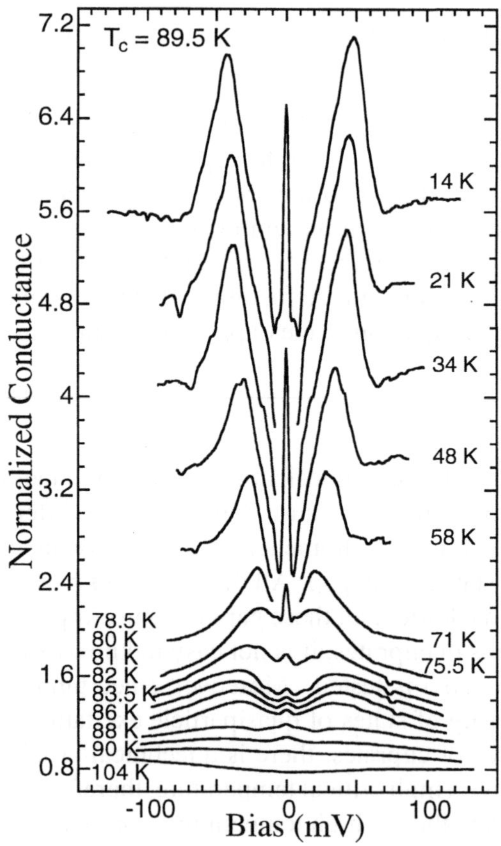

Figure 12.17 Temperature dependence of the conductance from Fig. 12.1, obtained in a Bi2212 single crystal with T_c = 89.5 K ($p \simeq 0.19$) by an SIS junction [1]. At 14 K the normal resistance of this junction was about 90 Ω. The spectra are offset vertically for clarity. In the spectra obtained at 21 K, 34 K, 58 K and 71 K, the zero-bias peak due to the Josephson current is removed for clarity.

In Fig. 12.17, one can see that the quasiparticle peaks indeed vanish at T_c. However, at high temperatures, there appears another gap: Figure 12.18 shows two curves from Fig. 12.17, enlarged vertically. The dashed lines in Fig.12.18 indicate the evolution of these two gaps. In Fig. 12.18, one can clearly see that, upon heating, one gap closes and the other expands. Figure

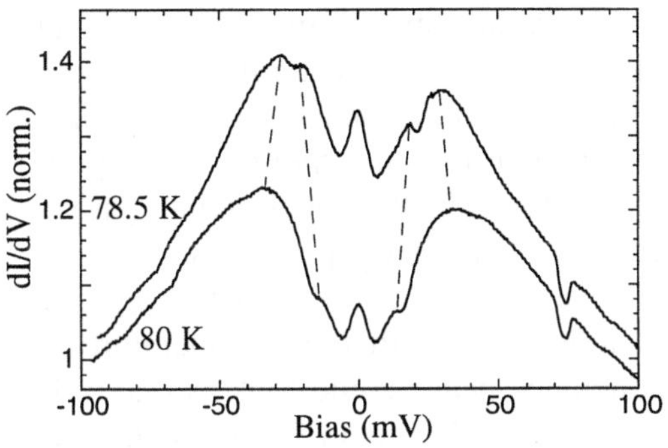

Figure 12.18. Two conductances from Fig. 12.17 enlarged vertically. The dashed lines indicate the evolution of two gaps.

12.19 shows the temperature dependence of the distance between conductance peaks in Fig. 12.17. In Fig. 12.19, above $T/T_c \approx 0.9$, there are two humps in the conductances, as illustrated in Fig. 12.18.

In Fig. 12.17, the gap magnitude at low temperature is about 23 meV, which is in good agreement with the magnitude of Δ_c in Bi2212. From Fig. 12.19, the magnitude of the other gap above T_c, 1.4×23 meV $\simeq$ 32 meV, is in good agreement with the magnitude of the pairing energy gap in Bi2212 having a doping level of 0.19, Δ_p = 30.7 meV.

We are now in a position to explain the data presented in Figs. 12.17–12.19. In Fig. 12.19, as the phase-coherence gap closes, the Cooper-pair excitations vanish at T_c. The other gap "appearing" near T_c corresponds to the pairing gap Δ_p. It is obvious that the magnitude of Δ_p in reality remains unchanged on

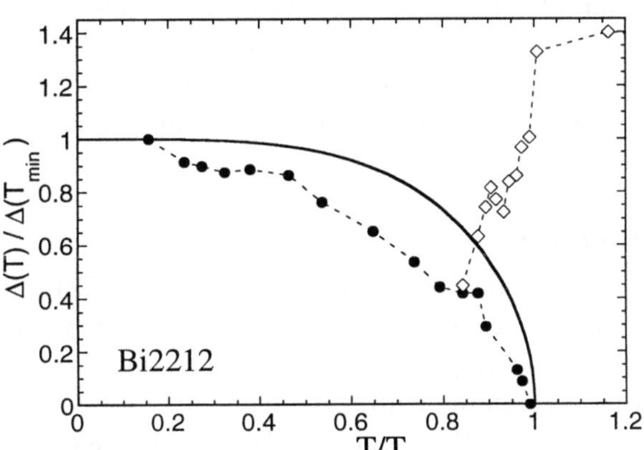

Figure 12.19. Temperature dependence of the distance between conductance peaks in Fig. 12.17. The dots and diamonds indicate two different gaps. The solid line shows the BCS temperature dependence.

crossing T_c. In Fig. 12.19, the apparent increase of the pairing gap, upon heating, is an artifact. In Fig. 12.17, at low temperature the junction was initially "tuned" to Cooper-pair excitations vanishing at T_c. This leads to a deepening of the conductance curves at low bias and revealing parts of conductances at higher bias, as shown in Fig. 12.18. This deepening at low bias creates a false effect that the magnitude of another gap increases.

If, for some reason we would not observe the closing of the phase-coherence gap at temperatures above $0.85T_c$, then the temperature dependence of the distance between conductance peaks could look as that in Fig. 12.20. The temperature dependence in Fig. 12.20 is in fact typical, at least, for Bi2212. The "jump" at T_c is larger in underdoped Bi2212, and decreases as the doping level increases.

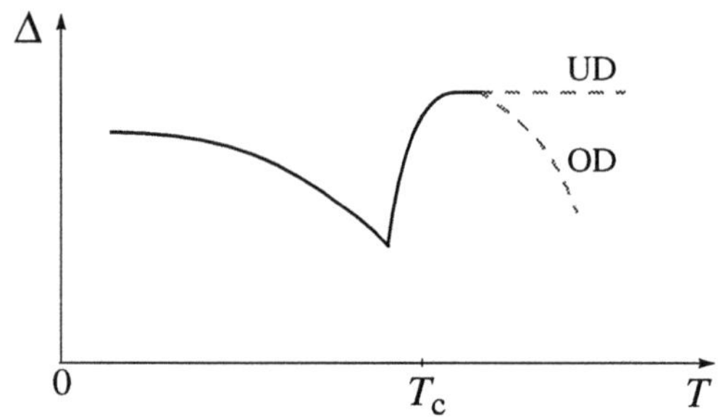

Figure 12.20. Sketch of the typical temperature dependence of the distance between conductance peaks in Bi2212. The grey dashed lines mark the gap evolution in underdoped (UD) and overdoped (OD) Bi2212 at high temperatures.

As the last example, Figure 12.21 shows the temperature dependence of two gaps in Bi2212, observed simultaneously in one conductance by an SIN junction. In Fig. 12.21, one can see that, upon heating, the magnitude of the larger gap remains unchanged, whilst the temperature dependence of the smaller gap is similar to that of Δ_c.

7.2 Normal state

Here we concentrate on tunneling data obtained in Bi2212 above the critical temperature. Figure 12.22 shows the temperature dependence of a conductance obtained in an overdoped Bi2212 single crystal by an SIS junction. In Fig. 12.22, there is no sign indicating at what temperature the phase-coherence gap was closed. Across T_c, the spectra evolve continuously into a normal-state pseudogap. In Fig. 12.22, the conductance obtained at 290 K seems to be smooth, without a sign of the pseudogap. Figure 12.23 depicted the 290 K conductance enlarged vertically. In Fig.12.23, one can see the conductance exhibits the presence of a weak pseudogap. The magnitude of the pseudogap

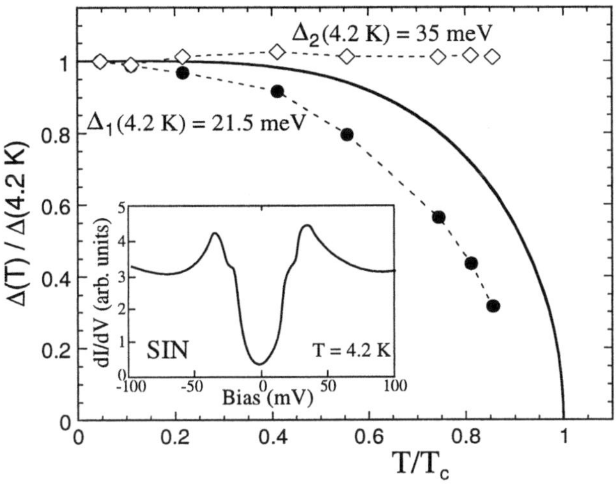

Figure 12.21. Temperature dependence of two gaps in an overdoped Bi2212 single crystal having T_c = 90 K, observed in one conductance by an SIN junction. The inset shows this conductance at 4.2 K (Fig. 3b in [11]). The magnitudes of the two gaps are given at 4.2 K.

is about 60 meV. One can then estimate the closing temperature of this pseudogap. From Eq. (10.2), the low-temperature magnitude of the pseudogap (charge gap) in Bi2212 with a doping level of p = 0.19 is about 92 meV. The pseudogap magnitude, upon heating, decreases to 60 meV at 290 K. Then, assuming that the pseudogap magnitude has a temperature dependence similar to the BCS temperature dependence, one can estimate that the closing temperature of this pseudogap is about 350 K. This value is in good agreement with T_{CO} = 360 K calculated from Eq. (10.1).

Figure 12.24 shows the temperature dependence of a conductance obtained in another Bi2212 single crystal above the critical temperature. In Fig. 12.24, one can see that the spectra are slightly asymmetrical about zero bias. This asymmetry is caused by an antiferromagnetic medium [10]. For convenience, it is easy to work with "more symmetrical" spectra, namely, with an even part of the conductances, defined as $G_e \equiv [G(V) + G(-V)]/2$. Figure 12.25 presents the even conductances of the spectra from Fig. 12.24, shown only at positive bias. In Fig. 12.25, one can observe that the conductances just above T_c have a hump which disappears at a temperature $T_0 > 100$ K. In Fig. 12.25, the spectra obtained above 103 K have another, small hump at high bias. Figure 12.26 depicts the temperature dependence of these two humps.

In Fig. 12.26, the temperature dependence of the low-bias hump is extrapolated: This gives an estimate of $T_0 \sim 115$ K. The temperature dependence of the high-bias hump indicates that, indeed, there are some changes in the system near 120 K. In addition, the temperature dependence of the high-bias

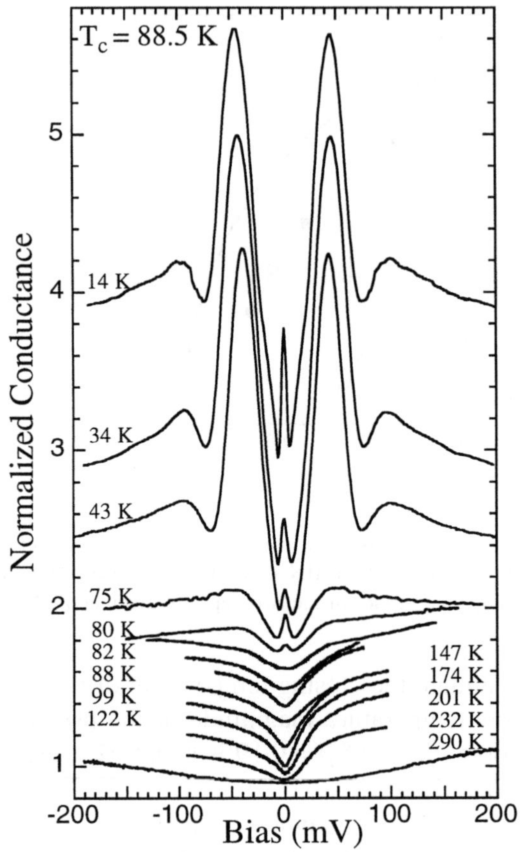

Figure 12.22 Temperature dependence of a conductance obtained in a Bi2212 single crystal with T_c = 88.5 K ($p \simeq$ 0.19) by an SIS junction. The spectra are offset vertically for clarity. The graph is taken from [12].

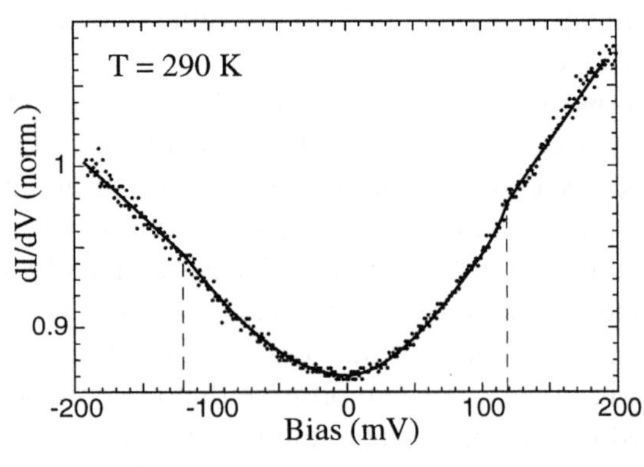

Figure 12.23. The 290 K conductance from Fig. 12.22 enlarged vertically. The lines are guides to the eye, and the dashed lines indicate the magnitude of a pseudogap.

hump shows that there are some changes near 210–220 K. Independently, the temperature dependence of zero-bias conductance (not shown) also indicates

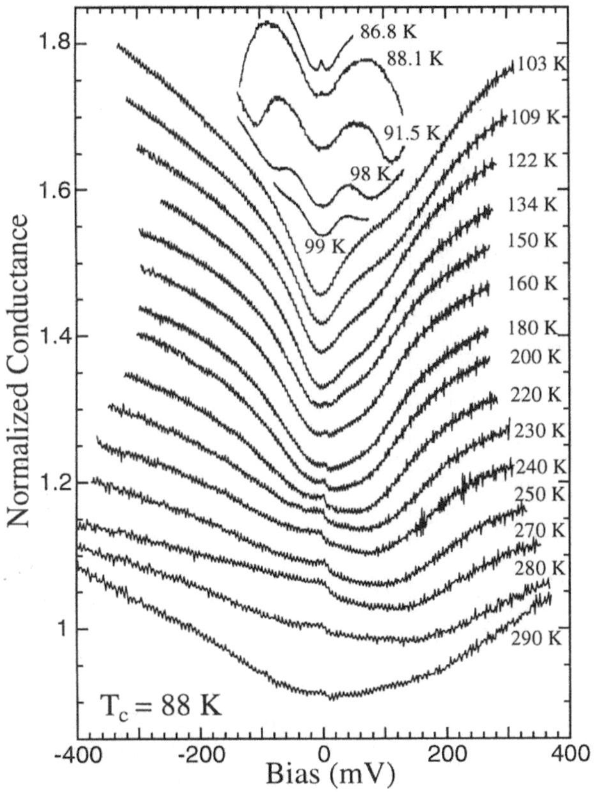

Figure 12.24 Temperature dependence of a conductance obtained in a Bi2212 single crystal with T_c = 88 K ($p \simeq$ 0.19) by an SIS junction. The spectra are offset vertically for clarity. The graph is taken from [12].

that there are some changes in the system near 110 K and 210 K—its slope changes at these temperatures.

In Bi2212 having a doping level of $p = 0.19$, the temperature $T_0 \sim 110$–120 K can be associated with T_{pair} shown in Fig. 10.8. The temperature 210 K can be attributed to the onset of magnetic correlations, since the closing temperature of the pseudogap (*charge gap*) is $T_{CO} \simeq 360$ K. The data for T_{pair} and T_{CO} are consistent with each other since $3T_{pair} \simeq T_{CO}$. The latter expression is in good agreement with the empirical relation $\Delta_{pg} \simeq 3\Delta_p$.

8. The Josephson product

In SIS-junction conductances shown in Figs. 12.1, 12.4, 12.12, 12.16 and in some others, the zero-bias peak is caused by Cooper-pair tunneling, known as the *dc* Josephson effect. The *dc* Josephson effect is a manifestation of a quantum state on a macroscopic scale. The magnitude of the Josephson current depends on the phase difference between two superconductors as $I = I_c \sin(\varphi_2 - \varphi_1)$, where $(\varphi_2 - \varphi_1)$ is the phase difference, and I_c is the critical Josephson current. The Ambegaokar-Baratoff theory for conventional superconductors predicts that, at low temperature, the Josephson $I_c R_n$ product is

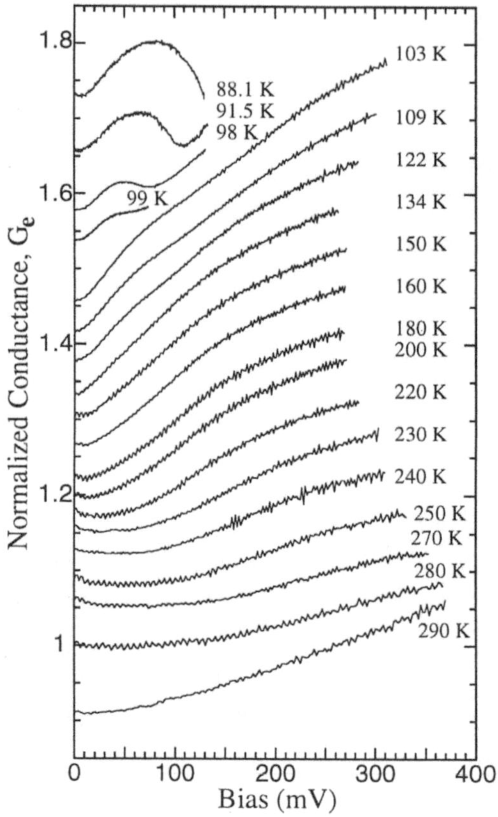

Figure 12.25 Even parts of the conductances from Fig. 12.24 shown exclusively at positive bias.

proportional to the gap magnitude (see Chapter 2), where R_n is the normal-state zero-bias resistance of the junction. In cuprates, the situation is more complicated because there two energy gaps.

Figure 12.27 shows the statistics of the Josephson I_cR_n product measured in Bi2212 single crystals having a doping level of 0.19. Figure 12.27a depicts the maximum and the average Josephson I_cR_n products as functions of the gap magnitude, observed in 110 tunneling spectra. The R_n value was estimated from high bias conductance which is relatively constant. Figure 12.27b shows the gap distribution in the same 110 tunneling spectra. It is worth recalling that, in Bi2212 with $p \simeq 0.19$, $2\Delta_c \simeq 42$ meV; $2\Delta_p \simeq 61$ meV, and $2\sqrt{\Delta_c^2 + \Delta_p^2} \simeq 74$ meV.

In Fig. 12.27a, the average Josephson I_cR_n product as a function of gap magnitude weakly correlates with the values of $2\Delta_c$, $2\Delta_p$, and $2\sqrt{\Delta_c^2 + \Delta_p^2}$ given above. Nevertheless, this correlation can be accidental because, in cuprates, the value of the Josephson I_cR_n product strongly depends on the quality of a junction. However, by doing a statistical study, the real value the Josephson

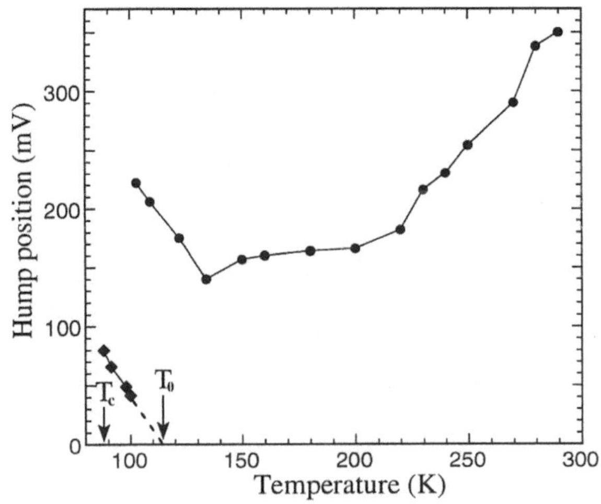

Figure 12.26. Temperature dependence of two humps in the conductances shown in Fig. 12.25. The dashed line extrapolates the closing temperature of the low-bias hump.

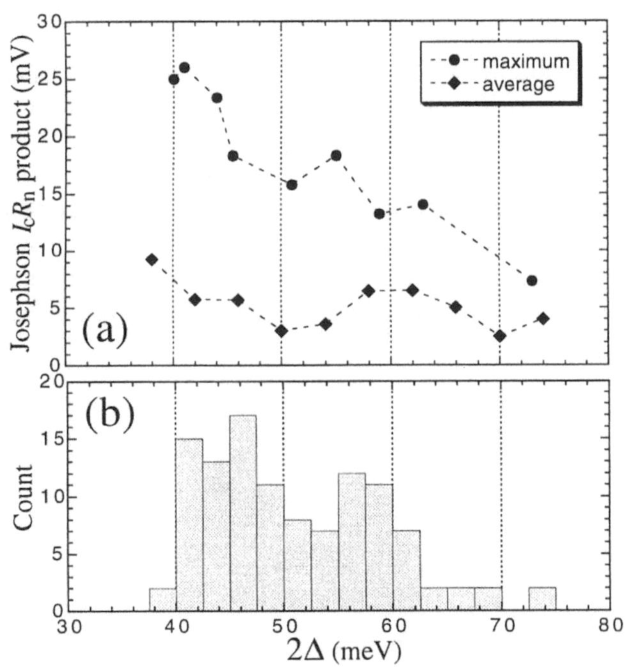

Figure 12.27. (a) Maximum and average Josephson I_cR_n products as functions of the gap magnitude in 110 Bi2212 break junctions. The doping level of Bi2212 is about 0.19. (b) The statistics of the gap magnitude for the same 110 junctions. The graph is taken from [5].

I_cR_n product can occasionally be observed, or at least, the value close to the actual one. In Fig. 12.27a, the maximum Josephson I_cR_n product almost linearly depends on the gap magnitude. The maximum value, 26 mV, is obtained at $2\Delta = 41$ meV. Thus, in Bi2212, the phase-coherence energy scale Δ_c has the maximum Josephson strength. Then, one may conclude that in cuprates, the Josephson product is proportional at low temperature to Δ_c, and not to Δ_p.

Generally speaking, the critical Josephson current I_c is determined by a phase stiffness, not by a pairing strength (if bosons are composite). For example, in underdoped cuprates, as the temperature increases, the Cooper pairs are still strongly coupled slightly above T_c, but the Josephson current is absent. The *dc* Josephson effect was recently observed in a Bose-Einstein condensate. In a Bose-Einstein condensate bosons are not composite, and I_c is exclusively determined by a phase stiffness. Thus, in cuprates, the strength of the phase-coherence gap Δ_c defines the value of the Josephson product.

Figure 12.28 shows the Josephson I_cR_n product as a function of doping in Bi2212. In Fig. 12.28, the grey line is the quadratic fit $I_cR_n = V_{max}[1 - 82.6(p-0.16)^2]$ proportional to Δ_c, where $V_{max} = 28$ mV. Tunneling measurements performed along the c axis in micron-size mesas suggest that, in Bi2212, $V_{max} \simeq 10$ mV. There are at least two different explanations to this discrepancy. First, in Bi2212, the value 28 mV is the in-plane value: in cuprates, the strength of Josephson I_cR_n product should be different in different directions. Secondly, it is possible that in break-junction measurements, the high value of I_cR_n is apparent and due to the presence of the zero-bias conductance peak (see the following Section). Thus, in Bi2212, the Josephson current at zero bias appears on top of this zero-bias conductance peak which is most likely caused by Andreev bound states confined to the surface. In the next Section, we shall see that a *dc* Josephson current can indeed appear on top of the zero-bias con-

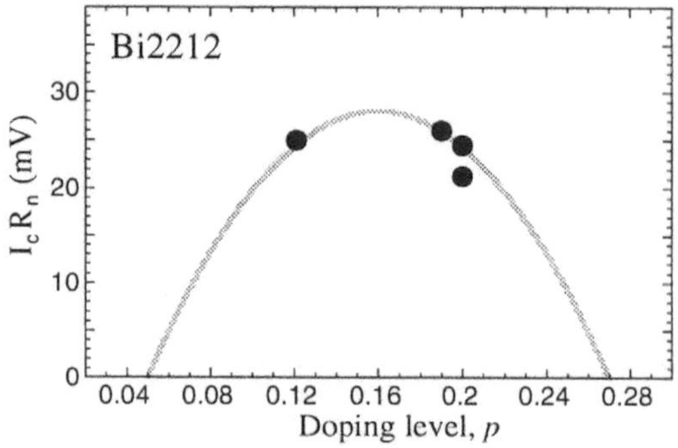

Figure 12.28. The Josephson I_cR_n product as a function of doping level. The data are taken from Figs. 12.27 and 6.4b, and from [12, 13]. The grey curve represents the quadratic fit with a maximum of 28 mV at $p = 0.16$.

ductance peak; however, it is possible, in first approximation, to separate these two contributions to tunneling spectra. The data presented in Fig. 12.28 seem to represent a predominant contribution from the Josephson current occurring at zero bias. So, the first explanation is most likely the cause of this discrepancy. It is also worth mentioning that, in YBCO, the value of the Josephson $I_c R_n$ product is 17 mV at 15 K [14] and 19 mV at 4.2 K [15].

9. Zero-bias conductance peak

In unconventional superconductors, the SIN-junction conductances often exhibit a zero-bias peak. First discovered in hole-doped YBCO [16, 17], the zero-bias conductance peak was later found in tunneling spectra of electron-doped NCCO (see Section 12 below), heavy fermions and organic superconductors. Figure 12.29 shows the $dI(V)/dV$ and $I(V)$ tunneling characteristics obtained in a Bi2212 single crystal by an SIN junction. In Fig. 12.29, the conductance has a peak at zero bias which is most likely caused by Andreev bound states [18, 19].

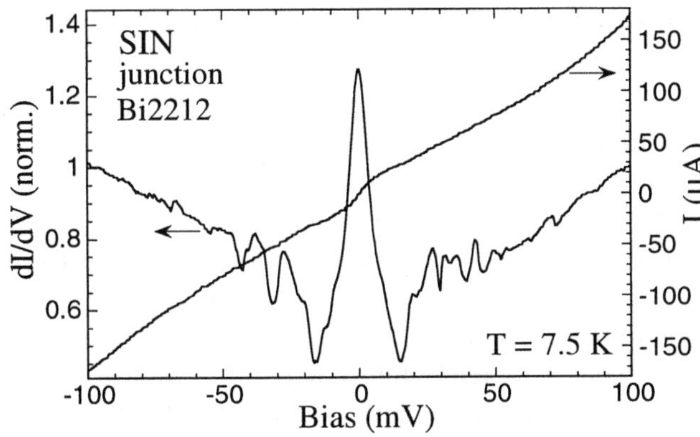

Figure 12.29. Tunneling spectra obtained in slightly overdoped Bi2212 single crystal ($p \simeq 0.19$) at 7.5 K by an SIN junction.

In superconductors having an order parameter with a d-wave symmetry, the midgap states theoretically exist for any specular surface except for the lobe direction of the d-wave gap perpendicular to the surface. The physical reason for the midgap states is that the neighboring lobes of the d-wave gap have different polarities: quasiparticles reflecting from the surface experience a change in the sign of the order parameter along their classical trajectory and subsequently undergo Andreev reflection, as shown in Fig. 12.30a. Constructive interference of incident and Andreev reflected quasiparticles result in bound states confined to the surface. That is, the Fermi-level states can exist due to Andreev reflections, where the quasiparticle changes from particle-like to hole-like and *vice*

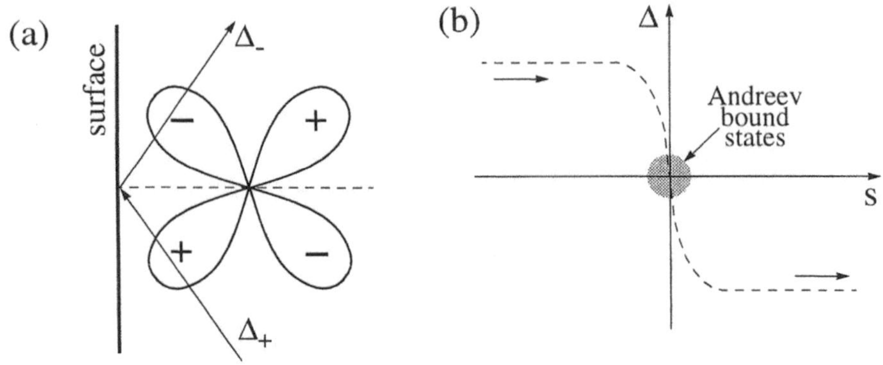

Figure 12.30. (a) Formation of Andreev bound states in a superconductor having a d-wave order parameter. The arrows indicate the propagation direction of a quasiparticle. (b) The order parameter is plotted versus the path s along the quasiparticle trajectory. If after the reflection, the order parameter changes sign, then the Andreev bound states are always formed at the surface ($s = 0$).

versa, and **k** changes sign. Therefore, the midgap states shown schematically in Fig. 12.30b are often denoted as Andreev bound states. If the tunneling is in-plane, the Andreev bound states result in a zero-bias conductance peak. As a consequence, tunneling spectroscopy is in fact a phase-sensitive technique enabling to identify unconventional superconductors.

In Fig. 12.29, the conductance does not have a gap structure. This suggests that it was measured near a node direction of the d-wave gap Δ_c. Figure 12.31 shows the temperature dependence of the conductance depicted in Fig. 12.29. Theoretically, a conductance peak caused by Andreev reflections should exhibit a linear T^{-1} temperature dependence [20]. Figure 12.32a depicts the relative amplitude of the zero-bias conductance peaks shown in Fig. 12.31, $(G_{max} - G_{ext})/G_{ext}$, as a function of T^{-1}, where G_{max} is the peak-conductance value at $V = 0$, and G_{ext} is the extrapolated conductance value at $V = 0$ if the zero-bias conductance peak *would* be absent. In Fig. 12.32a, one can see that the relative amplitude of zero-bias conductance peak linearly depends on T^{-1}. This indicates that the zero-bias conductance peak in Fig. 12.29 is most likely caused by Andreev bound states.

For comparison, Figure 12.32b presents the same dependence, $(G_{max} - G_{ext})/G_{ext}$ versus T^{-1}, for the zero-bias conductance peak caused by a *dc* Josephson current, shown in Fig. 6.5a. The data in Fig. 12.32b have no physical meaning and were presented exclusively for comparison. The zero-bias conductance peak due to a *dc* Josephson current, shown in Fig. 12.17, has a similar dependence to that in Fig. 12.32b. In Fig. 12.32b, one can see that at low temperature, the relative amplitude of Josephson current in Bi2212 tends to remain constant. Thus, the two different conductance peaks caused by An-

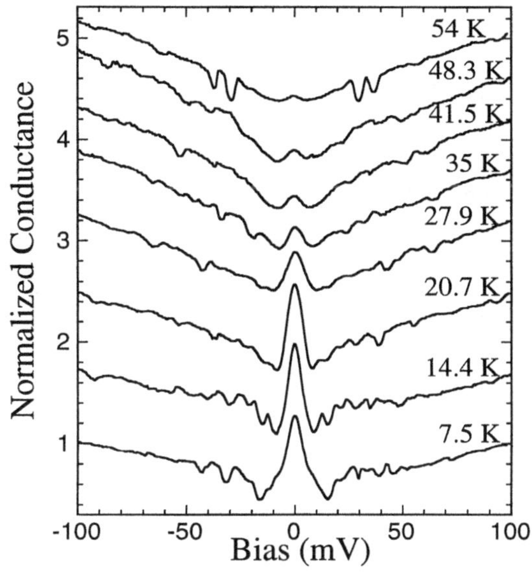

Figure 12.31. Temperature dependence of the conductance from Fig. 12.29. The spectra are offset vertically for clarity.

dreev bound states *and* by a *dc* Josephson current, both appearing at zero bias, have different tendencies at low temperature, as shown in Figs. 12.32a and 12.32b.

Initially found in SIN-junctions of cuprates, the zero-bias conductance peak

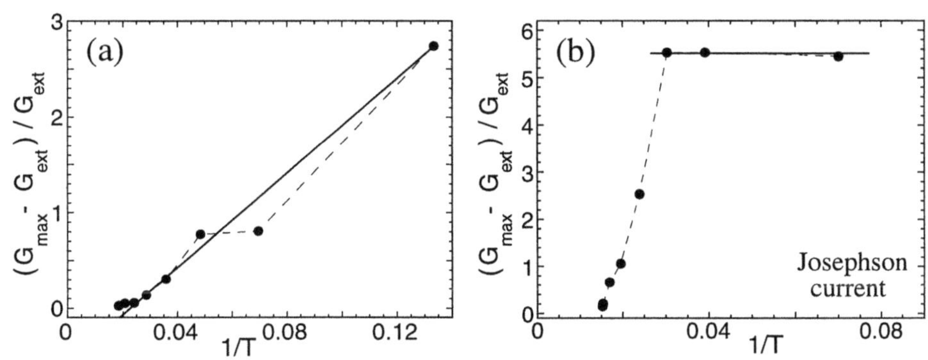

Figure 12.32. The relative amplitude of the zero-bias conductance peak $(G_{max} - G_{ext})/G_{ext}$ versus $1/T$ for the zero-conductance peak (a) in Fig. 12.31 and (b) in Fig. 6.5a. The latter is caused by a *dc* Josephson current. G_{max} is the conductance value at $V = 0$ and G_{ext} is the extrapolated conductance value at $V = 0$ if the zero-bias conductance peak *would* be absent. In both plots the solid lines indicate the data trend. The data in plot (b) have no physical meaning and are presented for comparison.

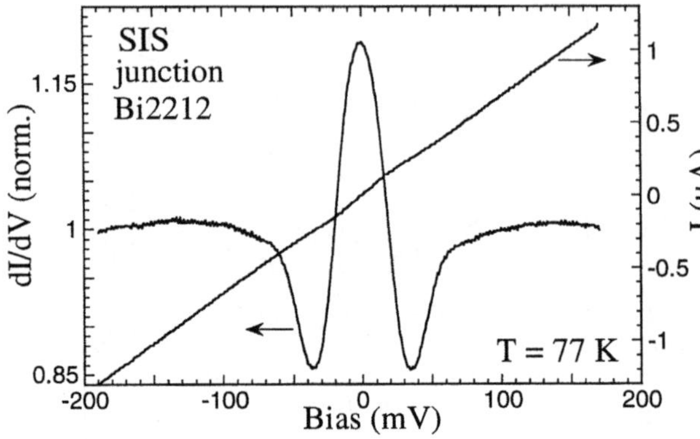

Figure 12.33. Tunneling spectra obtained in slightly overdoped Bi2212 single crystal ($p \simeq 0.19$) at 77 K by an SIS junction.

was later observed in SIS junctions too. As an example, Figure 12.33 shows the $dI(V)/dV$ and $I(V)$ tunneling characteristics obtained in a Bi2212 single crystal by an SIS junction. In Fig. 12.33, the conductance has a gap structure indicating that the spectra were obtained in a direction different from a nodal direction of the d-wave gap.

Single crystals of Bi2212 are extremely micaceous in nature. Performing point-contact measurements in Bi2212 single crystals, thus measurements by SIN junctions, a small piece of a Bi2212 single crystal can easily attach to the tip of metallic wire. In this case, the SIN junction becomes an SIS junction. Figure 12.34 shows the $dI(V)/dV$ and $I(V)$ tunneling characteristics obtained in a Bi2212 single crystal by a junction being initially the SIN junction. In Fig. 12.34, the shape of the zero-bias conductance peak is somewhat different

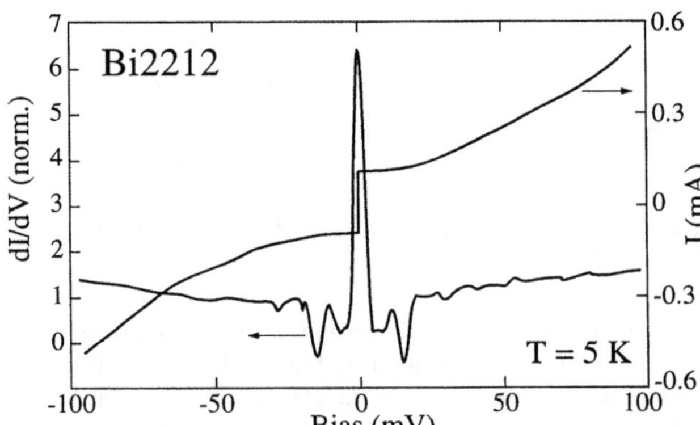

Figure 12.34. Tunneling spectra obtained in slightly overdoped Bi2212 single crystal ($p \simeq 0.19$) at 5 K by a junction being initially the SIN junction.

from those in Figs. 12.29 and 12.33—it is much narrower than the other two. Thus, one may conclude that the conductance peak at zero bias is not caused by Andreev bound states. The $I(V)$ characteristic in Fig. 12.34 indicates that this peak rather originates from a *dc* Josephson current (in addition, by changing the direction of the voltage sweep, the $I(V)$ characteristic exhibited a prominent hysteresis). This implies that this junction was in fact an SIS junction. Then, the estimated value of the Josephson I_cR_n product in this junction is about 41 mV. From Fig. 12.28, it is clear that, in Bi2212, the value 41 mV is too large for a Josephson I_cR_n product. One can then conclude that a *dc* Josephson current alone could not cause the occurrence of the conductance peak shown in Fig. 12.34.

Figure 12.35 depicts the relative amplitude of the zero-bias conductance peak shown in Fig. 12.34, $(G_{max} - G_{ext})/G_{ext}$, as a function of T^{-1}. Comparing Figs. 12.35 and 12.32, one can see that, at low temperature, the data in Fig. 12.35 exhibit a tendency which is different from those in Figs. 12.32a and 12.32b. In fact, the low-temperature behavior in Fig. 12.35 can be understood by assuming that the zero-bias conductance peak in Fig. 12.34 has two components: one is caused by a Josephson current and the other by Andreev bound states. Taking into account the two temperature dependences in Figs. 12.32a and 12.32b, one can estimate the relative ratio of these two components. In Fig. 12.35, therefore decrease at low temperature is most likely caused by Andreev bound states; then the slope of this decrease indicates the relative fraction of the component from Andreev bound states, as shown in Fig. 12.35. This estimate suggests that the conductance peak at zero bias in Fig. 12.34 is predominantly caused by a Josephson current and, in first approximation, the ratio of amplitudes of the two components is about 25/18 $\simeq$ 1.4.

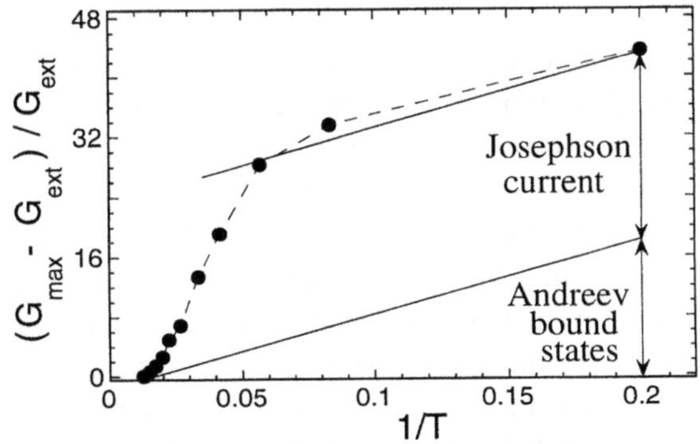

Figure 12.35. The relative amplitude of the zero-bias conductance peak (see Fig. 12.32) as a function of $1/T$ for the zero-conductance peak in Fig. 12.34. The upper solid line indicates the data trend at low T, and the lower line is parallel to the upper one. The arrows show the estimates of two components caused by a *dc* Josephson current and by Andreev bound states.

Comparing Figs. 12.29, 12.33 and 12.34, it seems that the dips in the conductance in Fig. 12.34 at about ± 15 mV are caused by Andreev bound states, while the small dips at about ± 7 mV by a *dc* Josephson current.

A few reports which can be found in the literature show that the zero-bias conductance peak is slightly split as the temperature decreases below 4 K. In addition, a few reports suggest that the zero-bias conductance peak can also be split by applying a magnetic field. However, the spitting in an applied magnetic field was not observed in all tunneling studies.

10. Zn and Ni doping in Bi2212

In Chapter 3, we already discussed how Ni and Zn atoms doped into CuO_2 planes affect locally superconductivity in Bi2212. Here we consider new features typical for superconductivity in Ni and Zn doped Bi2212, inferred from tunneling measurements.

Performing large amount of tunneling measurements in pure, Ni-doped and Zn-doped Bi2212 single crystals, I noticed that tunneling conductances obtained in Ni- and Zn-doped Bi2212 are slightly different from those measured in pure Bi2212. As an example, Figure 12.36 presents three conductances obtained in pure Ni- and Zn-doped Bi2212 single crystals. The conductances in Fig. 12.36 can be considered as typical for each type of Bi2212 samples.

Compare first the conductances measured in pure and Zn-doped Bi2212—the spectra A and C in Fig. 12.36, respectively. In the conductance C, the zero-bias peak due to a *dc* Josephson current is smaller than that in the conductance A. This fact indicates that Zn atoms doped into CuO_2 planes suppress the phase coherence. Since, in CuO_2 planes, Zn is a non-magnetic impurity (in contrast to Ni), this may indicate that, in Bi2212, the phase-coherence mechanism of superconductivity has the magnetic origin (see the following paragraph).

Consider now the conductances measured in pure and Ni-doped Bi2212—the spectra A and B in Fig. 12.36, respectively. In the conductance B, the distance between the humps is smaller than that in the conductance A. Since the humps in conductances obtained in cuprates are related to the pseudogap, this fact indicates that Ni atoms doped into CuO_2 planes destroy the pseudogap (in contrast to Zn). Since, in CuO_2 planes, Ni is a magnetic impurity, then this pseudogap has a non-magnetic origin (the same conclusion was made in Chapter 6). Indeed, the tunneling pseudogap is a charge gap. In Fig. 12.36, the second feature of the conductance B is that the zero-bias peak due to a *dc* Josephson current is slightly larger than that in the conductance A. This fact indicates that Ni atoms doped into CuO_2 planes promote the phase coherence (in contrast to Zn). Since Ni is magnetic in CuO_2 planes, then one can conclude that, in Bi2212, the phase-coherence mechanism of superconductivity is also magnetic.

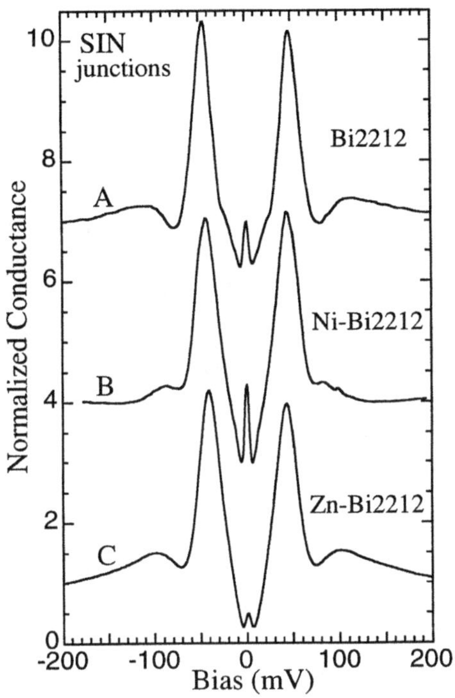

Figure 12.36 Normalized tunneling conductances obtained in different Bi2212 single crystals by SIS junctions: The spectra A, B and C are taken in pure ($p \simeq 0.19$), Ni-doped ($T_c \simeq 75$ K) and Zn-doped ($T_c \simeq 77$ K) Bi2212, respectively. The spectra A and B are offset vertically for clarity.

Comparing all the conductances in Fig. 12.36, one can see that the conductances A and B have weak subgaps, while the conductance C does not. This fact suggests that the subgap in the SIS-junction conductances has a magnetic origin.

Finally, let us discuss one observation more, common for Ni- and Zn-doped Bi2212—the "absence" of a pseudogap above T_c. In pure Bi2212, I observed the pseudogap above T_c from the first trial. However, I never found the pseudogap in Ni-doped Bi2212 samples and only once in Zn-doped Bi2212. This fact indicates that in CuO_2 planes, Ni and Zn ions "mask" the pseudogap above T_c. It is obvious that the pseudogap does exist above T_c in Ni- and Zn-doped Bi2212. However, it cannot be seen by tunneling measurements on the surface. Taking into account this observation, one can then understand the fact that the *tunneling* pseudogap was never observed in optimally doped and overdoped YBCO above T_c. In Tl2201, the pseudogap was never seen at any doping level by tunneling measurements.

11. Vortex-core states

Consider tunneling measurements performed in vortex cores. In conventional type-II superconductors, the Cooper pairs do not exist in the cores of quantized magnetic vortices. Thus, the superconducting order parameter is fully suppressed inside vortices. However, bound quasiparticle states can exist inside these cores. In cuprates the situation is different: the Cooper pairs are always present inside vortices. In addition, as in conventional superconductors, there are bound quasiparticle states in vortex cores; however, they are conceptually different from those in conventional superconductors.

Figure 12.37 shows two conductances obtained in an YBCO single crystal by SIN junctions: one was measured in zero magnetic field, and the other is an average spectrum of conductances obtained at the center of a vortex core along a 50 Å path. The zero-field conductance was already discussed (see Fig. 12.5b). The vortex-core conductance has humps at about ± 36 mV, corresponding to the pairing order parameter Δ_p in YBCO. These humps are similar to those in the zero-field curve. In Fig. 12.37, the humps in both conductances are caused by single-electron tunneling from the Cooper-pair states. The vortex-core conductance in Fig. 12.37 has two peaks at low bias ± 5.5 mV, which are in fact a superposition of two contributions: One is caused by an induced superconducting gap on chains, present also in the upper curve. The second originates from bound quasiparticle states in vortex cores. The magnitudes of these two contributions coincide, resulting in the appearance of a well-defined gap structure at low bias. In Fig. 12.37, the two conductances have different shapes because of the presence/absence of phase coherence: the phase coherence is suppressed in vortex cores. It is worth mentioning that the data in Fig. 12.37 are comparable with the data in Fig. 12.7.

In Bi2212 the results of tunneling measurements performed in vortex cores are similar to those for YBCO. Figure 12.38a depicts three conductances obtained

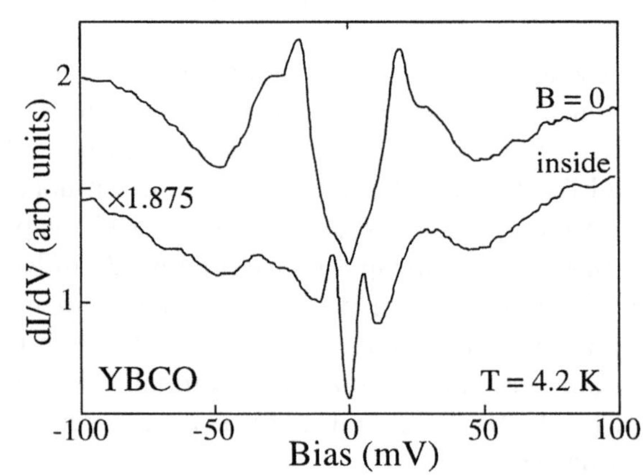

Figure 12.37. Tunneling conductances obtained in an YBCO single crystal having $T_c = 91$ K by SIN junctions at 4.2 K [21]. The upper curve was measured in zero magnetic field. The lower curve is the average spectrum of conductances obtained at the center of a vortex core along a 5 nm path and in $B = 6$ T applied along the c axis. The lower curve was magnified by a factor of 1.875. The upper curve is offset vertically for clarity.

in an underdoped Bi2212 single crystal by SIN junctions: one was measured between vortices, the second in a vortex core, and the last one above T_c. In Fig. 12.38a, one can see that all three conductances have gap structures with similar magnitudes. In Fig. 12.38a, the peaks and humps in the curves are caused by single-electron tunneling from the Cooper-pair states. The results of tunneling measurements performed in an overdoped Bi2212 are basically the same, as shown in Fig. 12.38b.

In Bi2212, the first results of vortex-core measurements did not display the presence of bound quasiparticle states in vortex cores—they were found later [23]. Moreover, at different doping levels, the energy of these vortex-core states shows an approximately linear scaling with the pairing energy gap Δ_p present between vortices [24]. This fact indicates that, in cuprates, the vortex-core states are directly connected with the Cooper-pair states present inside and outside vortices.

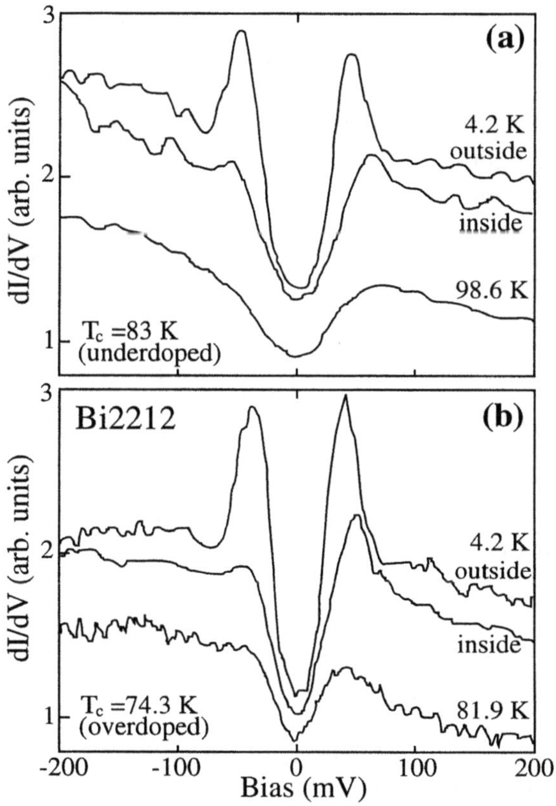

Figure 12.38 Tunneling conductances obtained by SIN junctions in Bi2212 single crystals: (a) underdoped (T_c = 83 K) and (b) overdoped (T_c = 74.3 K) [22]. In both plots, the upper and middle curves are measured at 4.2 K and at a magnetic field B = 6 T: the upper curve between vortices and the middle curve at the center of a vortex core. In both plots the lower conductances are taken in zero magnetic field above T_c: in plot (a) at 98.6 K and in plot (b) at 81.9 K. For clarity the 4.2 K curves are offset vertically.

12. NCCO

In this book, we mainly discussed hole-doped cuprates. In this Section, we concentrate on tunneling measurements performed in electron-doped NCCO. A single crystal of NCCO used in measurements is underdoped, having T_c = 14.2 ± 0.2 K [25]. The measurements were performed by using Ag and Pt-Ir normal tips and a Nb superconducting tip.

12.1 Symmetry of the order parameters

Soon after the discovery of electron-doped NCCO it was generally accepted that the order parameter in NCCO has an s-wave symmetry. Thus from the beginning, the electron-doped cuprates were thought to be different from hole-doped members of the cuprate family. What information can we extract to provide us tunneling measurements on this issue?

Figure 12.39 shows two normalized conductances obtained in underdoped NCCO. In Fig. 12.39 the upper curve is measured by using a Nb superconducting tip as a counterelectrode and the lower curve by an Ag normal tip. In Fig. 12.39, the upper conductance has a zero-bias peak due to a *dc* Josephson current. Since the order parameter in Nb has an s-wave symmetry, the presence of a *dc* Josephson current in this junction indicates the existence of an s-wave condensate in NCCO. In Fig. 12.39 the estimated Josephson I_cR_n product for this Nb-NCCO junction is about 200 μV.

In Fig. 12.39 the lower curve obtained by an SIN junction has a zero-bias conductance peak. As discussed above the zero-bias conductance peak is a manifestation of d-wave order parameter present in the system (see Fig. 12.30).

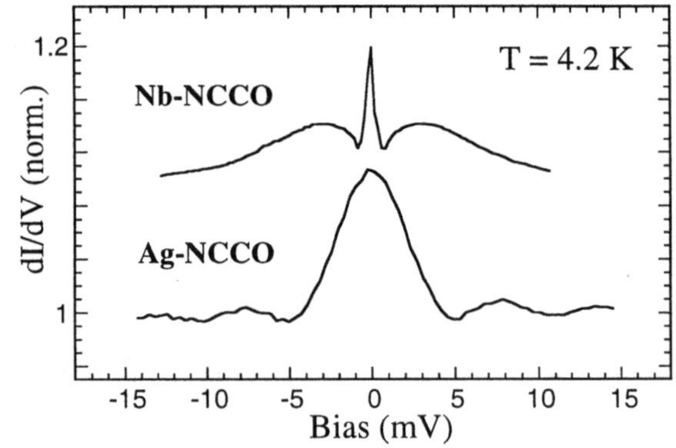

Figure 12.39. Normalized conductances obtained in a Nb-NCCO SIS junction (upper curve) and in an Ag-NCCO SIN junction (lower curve). The single crystal of NCCO is underdoped having T_c = 14.2 K. The Nb-NCCO spectrum is shifted up for clarity. The plot is taken from [25]. In these curves and in conductances presented in Figs. 12.40—12.44, the linear background, typical for point-contact junctions, is removed.

These results show that superconductivity in electron-doped NCCO is similar to that in hole-doped cuprates: the s-wave and d-wave order parameters are both present in NCCO. The question is why it took 10 years to find it out.

12.2 Two energy scales

From the beginning, tunneling data published in the literature did not show the distribution of gap magnitude in NCCO. In hole-doped cuprates (see e.g. Fig. 12.4), the gap distribution is due to the presence of two anisotropic energy gaps. What is the real situation in NCCO?

Figure 12.40 depicts two sets of normalized conductances obtained in NCCO by point contacts (a) along and (b) perpendicular to the c-axis (see Fig. 3.7). In Fig. 12.40a, one can see that indeed there is no gap distribution along the c-axis direction. In Fig. 12.40a, the dashed lines mark the magnitude of the c-axis gap, which is about 3.5 meV. In contrast, the magnitude of an in-plane tunneling gap varies approximately between 3.5 meV and 13 meV, as shown by

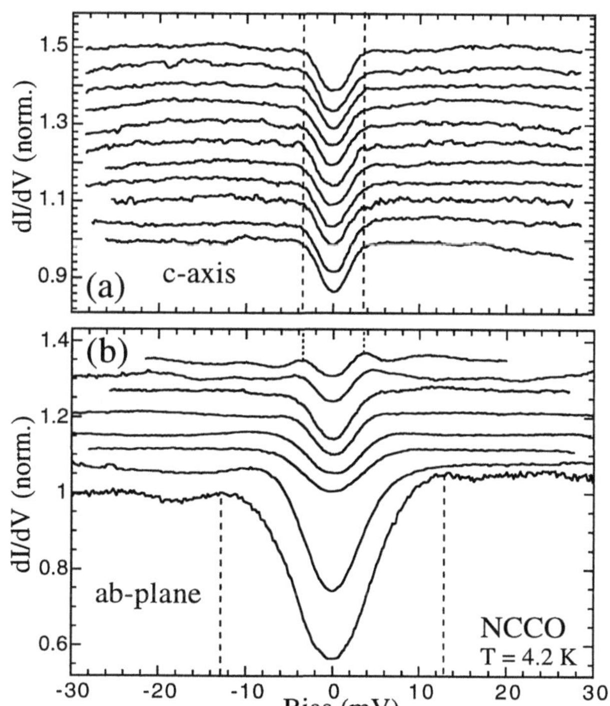

Figure 12.40. Normalized conductances obtained in an underdoped NCCO single crystal (same as in Fig. 12.39) by point contacts (a) along and (b) perpendicular to the c-axis. The dashed lines in plot (a) show the gap at ± 3.5 mV and in plot (b) the minimum and maximum magnitudes of the in-plane gap. In both plots the curves are offset vertically for clarity. The plot is taken from [25].

the dashed lines in Fig. 12.40b. Again there is a similarity between NCCO and hole-doped cuprates: there exists the distribution of in-plane gap magnitude in both types of cuprates.

In Fig. 12.40, the presence of two energy scales in NCCO is obvious. In addition, Andreev-reflection data supplement this picture: Figure 12.41 shows a normalized Andreev-reflection spectrum obtained in NCCO, and three tunneling conductances from Fig. 12.40. The dashed lines in Fig. 12.41 indicate the two energy scales in NCCO: One energy scale manifests itself in the in-plane energy gap having a maximum magnitude, while the other in the Andreev-reflection spectrum, in the *c*-axis energy gap and in the in-plane gap having a

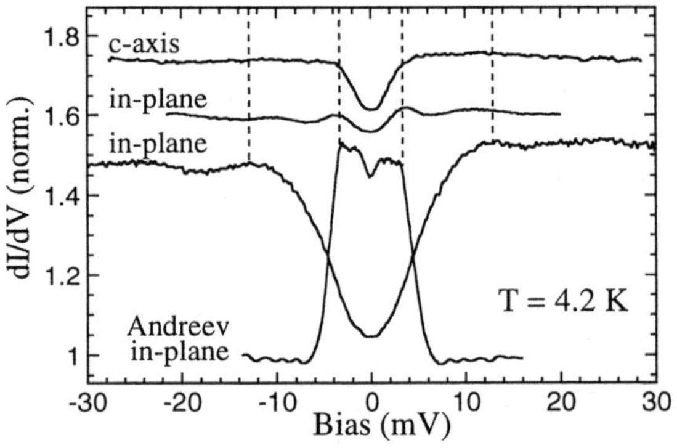

Figure 12.41. Normalized Andreev-reflection spectrum obtained in an underdoped NCCO single crystal (same as in Fig. 12.39) by a point contact and the tunneling conductances from Fig. 12.40. The dashed lines indicate two energy scales in NCCO. The tunneling curves are offset vertically for clarity [25].

minimum magnitude. What is even more interesting is that, when plotted in the phase diagram of hole-doped cuprates, these two energy scales in NCCO seem to mimic the two energy scales in hole-doped cuprates, as shown in Fig. 12.42. Then, by analogy, the smallest energy scale in NCCO can be associated with a d-wave symmetry, whilst the largest scale with an s-wave symmetry. In NCCO the presence of order parameters with s-wave and d-wave symmetries were discussed above.

12.3 Pseudogap

To complete the picture showing the similarity between the two mechanisms of superconductivity in electron- and hole-doped cuprates, it is necessary to demonstrate the existence of a tunneling pseudogap in NCCO.

Figure 12.43 depicts the temperature dependence of an Andreev-reflection spectrum obtained in NCCO by a point contact. In Fig. 12.41 the temperature dependence lies below the BCS temperature dependence as those in Fig. 9.7. The data in Fig. 12.43 do not provide new information; however, what is

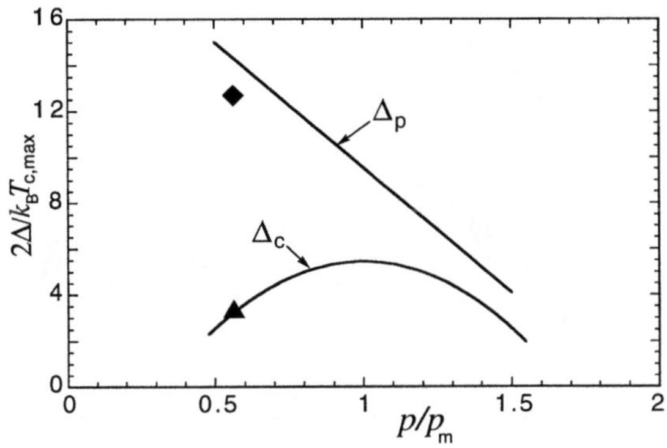

Figure 12.42. Phase diagram of hole-doped cuprates from Fig. 3.22. The triangle and diamond indicate the two energy scales in underdoped NCCO shown in Fig. 12.41. The plot is taken from [25].

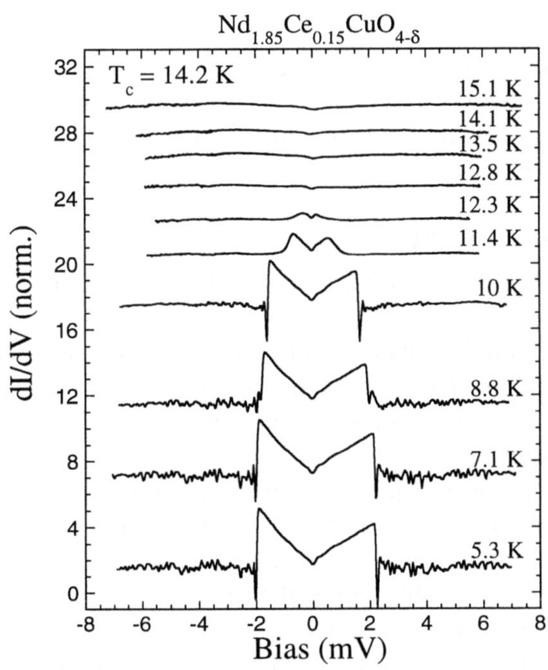

Figure 12.43. Temperature dependence of an Andreev-reflection spectrum obtained in an underdoped NCCO single crystal (same as in Fig. 12.39) by a point contact. The curves are offset vertically for clarity.

more important is that the spectra measured above T_c exhibit the presence of a pseudogap, as shown in Fig. 12.44. The conductances in Fig. 12.44 clearly

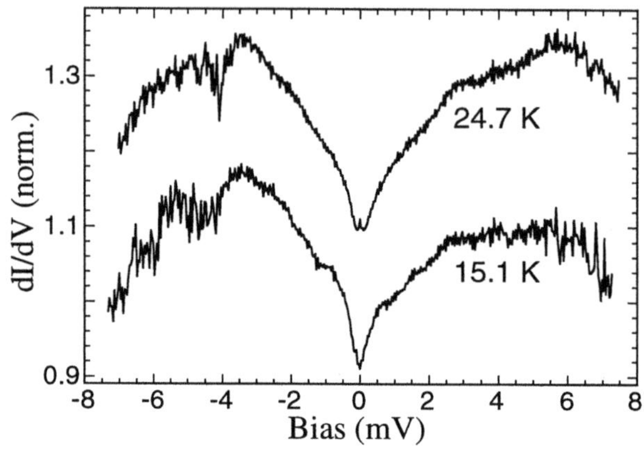

Figure 12.44. Two spectra from Fig. 12.43 enlarged vertically. Both curves exhibit a pseudogap above T_c. The upper curve is offset vertically for clarity.

display the presence of a pseudogap above T_c in NCCO.

Finally, from the tunneling data presented in this Section, one can conclude that the mechanism of superconductivity in electron-doped NCCO seems to be similar to that in hole-doped cuprates.

13. SIS-junction fit

In Chapter 2, we discussed tunneling measurements in conventional superconductors. In conventional superconductors, tunneling data obtained in SIN junctions can be well described by equations of the so-called BTK theory [26]. As an example, Figure 2.16 depicts the SIN-junction data and theoretical curve for Nb. In Fig. 2.16 one can notice that there is good agreement between the data and the theoretical fit.

At the same time in SIS junctions of conventional superconductors, a lack of correspondence has been noted between theoretical fits and tunneling data: Figure 2.17 demonstrates this disagreement. This issue has not been discussed in the literature to date. What can be the reason for the disagreement?

It is worth noting that we deal here exclusively with conventional superconductors. (I hope that after reading this book the reader will never use the BTK equations to fit tunneling characteristics of cuprates, obtained either in SIN or SIS junctions.) A visual inspection of Figs. 2.16 and 2.17b reveals that the conductance in Fig. 2.17b looks if it were obtained in an SIN junction formed between a normal metal and a conventional superconductor having an energy gap of 2Δ.

The BTK fits are very successful in describing SIN-junction data; thus it is most likely that the concept of Eq. (2.18) is to blame. We rewrite Eq. (2.18) as follows

$$I(V) = \int_0^\infty K(V)N(E,\Gamma)N(E-eV,\Gamma)[f(E,T)-f(E-eV,T)]dE, \quad (12.1)$$

where $N(E)$ is the density of states of quasiparticle excitations in the superconducting state, Γ is the phenomenological smearing parameter, and $f(E,T)$ is the Fermi function $f(E,T) = [\exp(E/k_BT) + 1]^{-1}$. In this equation, the constant K is now a matrix and a function of bias V.

It is possible that for an SIS junction, the matrix $K(V)$ in Eq. (12.1) is not diagonal, as generally assumed. Since this problem is not the primary focus of the book, we postpone this question for later study. This issue definitely needs further clarification.

14. Bisoliton fit

The purpose of this Section is to discuss tunneling characteristics obtained in Bi2212 in terms of the bisoliton fit. In Chapter 6, we already used this fit derived in Chapter 5 from the soliton theory to extrapolate tunneling characteristics of cuprates and manganites. For $dI(V)/dV$ and $I(V)$ characteristics, the SIN-junction fits are given by Eqs. (6.1) and (6.4), and the SIS-junction fits by Eqs. (6.2) and (6.3), respectively. Nevertheless, a few issues were left unconsidered which we shall discuss here.

14.1 Height of quasiparticle peaks

Consider the height of conductance peaks as a function of doping. Figure 12.45 depicts three tunneling conductances obtained in an underdoped, an optimally doped and an overdoped Bi2212 single crystal [27]. In Fig. 12.45 one can see that the gap magnitude decreases as the doping level increases. This dependence was already discussed a few times in this book. In addition, in Fig. 12.45 one can observe that the ratio of heights of quasiparticle peaks and humps increases as the doping level increases. Figure 12.46 schematically depicts this doping dependence. This fact is in good agreement with the soliton theory [28].

In Fig. 12.45 the data, obtained by another group, are presented intentionally: they were not selected according to the doping dependence shown in Fig. 12.46. The same doping dependence can also be found in Figs. 6.1a and 6.3a and in some other plots in this work.

It is worth noting that in the *highly* overdoped region, the *absolute* heights of quasiparticle peaks and humps in tunneling conductances both decrease because the fraction of charge-stripe domains diminishes. The pseudogap (charge gap) collapses at about $p \simeq 0.3$. Above this doping level cuprates are completely metallic. Superconductivity however disappears at about 0.27 due to a

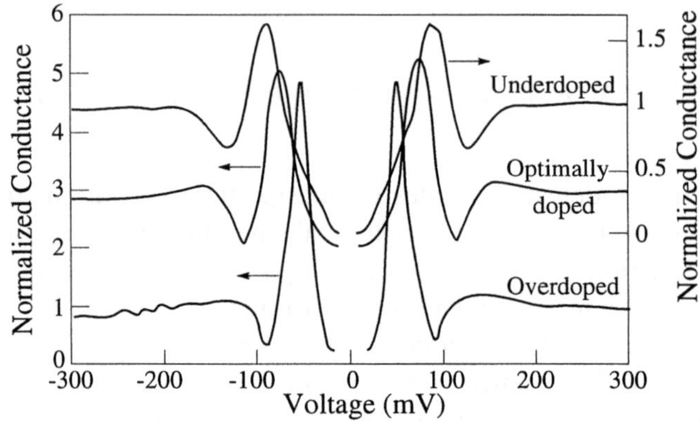

Figure 12.45. Tunneling conductances obtained in three Bi2212 single crystals by SIS junctions [27]. The upper curve is measured in an underdoped Bi2212 having $T_c = 83$ K, the middle curve in an optimally doped Bi2212 with $T_c = 95$ K, and the lower curve in an overdoped Bi2212 having $T_c = 82$ K. In all curves the zero-bias peak from the Josephson current has been removed. The middle spectrum is offset vertically for clarity.

lack of long-range phase coherence. Between 0.27 and 0.3 cuprates are quasi-metallic.

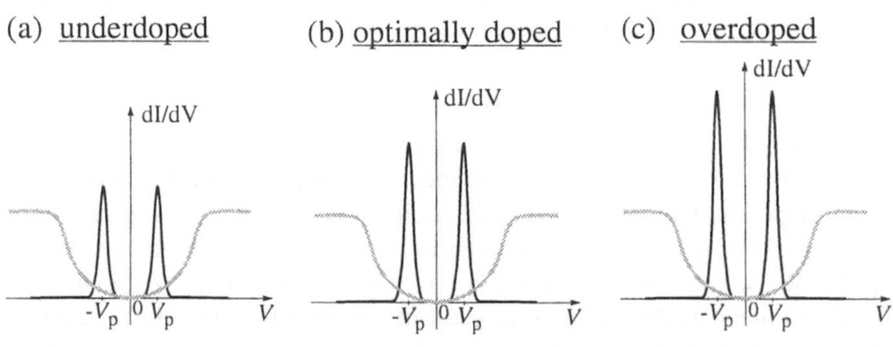

Figure 12.46. Sketch of tunneling pseudogap and quasiparticle peaks obtained in (a) underdoped, (b) optimally doped, and (c) overdoped Bi2212, demonstrating the relative ratio of heights of quasiparticle peaks and pseudogap as a function of doping. The pseudogap is schematically shown in grey.

14.2 Bisoliton fit in numbers

In Chapter 6 we used the bisoliton fit to extrapolate tunneling characteristics obtained in Bi2212 and in the manganite $La_{1.4}Sr_{1.6}Mn_2O_7$ (LSMO). The fit was applied to $dI(V)/dV$ and $I(V)$ characteristics measured in SIN and SIS junctions. In Chapter 6 however, the fits were not given explicitly: Table 12.1

lists the values of A_i (i = q, c), 2Δ and V_0 used in SIS-junction fits given by Eqs. (6.2) and (6.3).

Table 12.1. The values used in the SIS-junction bisoliton fits given by Eqs. (6.2) and (6.3), which were used in Chapter 6.

Figure	*compound*	A_q	A_c	2Δ	V_0
6.7a & 6.8a	Bi2212	1.05	0.181	118	35
6.7b & 6.8c	Bi2212	3.68	0.752	46	15
6.7c & 6.9a	Ni-Bi2212	5.9	0.78	35	15
6.8b	Bi2212	-	26	83	27
6.14b	LSMO	62	-	1.22	0.43

In Figs. 6.7d and 6.8d, the $dI(V)/dV$ and $I(V)$ curves were fitted by the SIN-junction fits given in Eqs. (6.1) and (6.4), respectively. In these SIN-junction fits the values $\Delta = 20$ and $V_0 = 6.6$ were used.

As noted in Chapter 10, one can mention that in the SIS-junction fits the ratio $\frac{2\Delta}{eV_0} \approx 3$, as well as in the SIN-junction fit, $\frac{\Delta}{eV_0} \approx 3$. Therefore, *in first approximation*, knowledge of gap magnitude to fit the data is not necessary, and the functions presented in Eqs. (6.1) and (6.2) and Eqs. (6.3) and (6.4) can be expressed by Eqs. (10.10) and (10.11), respectively. This fact indicates the existence of one important relationship. Since in the bisoliton fit the voltage V_0 is responsible for the width of conductance peaks, then consequently at different dopings, the peak width scales with the magnitude of the pairing gap.

In Table 12.1 there is one exception from this rule: the spectra obtained in Ni-doped Bi2212, for which $\frac{2\Delta}{eV_0} \approx 2.3$. In Fig. 6.6c one can notice that in Ni-doped Bi2212, the conductance peaks are slightly wider than those in Figs. 6.6a and 6.6b. In fact this is not surprising because in Ni-doped Bi2212 the conductance peaks contain two independent contributions shown in Fig. 6.9a. The predominant contribution is caused by bisolitons and the other by a magnetic resonance peak.

To visualize the relationship between Δ and V_0 Figure 12.47 presents seven normalized SIS-junction conductances obtained in Bi2212. In Fig. 12.47, the doping level of Bi2212 varies between 0.085 and 0.2: This interval represents more than a half of the superconducting phase of Bi2212. Figure 12.47 also depicts the SIN-junction fit for Bi2212 from Fig. 6.7d and the "SIS"-junction fit for LSMO manganite from Fig. 6.14b. In Fig. 12.47, one can see that all the conductance peaks coincide when normalized. It is worth recalling that the manganite LSMO never becomes superconducting. This fact is direct evidence for the presence of bisolitons in LSMO, residing on quasi-static charge stripes.

In first approximation the dependence $V_0(p)$ is linear: $V_0(p) \approx 167 \times [1 - p/0.3]$ (in mV) for SIS junctions and $V_0(p) \approx 84 \times [1 - p/0.3]$ (in mV) for SIN junctions.

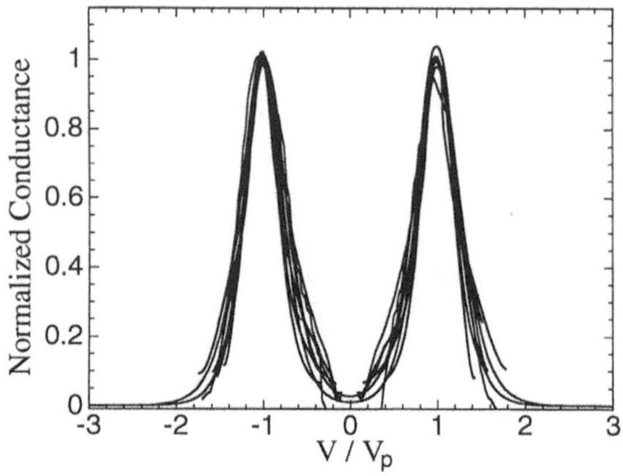

Figure 12.47. Seven normalized conductances and two normalized bisoliton fits. The conductances are taken from Figs. 4.2a, 6.2, 6.3a, 6.3b and 12.45, obtained in Bi2212 by SIS junctions. The bisoliton fits are the SIN-junction fit for Bi2212 from Fig. 6.7d and the fit from Fig. 6.14b. The latter was applied to a conductance obtained in a single crystal of the manganite LSMO. In this plot each curve is normalized twice—to its peak value and to its peak bias V_p. The humps at high bias and the zero-bias peaks from the Josephson current are not shown. In the conductance from Fig. 12.45, obtained in an underdoped Bi2212, the background from the humps is subtracted as that in Fig. 6.1.

14.3 SIS-junction fit

The bisoliton density of states near the gap energy $\pm\Delta_p$ is proportional to $sech^2[(E \pm \Delta_p)/E_0]$, where E_0 is the constant. As a consequence, the conductance peaks obtained by an SIN junction in a system having bisolitons can be fitted by the following function

$$\frac{dI(V)}{dV} \simeq A \times \left[sech^2\left(\frac{V+V_p}{V_0}\right) + sech^2\left(\frac{V-V_p}{V_0}\right)\right], \qquad (12.2)$$

derived in Chapter 5, where $V_p = \Delta_p/e$ is the peak bias. We are interested in to know the shape of SIS-junction conductance peaks caused by bisoliton states. As noted in Chapter 5, it is not possible to obtain analytically the exact solution. Even, for SIS junctions of conventional superconductors, the theoretical predictions disagree with data (see the previous Section). In cuprates as well as in conventional superconductors, the tunneling matrix $K(V)$ in Eq. (12.1) is unknown. The purpose of this subsection is to *estimate* the SIS-junction bisoliton fits for $dI(V)/V$ and $I(V)$ tunneling characteristics.

In Fig. 12.47, one can see that, *experimentally*, the SIS-junction data have the same pattern and can be well described by the hyperbolic $sech^2$ function given by Eq. (12.2), where V_p is substituted for $2V_p$. The agreement of SIN- and SIS-junction data in Fig. 12.47 indicates that the width of quasiparticle peaks in conductances obtained in SIS junctions is about two times wider than the width of quasiparticle peaks in conductances obtained in SIN junctions. This fact is more or less obvious: Consider a peak-to-peak tunneling in an SIS junction, as shown in Fig. 12.15. By varying bias near $\pm 2V_p$, the peak-to-peak tunneling will result in a conductance peak which is approximately two times wider than the bisoliton peak. Near peak bias, the presence of a pseudogap can be ignored.

Let us estimate shapes of SIS-junction fits. Take the bisoliton density of states as $N_b(E) = A \times (sech^2[(E - \Delta_p)/E_0] + sech^2[(E + \Delta_p)/E_0])$. To estimate exclusively the shape of SIS-junction conductance peaks, we neglect the presence of pseudogap. Then, substituting $N_b(E)$ into Eq. (12.1), the following integral should be solved

$$I(V) \propto \int_0^\infty sech^2\left(\frac{E - \Delta_p}{E_0}\right) sech^2\left(\frac{E - eV + \Delta_p}{E_0}\right) dE, \qquad (12.3)$$

where we neglect the smearing parameter Γ, the Fermi function and we assume that the matrix $K(V)$ is constant. The latter integral can be estimated by using the following result

$$\int sech^4(x)dx = \frac{\tanh x}{3} \times [2 + sech^2(x)] + C, \qquad (12.4)$$

where C is a constant. To visualize this result, Figure 12.48a depicts the latter function. In Fig. 12.48a, one can see that the hyperbolic function $\tanh(x)\,[2 + sech^2(x)]/3$ still has a $tanh$-function shape. Figure 12.48b presents for comparison the hyperbolic functions $sech(x)$, $sech^2(x)$ and $sech^4(x)$. In Fig. 12.48b, one can observe that the difference between $sech^2(x)$ and $sech^4(x)$ functions is small. It is possible that, for SIS-junction tunneling conductances, the exact solution of the bisoliton fit is not the $sech^2$ function but something

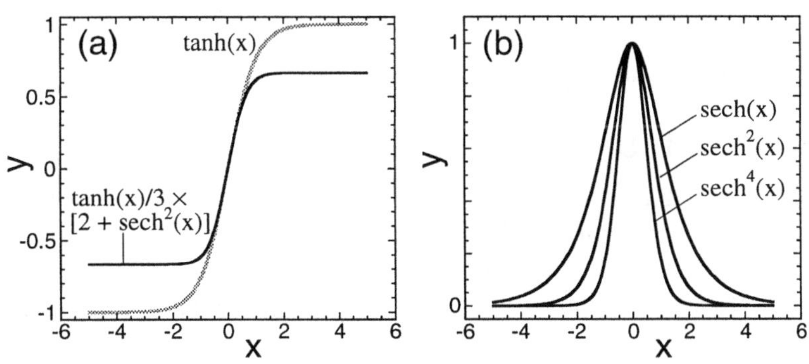

Figure 12.48. Hyperbolic functions: (a) $\tanh x$ and $\tanh x[2 + sech^2 x]/3$, and (b) $sech\ x$, $sech^2 x$ and $sech^4 x$.

between the two hyperbolic functions $sech^2$ and $sech^4$.

In the absence of exact analytical solutions for SIS-junction bisoliton fits, we must use either the $sech^2$ (and $tanh$) or $sech^4$ (and $tanh \times [2+sech^2]/3$) functions as first estimates of the fits.

APPENDIX A: Bi2212 single crystals

Overdoped Bi2212 single crystals were grown by the self-flux method in Al_2O_3 and ZrO_2 crucibles and then mechanically separated from the flux. The dimensions of these crystals are typically 2–3$\times$1$\times$ 0.1 mm^3. The chemical composition of the Bi-2:2:1:2 phase in these single crystals corresponds to the formula $Bi_2Sr_{1.9}CaCu_{1.8}O_{8+x}$ as measured by energy dispersive X-ray fluorescence (EDX). The crystallographic a, b, c values are of 5.41, 5.50 and 30.81 Å, respectively. The T_c value was determined by the four-contact method yielding T_c = 87–90 K with the transition width less than 1 K. To be sure that the Bi2212 single crystals are indeed overdoped, a few of them were carefully checked: the T_c value of these samples was increasing up to 95 K when some oxygen was chemically removed.

The underdoped Bi2212 single crystals were obtained by annealing the overdoped crystals in vacuum. The results presented here are obtained in an underdoped Bi2212 single crystal having $T_c \simeq 51$ K and the transition width of a few degrees. The estimated doping level in this Bi2212 single crystal is about 0.085.

Bi2212 single crystals in which Cu is partially substituted for Ni or Zn were also grown by the self-flux method. As measured by EDX the chemical composition of the Bi-2:2:1:2 phase in Ni-doped and Zn-doped Bi2212 single crystals corresponds to the formula $Bi_2Sr_{1.95}Ca_{0.95}(CuNi)_{2.05}O_{8+x}$ and $Bi_2Sr_{1.98}Ca_{0.83}(CuNi)_2O_{8+x}$, respectively. In these single crystals the Ni and Zn content with respect to Cu was approximately 1.5% and 1%, respectively. The T_c value was determined by the four-contact method: T_c = 75–77 K in Ni-doped Bi2212 and T_c = 76–78 K in Zn-doped Bi2212. In both types of doped Bi2212 single crystals, the transition width is a few degrees.

APPENDIX B: Details of test setups

Most tunneling data presented in this book were performed by the break-junction technique. Figure 12.49 schematically depicts the break-junction setup. A single crystal is glued to a flexible insulating substrate, as shown in Fig. 12.49, such that the ab plane of the crystal is perpendicular to the substrate surface. Four electrical contacts (typically with a resistance of a few Ohms) were made by attaching gold wires to the crystal with silver paint. The diameter of golden wires is 25 μm. Two contacts situated diagonally are used for bias input, which is slightly modulated by a lock-in, and the other two for measuring

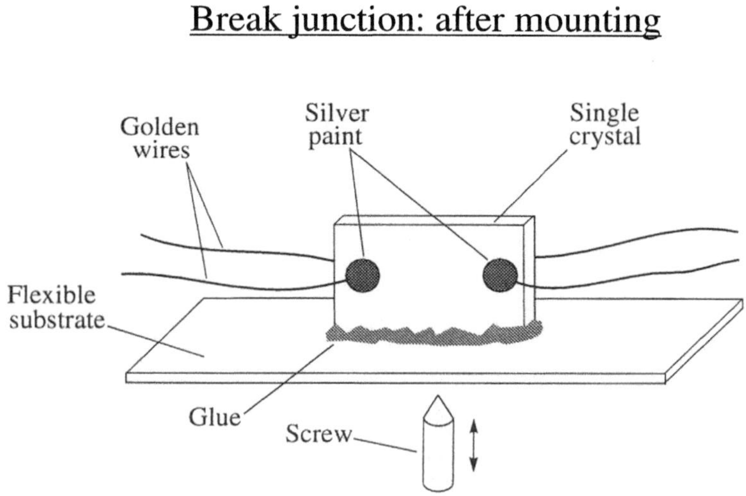

Figure 12.49. Schematic of a break junction after mounting: a single crystal with four electrical contacts is glued to a flexible substrate. The crystal is broken by bending the substrate using a differential screw. For more details, see text.

$dI(V)/dV$ and $I(V)$ tunneling characteristics. At low temperature, the crystal is broken in an He ambient in the ab plane by bending the flexible substrate. The bending force is applied by the differential screw shown schematically in Fig. 12.49. The differential screw has a precision of 10 μm per turn. Tunneling is achieved by changing the distance between broken parts of the crystal. The $dI(V)/dV$ and $I(V)$ tunneling characteristics are determined by a standard lock-in modulation technique.

In one junction a few tunneling spectra are usually obtained at low (constant) temperature by changing the distance between broken parts of a crystal, going back and forth etc., and each time the tunneling occurs most likely in different places.

In addition to SIS-junction measurements tunneling tests have also been carried out in Bi2212 and NCCO single crystals by forming SIN junctions. In point-contact measurements, the differential screw shown in Fig. 12.49 is used to push a normal tip against a fixed single crystal. Pt-Ir and Ag wires sharpened mechanically were used as normal tips. The point-contact measurements were performed either along or perpendicular to the c crystal axis.

References

Chapter 1

1 J. S. Russell, *Report on Waves*, in *Rep. 14th Meet. British Assoc. Adv. Sci.* (John Murray, 1844) p. 311.
2 H. Kamerlingh Onnes, *Commun. Phys. Lab. Univ. Leiden* **124c** (1911).
3 W. Meissner and R. Ochsenfeld, *Naturwissenschaften* **21**, 787 (1933).
4 C. Gorter and Casimir, *Physica* **1**, 306 (1934).
5 L. W. Schubnikow and I. E. Nakhutin, *Nature* **139**, 589 (1937).
6 E. Maxwell, *Phys. Rev.* **78**, 477 (1950).
7 C. A. Reynolds, B. Serin, W. H. Wright, and L.B. Nesbitt, *Phys. Rev.* **78**, 487 (1950).
8 V. Ginsburg and L. Landau, *Zh. Eksp. Teor. Fiz.* **18**, 241 (1950).
9 A. A. Abrokosov, *Zh. Eksp Teor Fiz.* **32**, 1442 (1957).
10 J. Bardeen, L. N. Cooper, and J. R. Schrieffer, *Phys. Rev.* **108**, 1175 (1957).
11 B. D. Josephson, *Phys. Lett.* **1**, 251 (1962).
12 J. Bardeen, *Phys. Rev. Lett.* **9**, 147 (1962).
13 L. S. Brizhik and A. S. Davydov, *Fiz. Nizk. Temperatur*, **10** 358 (1984).
14 D. Jérome, A. Mazaud, M. Ribault, and K. Bechgaard, *J. Phys. (Paris) Lett.* **41**, L195 (1980).
15 K. Miyake, S. Schmitt-Rink, and C. M. Varma, *Phys. Rev. B* **34**, 6554 (1986).
16 J. G. Bednorz and K. A. Müller, *Z. Phys. B* **64**, 189 (1986).
17 V. Kresin, *Solid State Comm.* **63**, 725 (1987); and V. Kresin and S. A. Wolf, *Solid State Comm.* **63**, 1141 (1987).
18 L. P. Gor'kov and A. V. Sokol, *JETP Lett.* **46**, 420 (1987).
19 Ph. W. Anderson, *Science* **235**, 1196 (1987).
20 A. S. Davydov, *Solitons in Molecular Systems* (Naukova Dumka, Kiev, 1988), in Russian.
21 J. C. Phillips, *Phys. Rev. Lett.* **59**, 1856 (1987).
22 W. W. Warren, Jr., R. E. Walstedt, G. F. Brennert, R. J. Cava, R. Tycko, R. F. Bell, and G. Dabbagh, *Phys. Rev. Lett.* **62**, 1193 (1989).
23 A. S. Davydov, *Phys. Rep.* **190**, 191 (1990).
24 A. S. Davydov, *Solitons in Molecular Systems* (Kluwer Academic, Dordrecht, 1991).
25 A. S. Alexandrov and N. F. Mott, *Rep. Prog. Phys.* **57**, 1197 (1994).
26 V. Emery and S. Kivelson, *Nature* **374**, 434 (1995).
27 J. M. Tranquada, B. J. Sternlieb, J. D. Axe, Y. Nakamura,and S. Uchida, *Nature* **375**, 561 (1995).
28 V. Emery, S. Kivelson, and O. Zachar, *Phys. Rev. B* **56**, 6120 (1997).
29 A. Mourachkine, *J. Low Temp. Phys.* **117**, 401 (1999), and references therein.
30 A. Mourachkine, *Europhys. Lett.* **55**, 86 (2001).
31 J. Rossat-Mignod, L. P. Regnault, C. Vettier, Ph. Bourges, P. Burlet, J. Bossy, J. Y. Henry, and G. Lapertot, *Physica C* **185-189**, 86 (1991).

32 A. Mourachkine, *Europhys. Lett.* **55**, 559 (2001).
33 A. Mourachkine, *Supercond. Sci. Technol.* **14**, 329 (2001).
34 A. Mourachkine, *J. Supercond.* **14**, 375 (2001).

Chapter 2

1 J. Bardeen, L. N. Cooper, and J. R. Schrieffer, *Phys. Rev.* **108**, 1175 (1957).
2 W. L. McMillan, *Phys. Rev.* **167**, 331 (1968).
3 V. Z. Kresin and S. A. Wolf, *Fundamentals of Superconductivity* (Plenum Press, New York, 1990).
4 V. Emery and S. Kivelson, *Nature* **374**, 434 (1995).
5 R. C. Dynes, V. Narayanamurti, and J. P. Mihaly, *Phys. Rev. Lett.* **41**, 1509 (1978).
6 S. H. Pan, E. W. Hudson, and J. C. Davis, *Appl. Phys. Lett.* **73**, 2992 (1998).
7 O. Naaman, W. Teizer, and R. C. Dynes, *Phys. Rev. Lett.* **87**, 097004 (2001).
8 V. Ambegaokar and A. Baratoff, *Phys. Rev. Lett.* **10**, 486 (1963).
9 A. F. Andreev, *Zh. Eksp. Teor. Fiz.* **46**, 1823 (1964); in English *Sov. Phys. JETP* **19**, 1228 (1964).

Chapter 3

1 J. G. Bednorz and K. A. Müller, *Z. Phys. B* **64**, 189 (1986).
2 N. Ichikawa, S. Uchida, J. M. Tranquada, T. Niemoller, P. M. Gehring, S.-H. Lee, and J. R. Schneider, *Phys. Rev. Lett.* **85**, 1738 (2000).
3 J. D. Axe, A. H. Moudden, D. Hohlwein, D. E. Cox, K. M. Mohanty, A. R. Moodenbough, and Y. Xu, *Phys. Rev. Lett.* **62**, 2751 (1989).
4 B. Nachumi *et al.*, *Phys. Rev. B* **58**, 8760 (1998).
5 Z. A. Xu, N. P. Ong, T. Noda, H. Eisaki, and S. Uchida, preprint cond-mat/9910123 (1999).
6 J. S. Zhou and J. B. Goodenough, *Phys. Rev. B* **56**, 6288 (1997).
7 W. Ting and K. Fossheim, *Phys. Rev. B* **48**, 16751 (1993).
8 M. Nohara, T. Suzuki, Y. Maeno, and T. Fujita, *Phys. Rev. B* **52**, 570 (1995).
9 R. P. Sharma, S. B. Ogale, Z. H. Zhang, J. R. Liu, W. K. Chu, B. Veal, A. Paulikas, H. Zheng, and T. Venkatesan, *Nature* **404**, 736 (2000).
10 Y.-N. Wang, J. Wu, H.-M. Shen, J.-S. Zhu, and X.-H. Chen, *Phys. Rev. B* **41**, 8981 (1990).
11 A. Fukuoka, A. Tokiwa-Yamamoto, M. Itoh, R. Usami, S. Adachi, and K. Tanabe, *Phys. Rev. B* **55**, 6612 (1997).
12 O. Chmaissem, J. D. Jorgensen, S, Short, A. Knizhnik, Y. Eckstein, and H. Shaked, *Nature* **397**, 45 (1999).
13 J. P. Attfield, A. L. Kharlamov, and J. A. McAllister, *Nature* **394**, 157 (1998).
14 M. R. Presland *et al.*, *Physica C* **176**, 95 (1991).
15 P. Ghigna, G. Spinolo, G. Flor, and N. Morgante, *Phys. Rev. B* **57**, 13426 (1998).
16 T. Fujii, T. Watanabe, and A. Matsuda, *Physica C* **357-360**, 173 (2001).
17 J. H. Schon, Ch. Kloc, and B. Batlogg, *Nature* **408**, 549 (2000).
18 T. Doderer, C. C. Tsuei, W. Hwang, and D. M. Newns, *Phys. Rev. B* **62**, 5984 (2000).
19 K. Yamada *et al.*, *Phys. Rev. B* **57**, 6165 (1998).
20 P. Imbert, G. Jehanno, P. Debray, C. Garcin, and J. A. Hodge, *J. Phys.* (Paris) **12**, 1405 (1992).
21 P. Imbert, J. A. Hodge, G. Jehano, P. Bonville, and C. Garcin, in *Phase Separation in Cuprate Superconductors*, K. A. Müller and G. Benedek (eds.) (World Scientific, Singapure, 1992), pp. 158 and 358.
22 G. V. M. Williams, S. Kramer, and M. Mehring, *Phys. Rev. B* **63**, 104514 (2001).
23 J. P. Frank, *Physica C* **282-287**, 198 (1997).
24 D. J. Pringle, G. V. M. Williams, and J. L. Tallon, *Phys. Rev. B* **62**, 12527 (2000).
25 G. V. M. Williams, D. J. Pringle, and J. L. Tallon, *Phys. Rev. B* **61**, R9257 (2000).

26 A. Lanzara, N. L. Saini, A. Bianconi, G.-M. Zhao, K. Conder, H. Keller, and K. A. Müller, preprint cond-mat/9812425 (1998).
27 J. Hofer, K. Conder, T. Sagawa, G.-M. Zhao, M. Willemin, H. Keller, and K. Kishio, *Phys. Rev. Lett.* **84**, 4192 (2000).
28 G. Zhao, V. Kirtikar, and D. E. Morris, *Phys. Rev. B* **63**, 220506R (2001).
29 D. Rubio Temprano, J. Mesot, S. Jansen, K. Conder, and A. Furrer, *Phys. Rev. Lett.* **84**, 1990 (2000).
30 A. Mourachkine, *Supercond. Sci. Technol.* **13**, 1378 (1999).
31 S. I. Vedeneev, A. G. M. Jansen, P. Samuely, V. A. Stepanov, A. A. Tsvetkov, and P. Wyder, *Phys. Rev. B* **49**, 9823 (1994).
32 Y. Ando, K. Segawa, S. Komiya, and A. N. Lavrov, *Phys. Rev. Lett.* **88**, 137005 (2002).
33 Y. J. Uemura *et al.*, *Phys. Rev. Lett.* **62**, 2317 (1989), and **66**, 2665 (1991).
34 C. Panagopoulos, J. R. Cooper, T. Xiang, Y. S. Wang, and C. W. Chu, *Phys. Rev. B* **61**, R3808 (2000).
35 J. L. Tallon, *Phys. Rev. B* **58**, R5956 (1998).
36 G. Villard, D. Pelloquin, and A. Maignan, *Phys. Rev B* **58**, 15231 (1998).
37 E. W. Hudson, K. M. Lang, V. Madhavan, S. H. Pan, H. Eisaki, S. Uchida, and J. C. Davis, *Nature* **411**, 920 (2001).
38 S. H. Pan, E. W. Hudson, K. M. Lang, H. Eisaki, S. Uchida, and J. C. Davis, *Nature* **403** 746 (2000).
39 B. von Hedt, W. Lisseck, K. Westerholt, and H. Bach, *Phys. Rev. B* **49**, 9898 (1994).
40 J. Corson, R. Mallozzi, J. Orenstein, J. N. Eckstein, and I. Bozovic, *Nature* **398**, 221 (1999).
41 J.-P. Salvetat, H. Berger, A. Halbritter, G. Mihaly, D. Pavuna, and L. Forro, *Europhys. Lett.*, **52**, 584 (2000).
42 See Fig. 2 in A. Mourachkine, preprint cond-mat/9812245 (1998).
43 G. Deutscher, *Nature* **397**, 410 (1999).
44 A. Mourachkine, *Europhys. Lett.* **50**, 663 (2000).
45 Q. Li, Y. N. Tsay, M. Suenaga, R. A. Klemm G. D. Gu, and N. Koshizuka, *Phys. Rev. Lett.* **83**, 4160 (1999).
46 A. Lanzara, P. V. Bogdanov, X. J. Zhou, S. A. Kellar, D. L. Feng, E. D. Lu, T. Yoshida, H. Eisaki, A. Fujimori, K. Kishlo, J.-I. Shimoyama, T. Noda, S. Uchida, Z. Hussain, and Z.-X. Sheb, *Nature* **412**, 510 (2001).
47 R. J. McQueeney, J. L. Sarrao, P. G. Pagliuso, P. W. Stephens, and R. Osborn, Phys. Rev. Lett. **87**, 077001 (2001).
48 G. M. Zhao, M. B. Hunt, H. Keller, and K. A. Müller, *Nature*, **385**, 236 (1997).
49 Y. Petrov, T. Egmi, R. J. McQueeney, M. Yethiraj, H. A. Mook, and F. Dogan, preprint cond-mat/0003414 (2000).
50 N. L. Saini, H. Oyanagi, and A. Bianconi, *Physica C* **357-360**, 117 (2001).
51 D. I. Khomskii and K. I. Kugel, *Europhys. Lett.* **55**, 208 (20001).
52 E. Kaneshita, M. Ichioka, and K. Machida, preprint cond-mat/0109466 (2001).
53 H. Ledbetter, *Physica C* **235-240**, 1325 (1994).
54 V. Z. Kresin and S. A. Wolf, *Phys. Rev. B* **46**, 6458 (1992).
55 L. Pintschovius, W. Reichardt, M. Braden, G. Dhalenne, and A. Revcolevschi, *Phys. Rev. B* **64**, 094510 (2001).
56 R. Osborn, E. A. Goremychkin, A. I. Kolesnokov, and D. G. Hinks, *Phys. Rev. Lett.* **87**, 017005 (2001).
57 H. He, Ph. Bourges, Y. Sidis, C. Ulrich, L. P. Regnault, S. Pailhes, N. S. Berzigiarova, N. N. Kolesnokov, and B. Keimer, *Science* **295**, 1045 (2002).
58 A. Mourachkine *Physica C* **341-348**, 917 (2000).
59 H. A. Mook and F. Dogan, preprint cond-mat/0103037 (2001).
60 T. Noda, H. Eisaki, and S. Uchida, *Science* **286**, 265 (1999).
61 A. Bianconi, N. L. Saini, A. Lanzara, M. Missori, T. Rossetti, H. Ogyanagi, H. Yamaguchi, K. Oka, and T. Ito, *Phys. Rev. Lett.* **76**, 3412 (1996).
62 M. Nohara, T. Suzuki, Y. Maeno, and T. Fujita, *Phys. Rev. B* **70**, 3447 (1993).
63 T. Fujita and T. Suzuki, in *Physics of High-Temperature Superconductors*, S. Maekawa and M. Sato (eds.) (Springer-Verlag, Berlin, 1992), p. 333.

64 A. Migliori, W. M. Visscher, S. Wong, S. E. Brown, I. Tanaka, H. Kojima, and P. B. Allen, *Phys. Rev. Lett.* **64**, 2458 (1990).
65 L. P. Gor'kov, *Pis'ma Zh. Eksp. Teor. Fiz.* **17**, 525 (1973), and *Zh. Eksp. Teor. Fiz.* **65**, 1558 (1973).
66 V. N. Naumov, G. I. Frolova, V. V. Nogteva, A. I. Romanenko, and T. Atake, preprint cond-mat/0109404 (2001).
67 A. Mourachkine, *Europhys. Lett.* **55**, 86 (2001).
68 *Science* **278**, 1879 (1997).
69 E. Dagotto, *Rev. Mod. Phys.* **66**, 763 (1994).
70 P. Brusov, *Mechanisms of High Temperature Superconductivity*, (Rostov State University Publishing, Rostov-on-Don, 1999), and references therein.
71 R. S. Markiewicz, *J. Phys. Chem. Solids* **58**, 1179 (1997).
72 S. Fujita and S. Godoy, *Theory of High Temperature Superconductivity* (Kluwer Academic, Dordrecht, 2001).
73 K. Miyake, S. Schmitt-Rink, and C. M. Varma, *Phys. Rev. B* **34**, R6554 (1986).

Chapter 4

1 A. Mourachkine, *Europhys. Lett.* **55**, 559 (2001).
2 A. Mourachkine, *Supercond. Sci. Technol.* **14**, 329 (2001).
3 G. E. Blonder, M. Tinkham, and T. M. Klapwijk, *Phys. Rev. B* **25**, 4515 (1982).
4 N. Miyakawa, P. Guptasarma, J. F. Zasadzinski, D. G. Hinks, and K. E. Gray, *Phys. Rev. Lett.* **80**, 157 (1998).
5 V. M. Krasnov, A. E. Kovalev, A. Yurgens, and D. Winkler, *Phys. Rev. Lett.* **86**, 2657 (2001).
6 A. G. Sun, A. Truscott, A. S. Katz, R. C. Dynes, B. W. Veal, ans C. Gu, Phys. Rev. B **54**, 6734 (1996).

Chapter 5

1 J. S. Russell, *Report on Waves*, in *Rep. 14th Meet. British Assoc. Adv. Sci.* (John Murray, 1844) p. 311.
2 D. J. Korteweg and G. de Vries, *On the Change of Form of Long Waves Advancing in a Rectangular Canal, and on a New Type of Long Stationary Waves*, *Phil. Mag.* **39**, 442 (1895).
3 E. Fermi, J. Pasta, and S. Ulam, *Studies of Nonlinear Problems* (Los Alamos National Laboratory, Los Alamos, 1955) Report No. LA1940.
4 N. J. Zabusky and M. D. Kruskal, *Phys. Rev. Lett.* **15**, 240 (1965).
5 A. T. Filippov, *The Versatile Soliton* (Birkhäuser, Boston, 2000).
6 Yu. I. Frenkel and T. A. Kontorova, *J. Phys. Moscow* **1**, 137 (1939).
7 M. Remoissenet, *Waves Called Solitons* (Springer-Verlag, Berlin, 1999).
8 W. P. Su, J. R. Schrieffer, and A. J. Heeger, *Phys. Rev. B* **22**, 2099 (1980).
9 E. J. Mele and M. J. Rice, *Phys. Rev. B* **23**, 5397 (1981).
10 A. S. Davydov, *Phys. Rep.* **190**, 191 (1990).
11 A. S. Davydov, *Solitons in Molecular Systems* (Kluwer Academic, Dordrecht, 1991).
12 G. A. Cambell and R. M. Foster, *Fourier Integrals for Practical Applications* (D. Van Nostrand Company, New York, 1951) p. 73.
13 M. Peyrard, (ed.), *Nonlinear Excitations in Biomolecules* (Springer-Verlag, Berlin, 1995).

Chapter 6

1 A. Mourachkine, *Europhys. Lett.* **55**, 559 (2001).
2 A. Mourachkine, *Supercond. Sci. Technol.* **14**, 329 (2001).
3 N. Miyakawa, P. Guptasarma, J. F. Zasadzinski, D. G. Hinks, and K. E. Gray, *Phys. Rev. Lett.* **80**, 157 (1998).

4 N. Miyakawa, J. F. Zasadzinski, L. Ozyuzer, P. Guptasarma, D. G. Hinks, C. Kendziora, and K. E. Gray, *Phys. Rev. Lett.* **83**, 1018 (1999).
5 L. Ozyuzer, J. F. Zasadzinski, C. Kendziora, and K. E. Gray, *Phys. Rev. B* **61**, 3629 (2000).
6 H. L. Edwards, A. L. Barr, J. T. Markert, and A. L. de Lozanne, *Phys. Rev. Lett.* **73**, 1154 (2001).
7 B. Grévin, B. Y. Berthier, and G. Collin, *Phys. Rev. Lett.* **85**, 1310 (2000).
8 A. L. de Lozanne, D. J. Derro, E. W. Hudson, K. M. Lang, S. H. Pan, J. C. Davis, and J. T. Markert, oral presentation at the "Stripes 2000" Conference in Rome 25-30 September 2000.
9 A. Miglori *et al.*, *Phys. Rev. B* **41**, 2098 (1990).
10 W.-K. Lee, M. Lew, and A. S. Nowick, *Phys. Rev. B* **41**, 149 (1990).
11 S. Bhattacharya, in *Ultrasonics of High-T_c and Other Unconventional Superconductors*, M. Levy (ed.) (Academic Press, San Diego,1992) pp. 303–348.
12 J. S. Zhou and J. B. Goodenough, *Phys. Rev. B* **56**, 6288 (1997).
13 I. M. Abu-Shiekah, O. Bakharev, H. B. Brom, and J. Zaanen, *Phys. Rev. Lett.* **87**, 237201 (2001).
14 T. Kimura, Y. Tomioka, H. Kuwahara, A. Asamitsu, M. Tamura, Y. Tokura, *Science* **274**, 1698 (1996).
15 T. Nachtrab, S. Heim, M. Mössle, R. Kleiner, O. Waldmann, R. Koch, P. Müller, T. Kimura, and Y. Tokura, *Phys. Rev. B* **65**, 012410 (2002), and cond-mat/0107463.
16 T. Nachtrab and P. Müller, private communications.

Chapter 7

1 A. S. Davydov, *Phys. Rep.* **190**, 191 (1990).
2 A. S. Davydov, *Solitons in Molecular Systems* (Kluwer Academic, Dordrecht, 1991).
3 L. S. Brizhik and A. S. Davydov, *Fiz. Nizk. Temperatur* **10**, 358 (1984), and 748 (1984).

Chapter 8

1 A. S. Davydov, *Phys. Rep.* **190**, 191 (1990).
2 A. S. Davydov, *Solitons in Molecular Systems* (Kluwer Academic, Dordrecht, 1991).
3 J. H. Schön, Ch. Kloc, and B. Batlogg. *Nature* **549**, 549 (2000).
4 M. V. Klein, *Physica C* **341-348**, 2173 (2000).
5 M. Nohara, T. Suzuki, Y. Maeno, and T. Fujita, *Phys. Rev. B* **52**, 570 (1995).
6 B. Renker, F. Gompf, D. Ewert, P. Adelmann, H. Schmidt, E. Gering, and H. Mutka, *Z. Phys. B* **77**, 65 (1990).
7 S. I. Vedeneev, A. G. M. Jansen, and P. Wyder, *Physica B*, **218**, 213 (1996).
8 R. S. Gonnelli, G. A. Ummarino, and V. A. Stepanov, *Physica C* **275**, 162 (1997).
9 A. Mourachkine, *Europhys. Lett.* **50**, 663 (2000).
10 Ph. Bourges, H. Casalta, A. S. Ivanov, and D. Petitgrand, *Phys. Rev. Lett.*, **79** (1997) 4906.
11 S. Kleefisch, B. Welter, A. Marx, L. Alff, R. Gross, and M. Naito, *Phys. Rev. B* **63**, 100507(R) (2001).

Chapter 9

1 J.-P. Salvetat, H. Berger, A. Halbritter, G. Mihaly, D. Pavuna, and L. Forro, *Europhys. Lett.* **52**, 584 (2000).
2 *Superconductivity in Ternary Compounds* II, O. Fischer and M. B. Maple (eds.) (Springer-Verlag, Berlin, 1982).
3 M. B. Malpe, *Physica C* **341-348**, 47 (2000), and *Physica B* **215**, 110 (1995).
4 A. M. Gabovich and A. I. Voitenko, *Fiz. Nizk. Temperatur* **26**, 419 (2000).
5 R. H. Heffner, D. E. MacLaughlin, J. E. Sonier, G. J. Nieuwenhuys, O. O. Bernal, B. Simovic, P. G. Pagliuso, J. L. Sarrao, and J. D. Thompson, cond-mat/0102137 (2001).
6 S. Uji, H. Shinagawa, T. Terashima, T. Yakabe, Y. Terai, M. Tokumoto, A. Kobayashi, *Nature* **410**, 908 (2001).

7 L. Balicas, J. S. Brooks, K. Storr, S. Uji, M. Tokumoto, H. Kabayashi, A. Kabayashi, V. Barzykin, and L. P. Gor'kov, cond-mat/0103463 (2001).

8 S. S. Saxena, P. Agarwal, K. Ahilan, F. M. Grosche, R. K. W. Haselwimmer, M. J. Steiner, E. Pugh, I. R. Walker, S. R. Julian, P. Monthoux, G. G. Lonzarich, A. Huxley, I. Sheikin, D. Braithwaite, and J. Flouquet, *Nature* **406**, 587 (2000).

9 C. Pflelderer, M. Uhlarz, S. M. Hyden, R. Vollmer, H. v. Löhneyesen, N. R. Bernhoeft, and G. G. Lonzarich, *Nature* **412**, 58 (2001).

10 D. Aoki, A. Huxley, E. Ressouche, D. Bralthwalte, J. Flouquet, J.-P. Briston, E. Lhotel, and C. Paulsen, *Nature* **413**, 613 (2001).

11 K. Shimizu, T. Kimura, S. Furomoto, K. Takeda, K. Kontani, Y. Onuki, and K. Amaya, *Nature* **412**, 316 (2001).

12 Y. Tokunaga, H. Kotegawa, K. Ishida, Y. Kitaoka, H. Takagiwa, and J. Akimitsu, *Phys. Rev. Lett.* **86**, 5767 (2001).

13 S. Y. Chen, J. Shulman, Y. S. Wang, D. H. Cao, C. Wang, Q. Y. Chen, T. H. Johansen, W. K. Chu, and C. W. Chu, cond-mat/0105510 (2001).

14 S.-M. Choi, J. W. Lynn, D. Lopez, P. L. Gammel, P. C. Canfield, and S. L. Bud'ko, *Phys. Rev. Lett.* **87**, 107001 (2001).

15 K. Miyake, S. Schmitt-Rink, and C. M. Varma, *Phys. Rev. B* **34**, 6554 (1986).

16 N. D. Martur, F. M. Grosche, S. R. Julian, I. R. Walker, D. M. Freye, R. K. W. Haselwimmer, and G. G. Lonzarich, *Nature* **394**, 39 (1998).

17 S. Sato, N. Aso, K. Miyake, R. Shlina, P. Thalmeier, G. Varelogiannis, G. Gelbel, F. Steglich, P. Fulde, and T. Komatsubara, *Nature* **410**, 340 (2001).

18 M. Hunt and M. Jordan, in *Advances in Solid State Physics*, Vol. 39, B. Kramer (ed.) (Vieweg, Braunschweig/Wiesbaden, 1999) p. 351.

19 M. Jordan, M. Hunt, and H. Adrian, *Nature* **398**, 47 (1999).

20 J. Singleton, *Rep. Prog. Phys.* **63**, 1111 (2000).

21 Y. De Wilde, J. Heil, A. G. M. Jansen, P. Wyder, R. Deltour, W. Assmus, A. Menovsky, W. Sun, and L. Taillefer, *Phys. Rev. Lett.* **72**, 2278 (1994).

22 N. Bernhoeft, N. Sato, B. Roessli, N. Aso, A. Hiess, G. H. Lander, Y. Endoh, and T. Komatsubara, *Phys. Rev. Lett.* **81**, 4244 (1998).

23 N. Metoki, Y. Haga, Y. Koike, and Y. Onuki, *Phys. Rev. Lett.* **80**, 5417 (1998).

24 A. Mourachkine, *Europhys. Lett.* **55**, 86 (2001).

25 C. Broholm, J. K. Kjems, W. J. L. Buyers, P. Matthews, T. T. M. Palstra, A. A. Menovsky, and J. A. Mydosh, *Phys. Rev. Lett.* **58**, 1467 (1987).

26 V. M. Krasnov, A. Yurgens, D. Winkler, P. Delsing, and T. Claeson, *Phys. Rev. Lett.* **84**, 5860 (2000).

27 H.-H. Klaus, W. Wagener, M. Hillberg, W. Kopmann, H. Walf, F. J. Litterst, M. Hücker, and B. Büchner, *Phys. Rev. Lett.* **85**, 5490 (2000).

28 K. Tamasaku, Y. Nakamura, and S. Uchida, *Phys. Rev. Lett.* **69**, 1455 (1992).

29 H. A. Mook, P. Dai, and F. Dogan, *Phys, Rev. B* **64**, 012502 (2001).

30 Y. Sidis, C. Ulrich, P. Bourges, C. Bernhard, C. Niedermayer, L. P. Regnault, N. H. Andresen, and B. Keimer, *Phys. Rev. Lett.* **86**, 4100 (2001).

31 J. A. Hodges, Y. Sidis, P. Bourges, I. Mirebeau, M. Hennion, and X. Chaud, cond-mat/0107218 (2001).

32 T. Uefuji, T. Kubo, K. Yamada, M. Fujita, K. Kurahashi, I. Watanabe, and K. Nagamine, *Physica C* **357-360**, 208 (2001).

33 J. E. Sonier, J. H. Brewer, R. F. Kiefl, R. I. Miller, G. D. Morris, C. E. Stronach, J. S. Gardner, S. R. Dusiger, D. A. Bonn, W. N. Hardy, R. Liang, and R. H. Heffner, *Science* **292**, 1692 (2001).

34 H. He, Ph. Bourges, Y. Sidis, C. Ulrich, L. P. Regnault, S. Pailhes, N. S. Berzigiarova, N. N. Kolesnokov, and B. Keimer, *Science* **295**, 1045 (2002).

35 See, for eample, D. K. Morr and D. Pines, *Phys. Rev. Lett.* **81**, 1086 (1998).

36 P. Dai, H. A. Mook, G. Aeppli, S. M. Hayden, and F. Dogan, *Nature* **406**, 965 (2000).

37 D. C. Tsui, R. E. Dietz, and L. R. Walker, *Phys. Rev. Lett.* **27**, 1729 (1971).

38 S. J. S. Lister, A. T. Boothroyd, N. H. Andersen, B. H. Lasen, A. A. Zhokhov, A. N. Christensen, and A. R. Wildes, *Phys. Rev. Lett.* **86**, 5994 (2001).

39 M. Apostu, R. Suryanarayanan, A. Revcolevschi, H. Ogasawara, M. Matsukawa, M. Yoshizawa, and N. Kobayashi, *Phys. Rev. B* **64**, 012407 (2001).

40 L. Ping, *Phys. Rev. B* **63**, 132503 (2001).
41 L. Ping, *J. Phys.: Condens. Matter* **11**, 7569 (1999).

Chapter 10

1 A. S. Davydov, *Solitons in Molecular Systems* (Kluwer Academic, Dordrecht, 1991).
2 K. Miyake, S. Schmitt-Rink, and C. M. Varma, *Phys. Rev. B* **34**, R6554 (1986).
3 J. M. Tranquada, B. J. Sternlieb, J. D. Axe, Y. Nakamura, and S. Uchida, *Nature* **375**, 561 (1995).
4 H. A. Mook, P. Dai, and F. Dogan, *Phys. Rev. Lett.* **88**, 097004 (2002).
5 C. Howald, H. Eisaki, N. Kaneko, and A. Kapitulnik, preprint cond-mat/0201546 (2002).
6 V. V. Moshchalkov, L. Trappeniers, and J. Vanacken, *Europhys. Lett.* **46**, 75 (1999).
7 A. Mourachkine, *Supercond. Sci. Technol.* **13**, 1378 (2000).
8 A. Bussmann-Holder, A. Simon, H. Büttner, and A. R. Bishop, *Philosoph. Mag. B* **80**, 1955 (2000).
9 M. Segev and D. N. Christodoulides, in *Spatial Solitons*, S. Trillo and W. Torruellas (eds.) (Springer, Belrin, 2001), pp. 87-125.
10 M. Bosch, W. van Saarloos, and J. Zaanen, *Phys. Rev. B* **63**, 092501 (2001).
11 H. L. Edwards, A. L. Barr, J. T. Markett, and A. L. de Lozanne, *Phys. Rev. Lett.* **73**, 1154 (1994).
12 M. G. Zacher, R. Eder, E. Arrigoni, and W. Hanke, *Phys. Rev. Lett.* **85**, 2585 (2000).
13 D. A. Bonn, J. C. Lynn, B. W. Gardner, Y.-J. Lin, R. Liang, W. N. Hardy, J. K. Kirtley, and K. A. Moler, *Nature* **414**, 887 (2001).
14 J. Zaanen, O. Y. Osman, and W. van Saarloos, *Phys. Rev. B* **58**, R11868 (1998).
15 Y. S. Lee, R. J. Birgeneau, M. A. Kastner, Y. Endoh, S. Wakimoto, K. Yamada, R. W. Erwin, S.-H. Lee, and G. Shirane, *Phys. Rev. B* **60**, 3643 (1999).
16 C.-M. Chang, A. H. Castro, and A. R. Bishop, *Philosoph. Mag. B* **81**, 827 (2001).
17 L. P. Gor'kov and A. V. Sokol, *JETP Lett.* **46**, 420 (1987).
18 J. A. Krumhansl and J. R. Schrieffer, *Phys. Rev. B* **11**, 3535 (1975).
19 J. W. Loram, J. L. Luo, J. R. Cooper, W. Y. Liang, and J. L. Tallon, *Physica C* **341-348**, 831 (2000).
20 J. L. Tallon, *Phys. Rev. B* **58**, R5956 (1998).
21 I. Watanabe, M. Akoshima, Y. Koike, S. Ohira, and K. Nagamine, *J. Low Temp. Phys.* **117**, 503 (1999).
22 N. Ichikawa, S. Uchida, J. M. Tranquada, T. Niemoller, P. M. Gehring, S.-H. Lee, and J. R. Schneider, *Phys. Rev. Lett.* **85**, 1738 (2000).
23 S. Tajima, T. Noda, H. Esaki, and S. Uchida, *Phys. Rev. Lett.* **86**, 500 (2001).
24 Y. Okuno and K. Miyake, *J. Phys. Soc. Jpn.* **67**, 3342 (1998).
25 P. Dai, H. A. Mook, and F. Dogan, *Phys. Rev. B* **63**, 054525 (2001).
26 N. K. Sato, N. Aso, K. Miyake, R. Shina, P. Thalmeler, G. Varelogiannis, C. Geibel, F. Steglich, P. Fulde, and T. Komatsubara, *Nature* **410**, 340 (2001).
27 J. P. Joshi, A. K. Sood, S. V. Bhat, A. R. Raju, and C. N. R. Rao, preprint, cond-mat/0201336 (2002).
28 B. Roessli, J. Schefer, G. A. Petrakovskii, B. Ouladdiaf, M. Boehm, U. Stuab, A. Vorotinov, and L. Bezmeternikh, *Phys.Rev. Lett.* **86**, 1885 (2001).
29 K. Frikach, M. Poirier, M. Castonguay, and K. D. Truong, *Phys. Rev. B* **61**, R6491 (2000).
30 B. Golding, in *Ultrasonics of High-T_c and Other Unconventional Superconductors*, M. Levy (ed.) (Academic Press, San Diego,1992) pp. 348–379.
31 Q. Li, Y. N. Tsay, M. Suenaga, R. A. Klemm G. D. Gu, and N. Koshizuka, *Phys. Rev. Lett.* **83**, 4160 (1999).
32 A. Mourachkine, *Europhys. Lett.* **50**, 663 (2000).
33 Ch. Walti, H. R. Ott, Z. Fisk, and J. L. Smith, *Phys. Rev. Lett.* **84**, 5616 (2000).
34 S. Shibata, A. Sumiyama, Y. Oda, Y. Haga, and Y. Onuki, *Phys. Rev. B* **60**, 3076 (2000).
35 A. Sumiyama, S. Shibata, Y. Oda, N. Kimura, E. Yamamoto, Y. Haga, and Y. Onuki, *Phys. Rev. Lett.* **81**, 5213 (1998).
36 A. S. Alexandrov, *Physica C* **305**, 46 (1998).

37 Y. Tokunaga, H. Kotegawa, K. Ishida, Y. Kitaoka, H. Takagiwa, and J. Akimitsu, *Phys. Rev. Lett.* **86**, 5767 (2001).
38 S. Y. Chen, J. Shulman, Y. S. Wang, D. H. Cao, C. Wang, Q. Y. Chen, T. H. Johansen, W. K. Chu, and C, W, Chu, preprint cond-mat/0105510 (2001).
39 S.-M. Choi, J. W. Lynn, D. Lopez, P. L. Gammel, P. C. Canfield, and S. L. Bud'ko, *Phys. Rev. Lett.* **87**, 107001 (2001).
40 E. Boaknin, R. W. Hill, C. Proust, C. Lupien, L. Taillefer, and P. C. Canfield, *Phys. Rev. Lett.* **87**, 237001 (2001).
41 M. Suzuki and M. Hikita, *Phys. Rev. B* **44**, 249 (1991).
42 Z. A. Xu, N. P. Ong, Y. Wang, T. Kakeshita, and S. Ushida, *Nature* **406**, 486 (2000).
43 K. Gorhy, O. M. Vyaselev, J. A. Martindale, V. A. Nandor, C. H. Pennington. P. C. Hammel, W. L. Hults, J. L. Smith, P. L. Kuhns, A. P. Reyers, and W. G. Moulton, *Phys. Rev. Lett.* **82**, 177 (1999).
44 W. Willemin, C. Rossel, J. Hofer, H. Keller, Z. F. Ren, and J. H. Wang, *Phys. Rev. B* **57**, 6137 (1998).
45 D. C. Tsui, R. E. Dietz, and L. R. Walker, *Phys. Rev. Lett.* **25**, 1729 (1971).
46 A. Mourachkine, *Physica C* **341-348**, 917 (2000).
47 J. Singleton, *Rep. Prog. Phys.* **63**, 1111 (2000).
48 J. Muller, M. Lang, F. Steglich, J. A. Schlueter, A. M. Kini, and T. Sasaki, preprint cond-mat/0107480 (2001).
49 L. Pintschovius, H. Rietschel, T. Sasaki, H. Mori, S. Tanaka, N. Toyota, M. Lang, and F. Steglich, *Europhys. Lett.* **37**, 627 (1997).
50 A. Girlando, M. Masino, A. Brillante, R. G. Della Valle, and E. Venuti, preprint cond-mat/0202141 (2002).
51 H. Elsinger, J. Wosnitza, S. Wanka, J. Hagel, D. Schweitzer, and W. Strunz, *Phys. Rev. Lett.* **84**, 6098 (2000).
52 J. H. Schön, M. Dorget, F. C. Beuran, X. Z. Xu, E. Arushanov, M. Lagues, and C. Deville Cavellin, *Science* **293**, 2430 (2001).
53 B. K. Sarma, M. Levy, S. Adenwalla, and J. B. Ketterson, in *Ultrasonics of High-T_c and Other Unconventional Superconductors*, M. Levy (ed.) (Academic Press, San Diego, 1992) pp. 108-189.

Chapter 11

1 M. Levy, M.-P. Xu, and B. Sarma, in *Ultrasonics of High-T_c and Other Unconventional Superconductors*, M. Levy (ed.) (Academic Press, San Diego, 1992) pp. 237–301.
2 D. A. Rudman and M. B. Beasley, *Phys. Rev. B* **30**, 2590 (1984).
3 L. R. Testardi, J. J. Hauser, and M. H. Read, *Solid State Commun.* **9**, 1829 (1971).
4 *Superconductivity in Ternary Compounds* I and II, O. Fischer and M. B. Maple (eds.) (Springer-Verlag, Berlin, 1982).
5 B. Wolf, J. Molter, G. Bruls, B. Luthi, and L. Jansen, *Phys. Rev. B* **54**, 348 (1996).
6 *Physics Today*, July issue, 11 (1994).
7 S. Zherlitsyn, B. Luthi, V, Gusakov, B. Wolf, F. Ritter, D. Wichert, S. Barilo, S. Shiryaev, C. Escribe-Filippini, and J. L. Tholence, *Europ. Phys. J. B* **16**, 59 (2000).
8 Ch. Niedermayer, C. Bernhard, T. Holden, R. K. Kremer, and K. Ahn, preprint cond-mat/0108431 (2001).
9 O. Gunnarsson, *Rev. Mod. Phys.* **69**, 575 (1997).
10 T. Takenobu, T. Muro, Y. Iwasa, and T. Mitani, *Phys. Rev. Lett.* **85**, 381 (2000).
11 J. H. Schön, Ch. Kloc, and B. Batlogg, *Nature* **408**, 549 (2000).
12 F. Carsey, R. Kagiwada, M. Levy, and K. Maki, *Phys. Rev. B* **4**, 854 (1971).
13 D. Djurek, Z. Medunic, A. Tonejc, and M. Paljevic, *Physica C* **351**, 78 (2001).
14 G.-m. Zhao and Y. S. Wang, preprint cond-mat/0111268 (2001).

Chapter 12

1 A. Mourachkine, preprint cond-mat/9811284 (1998).
2 G. Deutscher, *Nature* **397**, 410 (1999).

3 A. V. Fedorov, T. Valla, P. D. Johnson, Q. Li, G. D. Gu, and N. Koshizuka, *Phys. Rev. Lett.* **82**, 2179 (1999).
4 G. Allan, in *Magnetic Properties of Low-Demensional Systems*, L. M. Falicov and J. L. Moran-Lopez (eds.) (Springer, Berlin, 1986), p.2-9.
5 A. Mourachkine, *J. Supercond.* **14**, 375 (2001).
6 A. Mourachkine, *Physica C* **341-348**, 917 (2000).
7 W. Willemin, C. Rossel, J. Hofer, H. Keller, Z. F. Ren, and J. H. Wang, *Phys. Rev. B* **57**, 6137 (1998).
8 M. Gurvitch, J. M. Valles, Jr., A. M. Cucolo, R. C. Dynes, J. P. Garno, L. F. Schneemeyer, and J. V. Waszczak, *Phys. Rev. Lett.* **63**, 1008 (1989).
9 S. I. Vedeneev, A. G. M. Jansen, P. Samuely, V. A. Stepanov, A. A. Tsvetkov, and P. Wyder, *Phys. Rev. B* **49**, 9823 (1994).
10 D. C. Tsui, R. E. Dietz, and L. R. Walker, *Phys. Rev. Lett.* **27**, 1729 (1971).
11 H. J. Tao, A. Chang, F. Lu, and L. W. Wolf, *Phys. Rev. B* **45**, 10622 (1992).
12 A. Mourachkine, *Europhys. Lett.* **49**, 86 (2000).
13 N. Miyakawa, J. F. Zasadzinski, L. Ozyuzer, P. Guptasarma, D. G. Hinks, C. Kendziora, and K. E. Gray, *Phys. Rev. Lett.* **83**, 1018 (1999).
14 H. J. Tao, F. Lu, G. Zhang, and E. L. Wolf, *Physica C* **224**, 117 (1994).
15 A. Engelhardt, R. Dittmann, and A. I. Braginsky, *Phys. Rev. B* **59**, 3815 (1999).
16 M. A. Bari, F. Baudenbacher, J. Santiso, E. J. Tarte, J. E. Evetts, and M. G. Blamire, *Physica C* **256**, 227 (1996).
17 J. S. Tsai *et al.*, *Physica C* **153-155**, 1385 (1988).
18 J. Geerk, X. X. Xi, and G. Linker, *Z. Phys. B: Condens. Matter* **73**, 329 (1988).
19 C. R. Hu, *Phys. Rev. Lett.* **72**, 1526 (1994).
20 Y. Tanaka and S. Kashiwaya, *Phys. Rev. Lett.* **74**, 3451 (1995).
21 A. F. Andreev, *Zh. Eksp. Teor. Fiz.* **46**, 1823 (1964); in English *Sov. Phys. JETP* **19**, 1228 (1964).
22 I. Maggio-Aprile, Ch. Renner, A. Erb, E. Walker, and O. Fischer, *Phys. Rev. Lett.* **75**, 2754 (1995).
23 Ch. Renner, B. Revaz, K. Kadowaki, I. Maggio-Aprile, and O. Fischer, *Phys. Rev. Lett.* **80**, 3606 (1998).
24 S. H. Pan, E. W. Hudson, A. K. Gupta, K.-W. Ng, H. Esaki, S. Ushida, and J. C. Davis, *Phys. Rev. Lett.* **85**, 1536 (2000).
25 B. W. Hoogenboom, K. Kadowaki, B. Revaz, M. Li, Ch. Renner, and O. Fischer, *Phys. Rev. Lett.* **87**, 267001 (2001).
26 A. Mourachkine, *Europhys. Lett.* **50**, 663 (2000).
27 G. E. Blonder, M. Tinkham, and T. M. Klapwijk, *Phys. Rev. B* **25**, 4515 (1982).
28 N. Miyakawa, P. Guptasarma, J. F. Zasadzinski, D. G. Hinks, and K. E. Gray, *Phys. Rev. Lett.* **80**, 157 (1998).
29 E. J. Mele and M. J. Rice, *Phys. Rev. B* **23**, 5397 (1981).

Index

MIX
Papier aus verantwortungsvollen Quellen
Paper from responsible sources
FSC® C105338

If you have any concerns about our products,
you can contact us on
ProductSafety@springernature.com

In case Publisher is established outside the EU,
the EU authorized representative is:
Springer Nature Customer Service Center GmbH
Europaplatz 3, 69115 Heidelberg, Germany

Printed by Libri Plureos GmbH
in Hamburg, Germany